# COMPUTER-AIDED MODELING OF REACTIVE SYSTEMS

# COMPUTER-AIDED MODELING OF REACTIVE SYSTEMS

**WARREN E. STEWART**
Department of Chemical and Biological Engineering
University of Wisconsin
Madison, Wisconsin

**MICHAEL CARACOTSIOS**
Research and Development
UOP LLC (An affiliate of Honeywell International, Inc.)
Des Plaines, Illinois

A JOHN WILEY & SONS, INC., PUBLICATION

*Library of Congress Cataloging-in-Publication Data:*

Stewart, Warren E.
  Computer-aided modeling of reactive systems / Warren E. Stewart, Michael
Caracotsios.
      p. cm.
  Includes index.
  ISBN 978-0-470-27495-8 (cloth)
  1. Chemical reactions—Mathematical models.   2. Chemical reactions—Computer
simulation.   I. Caracotsios, Michael.   II. Title.
  QD503.S848 2008
  541'.39615118—dc22                                                    2007044566

Printed in the United States of America.

 9  8  7  6  5  4  3  2  1

*To our wives,*
*Jean Stewart and Joan Caracotsios,*
*for their patience, encouragement, and love.*

# Contents

Chapter 1. Overview .............................................. 1
    *REFERENCES and FURTHER READING* .................... 2

Chapter 2. Chemical Reaction Models ......................... **3**
  *2.1*  *STOICHIOMETRY OF REACTION SCHEMES* .............. 3
  *2.2*  *COMPUTABILITY OF REACTION RATES FROM DATA* ... 6
  *2.3*  *EQUILIBRIA OF CHEMICAL REACTIONS* ................. 7
  *2.4*  *KINETICS OF ELEMENTARY STEPS* ..................... 11
  *2.5*  *PROPERTIES OF REACTION NETWORKS* ................ 15
  *2.6*  *EVIDENCE FOR REACTION STEPS* ...................... 25
    *PROBLEMS* ..................................................... 28
    *REFERENCES and FURTHER READING* .................. 29

Chapter 3. Chemical Reactor Models ......................... **39**
  *3.1*  *MACROSCOPIC CONSERVATION EQUATIONS* ........... 39
    *3.1.1 Material Balances* ........................................ 39
    *3.1.2 Total Energy Balance* .................................. 41
    *3.1.3 Momentum Balance* .................................... 41
    *3.1.4 Mechanical Energy Balance* ........................... 42
  *3.2*  *HEAT AND MASS TRANSFER IN FIXED BEDS* .......... 47
  *3.3*  *INTERFACIAL STATES IN FIXED-BED REACTORS* ....... 48
  *3.4*  *MATERIAL TRANSPORT IN POROUS CATALYSTS* ...... 54
    *3.4.1 Material Transport in a Cylindrical Pore Segment* ....... 55
    *3.4.2 Material Transport in a Pore Network* .................. 56
    *3.4.3 Working Models of Flow and Diffusion in Isotropic Media* 57
    *3.4.4 Discussion* .............................................. 58
    *3.4.5 Transport and Reaction in Porous Catalysts* ............. 58
  *3.5*  *GAS PROPERTIES AT LOW PRESSURES* .................. 59
  *3.6*  *NOTATION* ................................................... 60
    *REFERENCES and FURTHER READING* .................... 62

Chapter 4. Introduction to Probability and Statistics ........ **65**
  *4.1*  *STRATEGY OF DATA-BASED INVESTIGATION* .......... 65
  *4.2*  *BASIC CONCEPTS IN PROBABILITY THEORY* .......... 66
  *4.3*  *DISTRIBUTIONS OF SUMS OF RANDOM VARIABLES* .... 69
  *4.4*  *MULTIRESPONSE NORMAL ERROR DISTRIBUTIONS* ... 72

4.5 *STATISTICAL INFERENCE AND CRITICISM* ............. 73
   *PROBLEMS* ................................................. 75
   *REFERENCES and FURTHER READING* .................. 76

**Chapter 5. Introduction to Bayesian Estimation** ............. **77**
5.1 *THE THEOREM* ........................................... 77
5.2 *BAYESIAN ESTIMATION WITH INFORMATIVE PRIORS* . 80
5.3 *INTRODUCTION TO NONINFORMATIVE PRIORS* ....... 84
5.4 *JEFFREYS PRIOR FOR ONE-PARAMETER MODELS* ..... 86
5.5 *JEFFREYS PRIOR FOR MULTIPARAMETER MODELS* ... 88
5.6 *SUMMARY* ................................................ 91
   *PROBLEMS* ................................................. 91
   *REFERENCES and FURTHER READING* .................. 92

**Chapter 6. Process Modeling with Single-Response Data** .... **95**
6.1 *THE OBJECTIVE FUNCTION $S(\theta)$* ........................ 96
6.2 *WEIGHTING AND OBSERVATION FORMS* ................ 98
6.3 *PARAMETRIC SENSITIVITIES; NORMAL EQUATIONS* ... 99
6.4 *CONSTRAINED MINIMIZATION OF $S(\theta)$* ................. 102
   *6.4.1 The Quadratic Programming Algorithm GRQP* ......... 102
   *6.4.2 The Line Search Algorithm GRS1* ..................... 104
   *6.4.3 Final Expansions Around $\hat{\theta}$* ........................... 105
6.5 *TESTING THE RESIDUALS* ............................... 106
6.6 *INFERENCES FROM THE POSTERIOR DENSITY* ....... 107
   *6.6.1 Inferences for the Parameters* .......................... 108
   *6.6.2 Inferences for Predicted Functions* ..................... 110
   *6.6.3 Discrimination of Rival Models by Posterior Probability* 112
6.7 *SEQUENTIAL PLANNING OF EXPERIMENTS* ........... 115
   *6.7.1 Planning for Parameter Estimation* ..................... 115
   *6.7.2 Planning for Auxiliary Function Estimation* ............ 116
   *6.7.3 Planning for Model Discrimination* ..................... 116
   *6.7.4 Combined Discrimination and Estimation* .............. 118
   *6.7.5 Planning for Model Building* ........................... 119
6.8 *EXAMPLES* ............................................... 119
6.9 *SUMMARY* ................................................ 122
6.10 *NOTATION* ............................................... 125
   *PROBLEMS* ................................................. 127
   *REFERENCES and FURTHER READING* .................. 133

**Chapter 7. Process Modeling with Multiresponse Data** ..... **141**
7.1 *PROBLEM TYPES* ......................................... 142
7.2 *OBJECTIVE FUNCTION* ................................... 148
   *7.2.1 Selection of Working Responses* ........................ 149
   *7.2.2 Derivatives of Eqs. (7.2-1) and (7.2-3)* .................. 150
   *7.2.3 Quadratic Expansions; Normal Equations* .............. 151

7.3  CONSTRAINED MINIMIZATION OF $S(\boldsymbol{\theta})$ ................. 152

    7.3.1 Final Expansions Around $\widehat{\boldsymbol{\theta}}$ ........................... 153

7.4  TESTING THE RESIDUALS .............................. 153

7.5  POSTERIOR PROBABILITIES AND REGIONS .......... 154

    7.5.1 Inferences Regarding Parameters ...................... 154

    7.5.2 Inferences Regarding Functions ......................... 156

    7.5.3 Discrimination Among Rival Models .................... 156

7.6  SEQUENTIAL PLANNING OF EXPERIMENTS .......... 158

7.7  EXAMPLES ............................................... 159

7.8  PROCESS INVESTIGATIONS ............................. 161

7.9  CONCLUSION ............................................ 165

7.10 NOTATION ............................................... 166

    ADDENDUM: PROOF OF EQS. (7.1-16) AND (7.1-17) ..... 168

    PROBLEMS ............................................... 169

    REFERENCES and FURTHER READING ................. 172

**Appendix A. Solution of Linear Algebraic Equations** ....... **177**

A.1  INTRODUCTORY CONCEPTS AND OPERATIONS ...... 177

A.2  OPERATIONS WITH PARTITIONED MATRICES ........ 180

A.3  GAUSS-JORDAN REDUCTION ........................... 182

A.4  GAUSSIAN ELIMINATION .............................. 184

A.5  LU FACTORIZATION ................................... 186

A.6  SOFTWARE ............................................. 187

    PROBLEMS ............................................... 187

    REFERENCES and FURTHER READING ................. 188

**Appendix B. DDAPLUS Documentation** ..................... **189**

B.1  WHAT DDAPLUS DOES ................................. 189

B.2  OBJECT CODE ......................................... 191

B.3  CALLING DDAPLUS ..................................... 191

B.4  DESCRIPTION OF THE CALLING ARGUMENTS ........ 191

B.5  EXIT CONDITIONS AND CONTINUATION CALLS ...... 200

B.6  THE SUBROUTINE fsub ................................. 203

B.7  THE SUBROUTINE Esub ................................ 203

B.8  THE SUBROUTINE Jac .................................. 205

B.9  THE SUBROUTINE Bsub ................................ 206

B.10 NUMERICAL EXAMPLES ............................... 208

    REFERENCES and FURTHER READING ................. 216

**Appendix C. GREGPLUS Documentation** ................... **217**

C.1  DESCRIPTION OF GREGPLUS .......................... 217

C.2  LEVELS OF GREGPLUS ................................ 219

C.3  CALLING GREGPLUS ................................... 219

C.4  WORK SPACE REQUIREMENTS FOR GREGPLUS ...... 226

C.5  SPECIFICATIONS FOR THE USER-PROVIDED MODEL . 226

*C.6 SINGLE-RESPONSE EXAMPLES* .......................... 228
*C.7 MULTIRESPONSE EXAMPLES* ........................... 244
     *REFERENCES and FURTHER READING* ................. 257

**Author Index** ................................................. 259

**Subject Index** ................................................. **265**

# Preface[†]

The senior author's interest in reaction and reactor modeling began in 1947 with an inspiring course from the text *Chemical Process Principles, Volume III* (Wiley 1947), coauthored by Olaf A. Hougen and Kenneth M. Watson. This interest was reawakened in the 1950s during Warren's work at Sinclair Research Labs and Baker & Kellogg, Inc., on the cooperative development of a catalytic reforming process, continued after he joined the Chemical Engineering Department of the University of Wisconsin in 1956, and was enriched by supportive interactions with Bob Bird, Ed Lightfoot, George Box, and Norman Draper in teaching and research on transport phenomena, statistics, and reaction modeling.

Michael Caracotsios's interest in mathematical modeling and parameter estimation began during his graduate years at the University of Wisconsin–Madison under the guidance of Warren Stewart. Upon completing his studies Michael joined Amoco Chemical, where he applied the fundamental principles of reactor modeling and parameter estimation to hydrocarbon polymerization and oxidation processes. He subsequently joined the kinetic modeling group at UOP LLC, where he continues to apply chemical engineering and rigorous statistical concepts to the modeling of complex refining and petrochemical processes.

Warren Stewart's graduate course on the Computer-Aided Modeling of Reactive Systems and his research with colleagues and graduate students, as well Michael Caracotsios's industrial experience and courses taught at Northwestern University and the Illinois Institute of Technology, have provided the framework for this book and for the software associated with it. Chapters 2 and 3 provide an overview of the basic modeling concepts, assumptions, and equations for multiple reactions involving multiple chemical species carried out in batch and continuous reactors. While not intended to be a comprehensive treatment of chemical reaction engineering, these chapters do provide a very general modeling framework and introduce specific chemical reaction problems that are used in later chapters

---

[†] Warren Stewart was working diligently to complete this book when he died unexpectedly on March 27, 2006, leaving this Preface, the Index, and a section of Chapter 3 unfinished. These remaining parts of the book were completed by Warren's colleagues Bob Bird, John Eaton, Dan Klingenberg, Tom Kuech, Jim Rawlings, and Harmon Ray, together with coauthor Michael Caracotsios.

to demonstrate fundamental parameter estimation techniques. Chapters 4 through 7 present the necessary theoretical background and computational methods for extracting reaction and process parameters from experimental data in the presence of measurement errors and incomplete data sets. These later chapters have benefited greatly from the writings and teachings of our colleague George Box, whose contributions to applied statistics are well known.

Because the computational procedures for simulating complex modeling problems and carrying out the parameter estimation are often extraordinarily complex and time consuming, a software package — Athena Visual Studio[1] — is available as an aid. Athena allows the modeler to solve model equations and estimate parameters for a wide variety of applications. The model may consist of ordinary and partial differential equations, implicit differential equations, differential-algebraic equations, or purely algebraic equations. The package DDAPLUS, described in Appendix B, is provided for solving models of these types. The program PDAPLUS solves models consisting of partial differential equation models with one time variable and one spatial variable. This class of models describes many applications involving simultaneous and coupled reaction, convection, and diffusion processes. Finally, the program GREGPLUS, described in Appendix C, provides the parameter estimation capability of Athena Visual Studio. GREGPLUS is a state-of-the-art parameter estimation tool that provides optimal parameter estimates and 95% confidence intervals for a wide variety of assumed measurement error structures. The error structures apply to both single-response systems presented in Chapter 6 and multiresponse systems presented in Chapter 7. The error structure may be completely known, or partially known with a block-diagonal structure, or completely unknown, in which case it is estimated from the data along with the parameters. GREGPLUS also provides the capability for performing optimal experimental design, which allows practitioners to design the most informative sequential experiments for estimating parameters of interest. Any prior information on parameter values may be included. If prior information is not available, GREGPLUS uses an appropriate noninformative prior, which is tailored to the model's measurement error structure. More details about this package may be found in Chapters 5 through 7 and Appendices B and C.

Some of the important research results that led to the material in this book were developed during the senior author's collaboration with former PhD student Jan Sørensen. This extensive research pioneered the development of orthogonal collocation methods for the solution of modeling and

---

[1]   Athena Visual Studio is a state-of-the-art software package that encapsulates the theoretical concepts presented in the chapters of this book in an integrated manner via a Windows interface. Athena Visual Studio can be downloaded for a free trial from www.AthenaVisual.com.

parameter sensitivity equations following Warren's initial discovery of orthogonal collocation while John Villadsen was a postdoctoral fellow at the University of Wisconsin in 1964. The authors are indebted to many others who influenced their work on process modeling and parameter estimation. Interactions with Linda Petzold concerning applications and modifications of her software for the solution of stiff differential-algebraic equations are greatly appreciated. Discussions with Gary Blau on parameter estimation application issues were very helpful. Some of the material on the development of experimental and computational methods for data analysis and parameter estimation reflects the senior author's research interactions with graduate students and research visitors over the years. These include Rod Bain, Tony Dixon, Chester Feng, Young Tai Lu, Sanjay Mehta, Dick Prober, Bob Shabaker, Dan Weidman, Tom Young, and others. We are grateful to Chemical Engineering Department staff Todd Ninman and Diane Peterson for their help throughout the project. Warren's children David Warren Stewart, Margaret Stewart Straus, and Mary Jean Stewart Glasgow deserve special thanks for their organizational and technical support in the latter stages of the writing.

The authors are especially grateful to their families for their constant support and gracious sacrifice during the writing of this book.

Warren Stewart
Michael Caracotsios

# List of Examples

2.1 Unimolecular Decomposition: Lindemann Scheme ............... 16
2.2 Thermal Reaction of Hydrogen with Bromine ................... 17
2.3 Homogeneous Catalysis by an Enzyme ......................... 19
2.4 Heterogeneous Catalysis ....................................... 20
2.5 Bifunctional Catalysis .......................................... 22
2.6 Composite Events; Stoichiometric Numbers ..................... 23

3.1 Balance Equations for Batch Reactors ........................... 42
3.2 Balance Equations for CSTR Reactors .......................... 44
3.3 Balances for Plug-Flow Homogeneous Reactors (PFR) ........... 45
3.4 Balances for Plug-Flow Fixed-Bed Reactors .................... 46
3.5 Calorimetric Measurement of Reaction Rate .................... 53

4.1 Tossing a Die .................................................. 68
4.2 The Univariate Normal Distribution ............................ 69

5.1 Likelihood Function for the Normal Mean ....................... 79
5.2 Estimation with Known Prior Probabilities ..................... 80
5.3 Estimation with Subjective Priors[1] ........................... 82
5.4 Prior for a Normal Mean ....................................... 84
5.5 Prior for a Normal Standard Deviation ......................... 84
5.6 Jeffreys Prior for the Normal Mean ............................ 87
5.7 Jeffreys Prior for the Normal Standard Deviation ............... 88
5.8 Jeffreys Prior for the Normal Mean and $\sigma$ ...................... 90

6.1 Discrimination Between Series and Parallel Models ............. 119
6.2 Discrimination of Models for Codimer Hydrogenation .......... 120

7.1 Kinetics of a Three-Component System ........................ 159
7.2 Multicomponent Diffusion in a Porous Catalyst ................ 160

A.1 Symbolic Inversion of a Partitioned Matrix .................... 181

B.1 Integration of State Equations ................................. 208
B.2 Integration with Two Stopping Criteria ........................ 210
B.3 Integration of a Differential-Algebraic Reactor Model .......... 211
B.4 Sensitivity Analysis of a Differential Equation ................. 214

C.1 Estimation with Numerical Derivatives ........................ 229
C.2 Nonlinear Estimation with Analytical Derivatives .............. 231
C.3 Analysis of Four Rival Models of a Catalytic Process ........... 233
C.4 Estimation with Numerical Derivatives ........................ 245
C.5 Multiresponse Estimation with Analytical Derivatives .......... 248
C.6 Multiresponse Discrimination and Goodness of Fit ............. 249

# List of Tables

6.1 Simulated Data for Example 6.1. ............................... 120

6.2 Estimates for Example 6.1. ..................................... 121

6.3 Analyses of Variance for Example 6.1. .......................... 121

6.4 Expectation Models for Response $y = \ln \mathcal{R}$ in Example 6.2. ...... 123

6.5 Testing of Kinetic Models for Data of Tschernitz et al. .......... 124

7.1 Full $Y$ and $\Sigma$ Structures. ....................................... 144

7.2 Block-Rectangular Data and Covariance Structures. ............. 146

7.3 An Irregular Data Structure and Its Covariances. ............... 148

7.4 Simulated Data, Adapted from Box and Draper (1965). ......... 159

7.5 Diffusion Model Discrimination; RD-150 Catalyst. .............. 161

7.6 Multiresponse Investigations with Unknown Covariance Matrix. . 163

B.1 Codes the User Provides to Go with DDAPLUS. ................ 192

# Chapter 1
# Overview

Great progress is being made in mathematical modeling of chemical and biochemical processes. Advances in computing, transport theory, chemical experimentation, and statistics all contribute to making process modeling more effective. Computerized laboratories are providing more and better data, and progress continues in statistical design and analysis of experiments. Modern computer hardware and software are making more realistic process models tractable, prompting fuller analyses of experimental data to build better models and establish probable process mechanisms.

This book will acquaint the reader with some of the important tools and strategies now available for investigating chemical processes and systems on personal computers. We present some useful mathematical models, numerical methods for solving them, and statistical methods for testing and discriminating candidate models with experimental data. The software package *Athena Visual Studio*, obtained from www.AthenaVisual.com, will enable you to do these computations and view the results on your personal computer screen, print them and transmit them. Familiarity with Microsoft Windows (1998 or later) is required, but little knowledge of Fortran, since Athena converts equation statements directly into Fortran code as described in the Athena tutorial on your computer's desktop.

Appendix A provides an introduction to linear algebra and to matrix methods for solving linear equation systems. A study or review of this appendix will prepare the reader for the applications of this material in Chapters 1 through 7. The other appendices describe Athena's solvers DDAPLUS and GREGPLUS, which are updated versions of our former subroutines DDASAC (Double-Precision Differential-Algebraic Sensitivity Analysis Code) and GREG (Generalized REGression package).

Chapter 2 deals with principles of chemical reaction modeling and gives some model forms for use in later chapters. A strong distinction is maintained between rates of reaction and of species production by defining reaction rates as frequencies of particular reaction events per unit reactor space. Various levels of reactor modeling are considered in Chapter 3 and implemented in Athena's samples of its solvers.

Chapter 4 begins the study of statistical methods and their role in process investigations. Concepts introduced include sample spaces, probability densities and distributions, the central limit theorems of DeMoivre and Lindeberg, the normal distribution, and its extension to multivariate

1

data. Chapter 5 presents Bayes' theorem and Jeffreys rule for unprejudiced inference of model parameters and auxiliary functions from experimental data.

Chapters 6 and 7 treat statistical aspects of process modeling, including parameter and function estimation, goodness-of-fit testing, model discrimination, and selection of conditions for additional experiments. Chapter 6 focuses on modeling with single-response data, and Chapter 7 deals with modeling with multiresponse data. Our software package GREGPLUS, included in Athena and described in Appendix C, performs these tasks for single or multiple models presented by the user.

Our aim in this book is to present modeling as a way of gaining insight, with or without detailed computations. With this in mind, we describe some diagnostic plots proposed by Yang and Hougen (1950) for reactions with a single controlling step, which give valuable clues for choosing good models and weeding out models that won't work.

Our students and colleagues have found it useful to examine the software descriptions and examples before choosing projects in their fields of interest. We suggest that the reader also begin by examining Appendices A, B, and C to learn the capabilities of Athena's solvers.

**REFERENCES and FURTHER READING**

Yang, K. H., and O. A. Hougen, Determination of mechanism of catalyzed gaseous reactions. *Chem. Eng. Prog.*, **46** (1950).

# Chapter 2
# Chemical Reaction Models

Chemical reaction models play a major role in chemical engineering as tools for process analysis, design, and discovery. This chapter provides an introduction to structures of chemical reaction models and to ways of formulating and investigating them. Each notation in this chapter is defined when introduced. Chapter 3 gives a complementary introduction to chemical **reactor** models, including their physical and chemical aspects.

A chemical reaction model consists of a list of chemical steps, each described by its stoichiometric equation and rate expression. The degree of detail included depends on the intended use of the model, the means available for solving it, and the range of phenomena to be described. Thus, the list of steps is always postulated and needs to be tested against experiments.

Chemical kinetics is an enormous subject; just a few basic principles will be treated in this chapter. Section 2.1 deals with reaction stoichiometry, the algebraic link between rates of reaction and of species production. Section 2.2 considers the computability of reaction rates from measurements of species production; the stoichiometric constraints on production rates are also treated there. The equilibrium and rate of a single reaction step are analyzed in Sections 2.3 and 2.4; then simple systems of reactions are considered in Section 2.5. Various kinds of evidence for reaction steps are discussed in Section 2.6; some of these will be analyzed statistically in Chapters 6 and 7.

## 2.1 STOICHIOMETRY OF REACTION SCHEMES

*Stoichiometry* is a compound of the Greek word $\sigma\tau o\iota\chi\epsilon\iota o$, for *element*, with $\mu\epsilon\tau\rho\varpi$ for *count*. The combination denotes "accounting of the elements." More particularly here, it denotes the balancing of chemical reactions with respect to each conserved entity therein. Thus, the reaction scheme

$$C_2H_4 + \tfrac{1}{2}O_2 \rightarrow C_2H_4O$$
$$C_2H_4O + \tfrac{5}{2}O_2 \rightarrow 2CO_2 + 2H_2O$$
$$C_2H_4 + 3O_2 \rightarrow 2CO_2 + 2H_2O$$

for ethylene oxidation is balanced for the elements C, H, and O and for electric charge, as the reader may verify.

3

Reaction schemes may be summarized as equation sets of the form

$$\sum_{i=1}^{\text{NS}} \nu_{ji} A_i = 0 \qquad j = 1, \ldots, \text{NR} \tag{2.1-1}$$

with NS denoting the number of chemical species in the reaction model, NR the number of reactions, and $A_i$ the formula for chemical species $i$. Then the coefficient $\nu_{ji}$ is the net number of particles of $A_i$ produced per occurrence of reaction $j$ in the forward direction (left to right as written). Thus, the stoichiometric coefficients for the foregoing reaction scheme form a matrix

$$
\nu = 
\begin{array}{ccccc}
C_2H_4 & O_2 & C_2H_4O & H_2O & CO_2 \\
\left(\begin{array}{ccccc}
-1 & -1/2 & 1 & 0 & 0 \\
0 & -5/2 & -1 & 2 & 2 \\
-1 & -3 & 0 & 2 & 2
\end{array}\right)
\end{array}
$$

of amounts produced per forward occurrence of each molecular reaction event. Here we have labeled each column with the corresponding chemical species.

Suppose that a set of chemical reactions occurs at total molar rates $\mathcal{R}_{1,\text{tot}}, \ldots, \mathcal{R}_{\text{NR,tot}}$ in a given macroscopic system. (The molar rate $\mathcal{R}_{j,\text{tot}}$ is the frequency of the molecular event $j$ in the chosen system minus that of the reverse event, divided by Avogadro's number: $\widetilde{N} = 6.023 \times 10^{23}$). Then the net rate of production of species $i$ in moles per second in the system is

$$R_{i,\text{tot}} = \sum_{j=1}^{\text{NR}} \nu_{ji} \mathcal{R}_{j,\text{tot}} \qquad i = 1, \ldots, \text{NS} \tag{2.1-2}$$

For the three-reaction scheme shown above, the resulting species production rates are

$$
\begin{aligned}
R_{C_2H_4,\text{tot}} &= -1\mathcal{R}_{1,\text{tot}} & +0\mathcal{R}_{2,\text{tot}} & \quad -1\mathcal{R}_{3,\text{tot}} \\
R_{O_2,\text{tot}} &= -1/2\mathcal{R}_{1,\text{tot}} & -5/2\mathcal{R}_{2,\text{tot}} & \quad -3\mathcal{R}_{3,\text{tot}} \\
R_{C_2H_4O,\text{tot}} &= 1\mathcal{R}_{1,\text{tot}} & -1\mathcal{R}_{2,\text{tot}} & \quad -0\mathcal{R}_{3,\text{tot}} \\
R_{H_2O,\text{tot}} &= 0\mathcal{R}_{1,\text{tot}} & +2\mathcal{R}_{2,\text{tot}} & \quad +2\mathcal{R}_{3,\text{tot}} \\
R_{CO_2,\text{tot}} &= 0\mathcal{R}_{1,\text{tot}} & +2\mathcal{R}_{2,\text{tot}} & \quad +2\mathcal{R}_{3,\text{tot}}
\end{aligned}
$$

The equation set (2.1-2) can be written in matrix form as

$$\mathbf{R}_{\text{tot}} = \nu^T \mathcal{R}_{\text{tot}} \tag{2.1-3}$$

with the production and reaction rates expressed as column vectors of lengths NS and NR, respectively. The matrix transposition here could have been avoided by defining $\nu$ differently (with the reactions written columnwise), but new transpositions would then appear in the expressions of Sections 2.3 and 2.4 for energies and entropies of reaction.

The two forms of rates introduced in Eq. (2.1-2) should be carefully distinguished. The net molar rate of production of chemical species $i$ is designated by a subscripted **Roman** letter $R_i$, whereas the net frequency of reaction event $j$ is designated by a subscripted **calligraphic** letter $\mathcal{R}_j$. Some practice in writing R and $\mathcal{R}$ is advised to express each rate type unambiguously.

It is important to distinguish *homogeneous* reactions (those that occur within a phase) from *heterogeneous* reactions (those that occur on interfaces). The presence or absence of heterogeneous reactions in a system is commonly investigated by experiments with various reactor surface/volume ratios or with various amounts or compositions of a dispersed phase. Labeling surface quantities with superscript $s$, one obtains the net rate of *production* of species $i$ [*mole/s*] in a system with interfacial surface $S$ and volume $V$ as

$$R_{i,\text{tot}} = \int_V R_i \, dV + \int_S R_i^s \, dS \qquad (2.1\text{-}4)$$

and the total rate of *reaction* $j$ [*mole/s*] in the system as

$$\mathcal{R}_{j,\text{tot}} = \int_V \mathcal{R}_j \, dV \qquad \text{if reaction } j \text{ is homogeneous} \qquad (2.1\text{-}5)$$

$$\mathcal{R}_{j,\text{tot}} = \int_S \mathcal{R}_j^s \, dS \qquad \text{if reaction } j \text{ is heterogeneous} \qquad (2.1\text{-}6)$$

Typical units for *local* rates of production or reaction are [*mole/cm³s*] for *homogeneous* reactions and [*mole/cm²s*] for *heterogeneous* ones. Various multiples of the SI unit mol (gram-mol) are common, such as kmol (kilogram-mol) and lb-mol (pound-mol), though in view of metrication, use of the latter unit is discouraged.

Local production rates are commonly used in reactor simulations. They can be expressed concisely as

$$R_i = \sum_{j=1}^{\text{NR}} \nu_{ji} \mathcal{R}_j \qquad i = 1, \ldots, \text{NS} \qquad (2.1\text{-}7)$$

if each reaction occurs only on spatial loci of a particular type, such as elements of volume in a particular fluid or solid phase or elements of area on a particular kind of surface. Then the subscript $j$ identifies both the reaction and its locus, and the superscript $s$ for surface reactions is no longer needed. Such models have been used in studies of metallurgical processes (Sohn and Wadsworth 1979), coal gasification reactors (Smoot and Smith 1979), atmospheric reactions (Seinfeld 1980), and heterogeneous polymerization processes (Ray 1983). Generic expressions for local reaction rates $\mathcal{R}_j$ will be considered beginning in Section 2.4.

Rates of reaction $\mathcal{R}_j$ have been defined here as frequencies of reaction events. Their computability from rates $R_i$ of species production is treated in the following section. Time derivatives are not needed here, but will arise in Chapter 3 for transient reactor operations.

## 2.2 COMPUTABILITY OF REACTION RATES FROM DATA

For a single-reaction system, Eq. (2.1-2) gives

$$\mathcal{R}_{1,\text{tot}} = R_{1,\text{tot}}/\nu_{11} = ,\ldots, = R_{\text{NS},\text{tot}}/\nu_{1,\text{NS}} \quad (\text{NR} = 1; \nu_{11} \neq 0) \quad (2.2\text{-}1)$$

The rate $\mathcal{R}_{1,\text{tot}}$ of a *single* reaction can thus be determined from a measurement of $R_{i,\text{tot}}$ for any species produced or consumed by that reaction. The species production rates are then directly proportional:

$$R_{i,\text{tot}} = (\nu_{1i}/\nu_{11})R_{1,\text{tot}} \qquad i = 2,\ldots,\text{NS} \quad (\text{NR} = 1; \nu_{11} \neq 0) \quad (2.2\text{-}2)$$

Corresponding relations hold for the local rates $R_i$ of Eq. (2.1-7) and also for the macroscopic extents of production $\xi_i = \int R_{i,\text{tot}}dt$, obtained from overall material balances on transient reactor operations with one reaction.

For multiple reactions, Eq. (2.1-3) is solvable for the full vector $\mathcal{R}_{\text{tot}}$ if and only if the matrix $\nu$ has full rank NR, i.e., if and only if the rows of $\nu$ are linearly independent. If any species production rates $R_{i,\text{tot}}$ are not measured, the corresponding columns of $\nu$ must be suppressed when testing for solvability of Eq. (2.1-3). Gaussian elimination (see Section A.4) is convenient for doing this test and for finding a full or partial solution for the reaction rates.

Consider the three-reaction scheme of Section 2.1. Suppose that an experiment is available with measured values of all the species production rates $R_{i,\text{tot}}$. Therefore, use all columns of the stoichiometric matrix

$$\nu = \begin{pmatrix} -1 & -1/2 & 1 & 0 & 0 \\ 0 & -5/2 & -1 & 2 & 2 \\ -1 & -3 & 0 & 2 & 2 \end{pmatrix}$$

to test the computability of the three reaction rates.

Dividing the first row of $\nu$ by its leading nonzero element (thus choosing $\nu_{11}$ as the first pivot) and adding this new row to row 3, we obtain the matrix

$$\nu' = \begin{pmatrix} 1 & 1/2 & -1 & 0 & 0 \\ 0 & -5/2 & -1 & 2 & 2 \\ 0 & -5/2 & -1 & 2 & 2 \end{pmatrix}$$

thus completing one elimination cycle. Dividing row 2 of this matrix by its leading nonzero element (thus choosing $\nu'_{22}$ as the second pivot) and adding $5/2$ times this new row to row 3, we obtain the matrix

$$\nu'' = \begin{pmatrix} 1 & 1/2 & -1 & 0 & 0 \\ 0 & 1 & 2/5 & -4/5 & -4/5 \\ 0 & 0 & 0 & 0 & 0 \end{pmatrix}$$

Since the remaining row is zero, the elimination ends here and the stoichiometric matrix $\boldsymbol{\nu}$ has rank 2. Thus, the three reaction rates in this scheme cannot be found from species production rates alone. However, a full solution becomes possible whenever one or two reaction rates are known to be negligible, as in experiments with negligible concentration of either ethylene ($C_2H_4$) or ethylene oxide ($C_2H_4O$).

A multireaction version of Eq. (2.2-2) is obtainable by continuing the elimination above as well as below the pivotal elements; this is the Gauss-Jordan algorithm of Section A.3. The final nonzero rows form a matrix $\boldsymbol{\alpha}$, here given by

$$\boldsymbol{\alpha} = \begin{pmatrix} 1 & 0 & -6/5 & 2/5 & 2/5 \\ 0 & 1 & 2/5 & -4/5 & -4/5 \end{pmatrix}$$

Each row of $\boldsymbol{\alpha}$ can be regarded as a lumped reaction that yields a unit amount of the corresponding pivotal species. Thus $\alpha_{1i}$ is the production of species $i$ per mole produced of the first pivotal species, and $\alpha_{2i}$ is the production of species $i$ per mole produced of the second pivotal species. Adding these contributions, we get

$$(R_1\ R_2\ R_3\ R_4\ R_5)_{\text{tot}} = (R_1\ R_2) \begin{pmatrix} 1 & 0 & -6/5 & 2/5 & 2/5 \\ 0 & 1 & 2.5 & -4/5 & -4/5 \end{pmatrix}$$

The general result for a system of stoichiometric rank NK is

$$R_{i,\text{tot}} = \sum_{K=1}^{NK} \alpha_{K,i} R_{K,\text{tot}} \tag{2.2-3}$$

Here the integers $K = 1, \ldots, NK$ denote the pivotal species, numbered in order of their selection.

## 2.3 EQUILIBRIA OF CHEMICAL REACTIONS

Before considering expressions for reaction rates, we pause to summarize the theoretical results derived by J. Willard Gibbs (1875–1878) for chemical reaction equilibria. His results, published in a monumental two-part treatise, have stood the test of time and are essential for construction of thermodynamically consistent reaction models.

The thermodynamic condition for equilibrium of reaction step $j$ at a point in a reactor system is

$$\sum_{i=1}^{NS} \nu_{ji}\mu_i = 0 \quad \text{(equilibrium of step } j) \tag{2.3-1}$$

At such a state the net rate $\mathcal{R}_j$ vanishes, as the molecular event $j$ and its reverse occur with equal frequency.

The chemical potentials $\mu_i$ are expressible in J/mol or kJ/kmol as

$$\mu_i = \mu_i^\circ + RT \ln a_i \qquad i = 1, \ldots, \text{NS} \qquad (2.3\text{-}2)$$

Here $\mu_i^\circ$ is the value of $\mu_i$ in a chosen standard state for species $i$ at the same temperature $T$, and $a_i$ is the thermodynamic activity (Poling, Prausnitz, and O'Connell, 2000) of species $i$ relative to that standard state. Several practical expressions for $a_i$ (with correspondingly different standard states) are listed here for future reference:

$$a_i = p_i \gamma_{ia} = p x_i \gamma_{ia} \quad \text{(gaseous constituents)} \qquad (2.3\text{-}3\text{a})$$
$$a_i = c_i \gamma_{ib} = c x_i \gamma_{ib} \quad \text{(fluid, solid, or surface constituents)} \qquad (2.3\text{-}3\text{b})$$
$$a_i = x_i \gamma_{ic} \quad \text{(fluid, solid, or surface constituents)} \qquad (2.3\text{-}3\text{c})$$
$$a_i = \theta_i \gamma_{id} \quad \text{(surface constituents)} \qquad (2.3\text{-}3\text{d})$$
$$a_i = 1 \quad \text{(pure phase or major constituent)} \qquad (2.3\text{-}3\text{e})$$

Here $p_i$ is a partial pressure, $c_i$ is a molar concentration (per unit volume or surface, as appropriate), $x_i$ is a mole fraction, and $\theta_i$ is a fractional surface coverage. The notation $[A_i]$ is often used for $c_i$. For convenience in later discussions, we summarize these expressions by

$$a_i = b_i \gamma_i \qquad i = 1, \ldots, \text{NS} \qquad (2.3\text{-}4)$$

in which $b_i$ represents $p_i$, $c_i$, $x_i$, $\theta_i$, or 1, as appropriate and $\gamma_i$ is the corresponding activity coefficient (1 for Eq. (2.3-3e)). In this book, we will usually take $\gamma_i = 1$; this is the ideal-solution approximation.

Substitution of Eq. (2.3-2) into (2.3-1) gives the equilibrium condition

$$\sum_{i=1}^{\text{NS}} \nu_{ji} \ln a_i = -\frac{\Delta \mu_j^\circ}{RT} \quad \text{(equilibrium of step } j) \qquad (2.3\text{-}5)$$

in which $\Delta \mu_j^\circ$ is the standard chemical potential change for reaction $j$:

$$\Delta \mu_j^\circ = \sum_{i=1}^{\text{NS}} \nu_{ji} \mu_i^\circ \qquad (2.3\text{-}6)$$

Taking the antilogarithm of Eq. (2.3-5), we get the equilibrium condition

$$\prod_{i=1}^{\text{NS}} (a_i)^{\nu_{ji}} = K_j \quad \text{(equilibrium of step } j) \qquad (2.3\text{-}7)$$

Here $K_j$ is the "equilibrium constant" for reaction step $j$ at temperature $T$, given by

$$K_j = \exp\left(-\frac{\Delta \mu_j^\circ}{RT}\right) \qquad (2.3\text{-}8)$$

Equations (2.3-5) to (2.3-8) are very important, not only for prediction of equilibrium compositions but also for determination of standard chemical potential changes $\Delta\mu_j^\circ$ and entropy changes $\Delta S_j^\circ$ from reactor data. Furthermore, they determine a locus of states at which the rate of step $j$ vanishes, whether or not the other reaction steps in the system are at equilibrium (Shinnar and Feng 1985).Solutions of Eq. (2.3-7) have nonzero activities $a_i$ for all reactants and products of step $j$, as long as no pure phase appears or disappears (as in the formation, combustion, or dissolution of a pure solid; see Eq. (2.3-3e). Under this condition, the progress of any reaction step can be reversed by suitable adjustment of the activities $a_i$. A step with very large $K_j$ is sometimes called "irreversible," meaning that the reverse step, $-j$, is much slower than the forward step, $j$, over the usual range of concentrations.

The chemical potential $\mu_i$ can be identified with the molar Gibbs energy $\widetilde{G}_i$ at the standard-state temperature and pressure or with the molar Helmholtz energy $\widetilde{A}_i$ $(=\widetilde{G}_i - p\widetilde{V}_i = \widetilde{U}_i - T\widetilde{S}_i)$ at the standard-state temperature and molar volume. The first alternative is usual in thermodynamic tables; it gives the relations

$$K_j = \exp\left(-\frac{\Delta\widetilde{G}_j^\circ}{RT}\right) \tag{2.3-9a}$$

$$= \exp\left(-\frac{\Delta\widetilde{H}_j^\circ}{RT} + \frac{\Delta\widetilde{S}_j^\circ}{R}\right) \tag{2.3-9b}$$

$$\frac{d\ln K_j}{dT} = \frac{\Delta\widetilde{H}_j^\circ}{RT^2} \tag{2.3-10}$$

with standard-state pressures $p_i^\circ$ independent of $T$.

The second alternative gives standard states $(T, c_i^\circ)$. This choice is common in chemical kinetics because of its convenience for analytic modeling of constant-volume systems. With this choice, one obtains

$$K_j = \exp\left(-\frac{\Delta\widetilde{A}_j^\circ}{RT}\right) \tag{2.3-11a}$$

$$= \exp\left(-\frac{\Delta\widetilde{U}_j^\circ}{RT} + \frac{\Delta\widetilde{S}_j^\circ}{R}\right) \tag{2.3-11b}$$

$$\frac{d\ln K_j}{dT} = \frac{\Delta\widetilde{U}_j^\circ}{RT^2} \tag{2.3-12}$$

with a correspondingly different equilibrium constant.

Changing the formulas according to the standard states is awkward, especially for heterogeneous reactions, so we prefer the more general equation (2.3-8). Defining a generalized energy function $\widetilde{\mathcal{E}} = \mu + T\widetilde{S}$ to represent the molar enthalpy $\widetilde{H}$ or internal energy $\widetilde{U}$ as needed, we can write Eq. (2.3-8) as

$$K_j = \exp\left(-\frac{\Delta\widetilde{\mathcal{E}}_j^\circ}{RT} + \frac{\Delta\widetilde{S}_j^\circ}{R}\right) \tag{2.3-13}$$

which gives

$$\frac{d\ln K_j}{dT} = \frac{\Delta\widetilde{\mathcal{E}}_j^\circ}{RT^2} \tag{2.3-14}$$

Equations (2.3-9a) to (2.3-12) are included in these two results. Here each species must have a constant standard-state pressure or concentration, but the standard states can be chosen independently for the various species, as is often done for gas-liquid and gas-solid reactions.

Analytic expressions for $\mu_i^\circ$ and $\widetilde{\mathcal{E}}_i^\circ$ are obtainable by use of a polynomial expansion of the standard-state heat capacity $\widetilde{C}_i^\circ = d\widetilde{\mathcal{E}}_i^\circ/dT$:

$$\widetilde{C}_i^\circ = C_{i0} + TC_{i1} + T^2 C_{i2} + \dots \tag{2.3-15}$$

Here $C_{i0}$, $C_{i1}$, etc. are coefficients chosen to give a good fit of $\widetilde{C}_i^\circ$ over the temperature range of interest. The standard state chosen for species $i$ determines whether the heat capacity at constant pressure or at constant volume is required. The corresponding expansions for the standard-state energy functions around a datum temperature $T_D$ are

$$\begin{aligned}
\widetilde{\mathcal{E}}_i^\circ(T) &= \widetilde{\mathcal{E}}_i^\circ(T_D) + \int_{T_D}^{T} \widetilde{C}_i^\circ(T')dT' \\
&= \widetilde{\mathcal{E}}_i^\circ(T_D) + (T - T_D)C_{i0} + (T^2 - T_D^2)C_{i1}/2 + \dots \\
&= \widetilde{\mathcal{E}}_i^\circ(T_D) + \sum_{k\geq 0}(T^{k+1} - T_D^{k+1})C_{ik}/(k+1)
\end{aligned} \tag{2.3-16}$$

$$\begin{aligned}
\frac{\mu_i^\circ(T)}{T} &= \frac{\mu_i^\circ(T_D)}{T_D} - \int_{T_D}^{T} \frac{\widetilde{\mathcal{E}}_i^\circ(T')}{T'^2}dT' \\
&= \frac{\mu_i^\circ(T_D)}{T_D} - \int_{T_D}^{T}\left\{\widetilde{\mathcal{E}}_i^\circ(T_D) + \sum_{k\geq 0}\left(T'^{k+1} - T_D^{k+1}\right)C_{ik}/(k+1)\right\}\frac{dT'}{T'^2} \\
&= \frac{\mu_i^\circ(T_D)}{T_D} + \widetilde{\mathcal{E}}_i^\circ(T_D)\left(\frac{1}{T} - \frac{1}{T_D}\right) + \sum_{k\geq 0}\frac{C_{ik}T_D^{k+1}}{k+1}\left(\frac{1}{T_D} - \frac{1}{T}\right) \\
&\quad - C_{i0}\ln(T/T_D) + \sum_{k\geq 1}\frac{C_{ik}}{k(k+1)}\left(T_D^k - T^k\right)
\end{aligned} \tag{2.3-17}$$

Corresponding expansions hold for the functions $\Delta\widetilde{\mathcal{E}}_j^\circ(T)$ and $\ln K_j(T) = -\Delta\mu_j^\circ(T)/RT$ in reaction models.

The accuracy of calculated equilibrium states depends critically on the data sources used. Accurate predictions of heat capacities are often available from statistical mechanics and spectroscopic data, but predictions of the constants $\Delta\widetilde{\mathcal{E}}_j^{\circ}(T_D)$ and $\ln K_j(T_D)$ are more difficult, especially when obtained by difference from measured heats of combustion. Much better values of these constants are available nowadays by parameter estimation from reactor data.

Phase equilibria of vaporization, sublimation, melting, extraction, adsorption, etc. can also be represented by the methods of this section within the accuracy of the expressions for the chemical potentials. One simply treats the phase transition as if it were an equilibrium reaction step and enlarges the list of species so that each member has a designated phase. Thus, if $A_1$ and $A_2$ denote liquid and gaseous species $i$, respectively, the vaporization of $A_1$ can be represented stoichiometrically as $-A_1 + A_2 = 0$; then Eq. (2.3-17) provides a vapor pressure equation for species $i$. The same can be done for fusion and sublimation equilibria and for solubilities in ideal solutions.

## 2.4 KINETICS OF ELEMENTARY STEPS

Elementary steps are molecular events in which "reactants are transformed into products directly, i.e., without passing through an intermediate that is susceptible of isolation" (Boudart 1968). We can represent such an event symbolically as

$$\sum_{i=1}^{NS} \epsilon_{ji}A_i \rightleftharpoons \sum_{i=1}^{NS} \epsilon_{-ji}A_i \qquad (2.4\text{-}1)$$

with nonnegative integer coefficients $\epsilon_{ji}$ and $\epsilon_{-ji}$. The total numbers of molecules on the left and on the right are known as the *molecularities* of the forward and reverse reactions, respectively, and are integers not normally greater than 3. The coefficients $\epsilon_{ji}$ and $\epsilon_{-ji}$ must satisfy the conservation relations for the event, just as the stoichiometric coefficients $\nu_{ji}$ must in Eq. (2.1-1).

Suppose, for the moment, that a postulated reaction scheme consists entirely of elementary steps with known coefficients $\epsilon_{ji}$ and $\epsilon_{-ji}$. Then it is natural to apply Eq. (2.1-1) to those steps, and set

$$\nu_{ji} = \epsilon_{-ji} - \epsilon_{ji} \qquad i = 1, \ldots, NS; \quad j = 1, \ldots, NR \qquad (2.4\text{-}2)$$

as the stoichiometric coefficients. We will follow this procedure whenever a reaction step is postulated to be elementary.

Probability considerations suggest that the frequency of the forward event $j$ in Eq. (2.4-1) at a given temperature should be proportional to each of the concentrations $b_i$ raised to the power $\epsilon_{ji}$, and a similar relation with exponents $\epsilon_{-ji}$ should hold for the reverse event. (This method of reasoning is restricted to ideal solutions.) The resulting net rate expression,

$$\mathcal{R}_j = k_j(T)\prod_{i=1}^{NS} b_i^{\epsilon_{ji}} - k_{-j}(T)\prod_{i=1}^{NS} b_i^{\epsilon_{-ji}} \qquad (2.4\text{-}3)$$

is the standard form for reactions in ideal solutions. The functions $k_j(T)$ and $k_{-j}(T)$ are the rate coefficients of the forward and reverse reactions. The generalized notation $b_i$ for all forms of concentrations was introduced in Eq. (2.3-4).

At thermodynamic equilibrium, $\mathcal{R}_j$ must vanish for every reaction; otherwise, the equilibrium could be shifted by adding catalysts or inhibitors to alter the nonzero rates $\mathcal{R}_j$. Formal proofs of this "detailed balance" principle are presented by de Groot and Mazur (1962). Therefore, setting $\mathcal{R}_j = 0$ at equilibrium and using Eq. (2.4-3), we get

$$\prod_{i=1}^{\text{NS}} \left( b_i^{\nu_{ji}} \right)_{\text{equil.}} = \frac{k_j(T)}{k_{-j}(T)} \qquad (2.4\text{-}4)$$

which is the ideal-solution approximation to Eq. (2.3-7). Hence, Eq. (2.4-3) can be written in the form

$$\mathcal{R}_j = k_j(T) \left[ \prod_{i=1}^{\text{NS}} b_i^{\epsilon_{ji}} - \frac{1}{K_j(T)} \prod_{i=1}^{\text{NS}} b_i^{\epsilon_{ji}+\nu_{ji}} \right] \qquad (2.4\text{-}5)$$

to satisfy the detailed-balance condition directly. Arrhenius (1889) explained the temperature dependence of reaction rates in terms of the temperature-dependent distribution of molecular velocities and the assumption that only sufficiently energetic collisions produce reaction. The result, commonly written as

$$k_j(T) = A_j \exp\left( -E_j/RT \right) \qquad (2.4\text{-}6)$$

remains a popular expression for rate coefficients. The coefficient $A_j$ is called the *frequency factor* because of its connection with the collision frequency used in early molecular theories; the quantity $E_j$ is the Arrhenius activation energy.

Equation (2.4-6) is rather awkward for representation of data, because $A_j$ is an extrapolated value of $k_j$ at infinite temperature. Arrhenius (1889) actually used the form

$$k_j(T) = k_j(T_j) \exp[(E_j/R)(1/T_j - 1/T)] \qquad (2.4\text{-}7)$$

which was also preferred by Box (1960) on statistical grounds. Here $T_j$ is a base temperature for reaction $j$, typically a temperature near the middle of the range in which $k_j$ was investigated. This simple reparameterization of Eq. (2.4-6) is very advantageous for estimation of $k_j(T)$ from data and will be used in Chapter 6.

A power of $T$ may appear in the expression for $k_j(T)$ when alternative standard states are introduced, or when a more detailed theoretical expression for $k_j(T)$ is used. For these situations, the expression

$$k_j(T) = k_j(T_j)(T/T_j)^{m_j} \exp[(E_j/R)(1/T_j - 1/T)] \qquad (2.4\text{-}8)$$

is more realistic. Values of $E_j$ appropriate to this equation differ somewhat from those for Eq. (2.4-7), except when $m_j = 0$.

More detailed expressions are available from absolute reaction rate theory (Pelzer and Wigner 1932; Evans and Polanyi 1935; Eyring 1935). This approach treats the forward and reverse reaction processes as crossings of molecular systems over a "mountain pass" on their potential energy surface. Systems at the pass are called *activated complexes,* denoted by $X^{\ddagger}_{jf}$ or $X^{\ddagger}_{jr}$ according to the direction from which they come. Equilibrium is assumed to hold among all configurations approaching the pass from a given side. The forward and reverse processes of Eq. (2.4-1) then take the forms

$$\sum_{i=1}^{NS} \epsilon_{ji} A_i \overset{\text{equil.}}{\rightleftharpoons} X^{\ddagger}_{jf} \longrightarrow \text{Products} \qquad (2.4\text{-}9a)$$

$$\text{Reactants} \longleftarrow X^{\ddagger}_{jr} \overset{\text{equil.}}{\rightleftharpoons} \sum_{i=1}^{NS} \epsilon_{-ji} A_i \qquad (2.4\text{-}9b)$$

and the concentrations of the activated complexes are given by

$$\left[X^{\ddagger}_{jf}\right] = \frac{1}{\gamma^{\ddagger}_j} \exp\left(-\frac{\Delta\mu^{\ddagger}_j}{RT}\right) \prod_{i=1}^{NS} (a_i)^{\epsilon_{ji}} \qquad (2.4\text{-}10a)$$

$$\left[X^{\ddagger}_{jr}\right] = \frac{1}{\gamma^{\ddagger}_j} \exp\left(-\frac{\Delta\mu^{\ddagger}_j - \Delta\mu^{\circ}_j}{RT}\right) \prod_{i=1}^{NS} (a_i)^{\epsilon_{ji} + \nu_{ji}} \qquad (2.4\text{-}10b)$$

Here Eqs. (2.3-3b), (2.3-7) and (2.3-8) have been used, with the activated complexes treated formally as additional species beyond the number NS.

Statistical mechanics (Horiuti 1938; Eyring, Walter, and Kimball 1944; Mahan 1974) gives the rates of passage over the potential-energy barrier as

$$\mathcal{R}_{jf} = \frac{\kappa T}{h} \left[X^{\ddagger}_{jf}\right] \qquad (2.4\text{-}11a)$$

$$\mathcal{R}_{jr} = \frac{\kappa T}{h} \left[X^{\ddagger}_{jr}\right] \qquad (2.4\text{-}11b)$$

on the assumption that no reflections back across the barrier occur. Here $\kappa$ $(= R/\tilde{N})$ is the Boltzmann constant and $h$ is Planck's constant. The resulting net rate of elementary reaction $j$, after use of Eq. (2.3-8), is

$$\mathcal{R}_j = \frac{\kappa T}{h\gamma^{\ddagger}_j} \exp\left(-\frac{\Delta\mu^{\ddagger}_j}{RT}\right) \left[\prod_{i=1}^{NS} (a_i)^{\epsilon_{ji}} - \frac{1}{K_j(T)} \prod_{i=1}^{NS} (a_i)^{\epsilon_{ji} + \nu_{ji}}\right] \qquad (2.4\text{-}12)$$

in agreement with Wynne-Jones and Eyring (1935). Here the detailed-balance condition $\mathcal{R}_j = 0$ at equilibrium is automatically fulfilled. The original version of the theory (Eyring 1935) gives predictions in terms of gas-phase partition functions; the version given here is applicable to all states of matter but leaves the transition-state values as thermodynamic parameters. The *transmission coefficient*, used in detailed theories to correct for tunneling through the barrier and reflections across it, is absorbed here into the adjustable free energy of activation, $\Delta\mu_j^{\ddagger}$.

The exponents on the activities in Eq. (2.4-12), and on the concentrations in Eq. (2.4-5), are known as the *orders* of the forward and reverse reaction $j$ with respect to the participating species. From these experimentally determinable exponents and Eq. (2.4-9a) or (2.4-2 and 2.4-9b), the elemental makeup of the activated complex can be calculated if the reaction is indeed an elementary one.

Use of Eq. (2.4-12) to evaluate the forward rate term in Eq. (2.4-5) gives

$$
\begin{aligned}
k_j &= \frac{\kappa T}{h} \exp\left(-\frac{\Delta\mu_j^{\ddagger}}{RT}\right) \left[\frac{1}{\gamma_j^{\ddagger}} \prod_{i=1}^{\mathrm{NS}} (\gamma_i)^{\epsilon_{ji}}\right] \\
&= \frac{\kappa T}{h} \exp\left(-\frac{\Delta\varepsilon_j^{\ddagger}}{RT} + \frac{\Delta S_j^{\ddagger}}{R}\right) \left[\frac{1}{\gamma_j^{\ddagger}} \prod_{i=1}^{\mathrm{NS}} (\gamma_i)^{\epsilon_{ji}}\right]
\end{aligned}
\qquad (2.4\text{-}13)
$$

in which the bracketed term is the correction for nonideal solution behavior, first proposed by Brønsted (1922). This nonideality correction has been closely confirmed for dilute solutions of electrolytes (Davies 1961) and also for moderately dense gas systems (Eckert and Boudart 1963); it is one of the reasons why $k_j$ may depend on composition. This theory is also able to describe the effects of total pressure (Eckert and Grieger 1970; Waissbluth and Grieger 1973) and electrical potential (Delahay 1965) on reaction rates.

Transition-state theory is an important conceptual aid in the study of reaction mechanisms. It has proved useful in selecting dominant reaction paths in many multireaction systems. Modern computational chemistry methods, employing various approximations of quantum mechanics, are beginning to provide accurate thermochemistry, along with activation energies and frequency factors, thus allowing semiquantitative rate predictions from first principles. A review of recent investigations along these lines at UW-Madison is given by Greeley, Nørskov, and Mavrikakis (2002). For reactor modeling we will use Eqs. (2.4-5) and (2.4-8), recognizing that these are ideal-solution approximations.

The principle of conservation of orbital symmetry, enunciated by Woodward and Hoffman (1970), has proved its worth as a guide to preferred routes for many organic syntheses. This principle is a promising one for modeling of reaction kinetics.

## 2.5 PROPERTIES OF REACTION NETWORKS

In this section we discuss some simple properties of reaction networks and give a few examples to show the consequences of various postulated network features. Section 2.6 and Chapters 6 and 7 will introduce the more difficult problem of reasoning from observations to realistic reaction models.

A thermodynamically consistent reaction model will have equilibrium states $(\boldsymbol{b}^*, T^*)$ at which the reaction rates $\mathcal{R}_j$ all vanish and the generalized concentrations $b_i$ are positive in each phase. Then a Taylor expansion of $\mathcal{R}_j$ gives

$$
\begin{aligned}
\mathcal{R}_j = 0 + \sum_{m=1}^{\mathrm{NS}} \left( \left. \frac{\partial \mathcal{R}_j}{\partial b_m} \right|_{\boldsymbol{b}^*, T^*} \right) (b_m - b_m^*) \\
+ \left( \left. \frac{\partial \mathcal{R}_j}{\partial T} \right|_{\boldsymbol{b}^*, T^*} \right) (T - T^*) + \dots \qquad j = 1, \dots, \mathrm{NR}
\end{aligned}
\tag{2.5-1}
$$

so that the kinetic scheme becomes linear at an equilibrium state. This linearity forms the basis of various techniques for measuring fast kinetics by perturbing an equilibrium system and observing its relaxation toward equilibrium (Eigen and Kustin 1963).

Multireaction systems often have some "quasi-equilibrium steps" whose forward and reverse rates greatly exceed the net rate $\mathcal{R}_j$ at all conditions of interest. For such a reaction, the approximation

$$
(\mathcal{R}_j / k_j) \approx 0 \qquad \text{(quasi-equilibrium step)}
\tag{2.5-2a}
$$

is permissible, and reduces Eq. (2.4-5) to the form

$$
\prod_{i=1}^{\mathrm{NS}} b_i^{\nu_{ji}} \approx K_j(T) \qquad \text{(quasi-equilibrium step)}
\tag{2.5-2b}
$$

Thus, the rate expression is replaced by an algebraic constraint on the concentrations. This treatment is customary for physical changes of state such as vaporization and is often applied to fast chemical steps as well. Note that the true equilibrium condition of vanishing $\mathcal{R}_j$ is replaced here by the less stringent condition of Eq. (2.5-2a), which permits step $j$ to proceed at a nonzero rate in response to concentration changes caused by other reactions.

Highly reactive intermediates, such as atoms and free radicals, occur at small concentrations in many reaction systems. Their net rates of production, $R_i$, are normally small relative to their total rates of formation and consumption, except for small regions near the start and end of the reaction interval of space or time. This behavior prompted the assumption

$$
R_i \approx 0 \qquad \text{(quasi-steady-state approximation)}
\tag{2.5-3}
$$

introduced by Bodenstein (1913), Christiansen (1919), Herzfeld (1919), and Polanyi (1920). This approximation has become a standard tool for analyzing reaction schemes. Several comparisons with more detailed approaches are available (Bowen, Acrivos, and Oppenheim 1963; Tsuchida and Mimashi 1965; Blakemore and Corcoran 1969; Farrow and Edelson 1974; Georgakis and Aris 1975).

A similar-looking condition, $R_i = 0$, holds for any immobile species in a true steady-state operation. For example, the surface species on a fixed bed of catalyst are immobile in those reactor models that neglect surface diffusion, and a mass balance then requires that $R_i$ vanish at steady state. When desired, one can include both $R_i = 0$ and $R_i \approx 0$ as possibilities by invoking a *stationarity condition* as in Example 2.4.

Another common property of multireaction networks is *stiffness*, that is, the presence of kinetic steps with widely different rate coefficients. This property was pointed out by Curtiss and Hirschfelder (1952), and has had a major impact on the development of numerical solvers such as DASSL (Petzold 1983) and DDAPLUS of Appendix B. Since stiff equations take added computational effort, there is some incentive to reduce the stiffness of a model at the formulation stage; this can be done by substituting Eq. (2.5-2b) or (2.5-3) for some of the reaction or production rate expressions. This strategy replaces some differential equations in the reactor model by algebraic ones to expedite numerical computations.

*Chain reactions* are recursive reaction cycles that regenerate their intermediates. Such cycles occur in combustion, atmospheric chemistry, pyrolysis, photolysis, polymerization, nuclear fusion and fission, and catalysis. Typical steps in these systems include initiation, propagation, and termination, often accompanied by chain branching and various side reactions. Examples 2.2 to 2.5 describe simple chain reaction schemes.

### Example 2.1. Unimolecular Decomposition: Lindemann Scheme

Observations of first-order kinetics for various gas decompositions, $A \to P$, seemed strange to early kineticists because the mechanism of activation of the reactant was unclear. If such a reaction were controlled by collisional activation it should be bimolecular, since the pair-collision frequency in a unit volume of pure A is proportional to $[A]^2$. Careful measurements for such reactions indeed showed a decrease in the apparent rate coefficient $\mathcal{R}/[A]$ with decreasing pressure, suggesting that collisional activation was beginning to limit the reaction rate.

Lindemann (1922) explained this behavior with a simple two-step scheme:

$$A + A \overset{1}{\rightleftharpoons} A^* + A \qquad (2.5\text{-}4)$$

$$A^* \overset{2}{\to} P \qquad (2.5\text{-}5)$$

Step (1) describes the formation and quenching of activated molecules by bimolecular gas-phase collisions. Step (2) is a postulated unimolecular decomposition of activated molecules to the ultimate reaction products. Application of Eqs. (2.1-7) and (2.5-3) to the activated entity $A^*$ gives

$$R_{A^*} = k_1[A]^2 - k_{-1}[A^*][A] - k_2[A^*] \approx 0 \qquad (2.5\text{-}6)$$

Hence,

$$[A^*] \approx k_1[A]^2/(k_{-1}[A] + k_2) \qquad (2.5\text{-}7)$$

and the rate of the final decomposition is

$$\mathcal{R}_2 \approx \frac{k_2 k_1 [A]^2}{k_{-1}[A] + k_2} \qquad (2.5\text{-}8)$$

The low- and high-pressure forms of this expression are

$$\mathcal{R}_2 \sim k_1[A]^2 \qquad \text{for } [A] << k_2/k_{-1} \qquad (2.5\text{-}9)$$

and

$$\mathcal{R}_2 \sim (k_2 k_1/k_{-1})[A] \qquad \text{for } [A] >> k_2/k_{-1} \qquad (2.5\text{-}10)$$

Thus, at low pressures, the reaction becomes second-order and is controlled by collisional activation, whereas at high pressures the reaction becomes first-order and is controlled by the vibrational dynamics of the energized species $A^*$.

### Example 2.2. Thermal Reaction of Hydrogen with Bromine

The reaction of hydrogen with bromine, with overall stoichiometry

$$H_2 + Br_2 \rightarrow 2HBr \qquad (2.5\text{-}11)$$

was investigated by Bodenstein and Lind (1907). From their experiments, they constructed the empirical rate expression

$$R_{HBr} = \frac{k[H_2][Br_2]^{1/2}}{1 + [HBr]/m[Br_2]} \qquad (2.5\text{-}12)$$

for the reaction in the absence of light. The term $[HBr]/m[Br_2]$ in the denominator indicates that the product HBr inhibits the reaction, despite the large equilibrium constant of the overall process.

A set of elementary steps consistent with these facts was proposed independently by Christiansen (1919), Herzfeld (1919), and Polanyi (1920). It is a chain reaction scheme with the following steps:

$$Br_2 + M \overset{1}{\rightleftharpoons} 2Br + M \qquad (2.5\text{-}13)$$

$$\text{Br} + \text{H}_2 \overset{2}{\rightleftharpoons} \text{HBr} + \text{H} \qquad (2.5\text{-}14)$$

$$\text{H} + \text{Br}_2 \overset{3}{\rightarrow} \text{HBr} + \text{Br} \qquad (2.5\text{-}15)$$

Initiation occurs by the forward reaction 1 and termination by the reverse reaction. A collider molecule M was included in reaction 1 after the work of Lindemann (1922) to provide energy for the dissociation of $\text{Br}_2$. Reactions 2 and 3 form the propagation sequence.

The net rates of production of the atomic species are

$$\text{R}_{\text{Br}} = 2\mathcal{R}_1 - \mathcal{R}_2 + \mathcal{R}_3 \qquad (2.5\text{-}16)$$
$$\text{R}_{\text{H}} = \mathcal{R}_2 - \mathcal{R}_3. \qquad (2.5\text{-}17)$$

Quasi-steady-state approximations for these species give

$$\mathcal{R}_1 \approx 0 \qquad (2.5\text{-}18)$$
$$\mathcal{R}_2 \approx \mathcal{R}_3 \qquad (2.5\text{-}19)$$

The first of these conditions gives a quasi-equilibrium,

$$[\text{Br}] = (K_1[\text{Br}_2])^{1/2} \qquad (2.5\text{-}20)$$

thus removing $k_1$ and $[\text{M}]$ from the analysis. Eq. (2.5-19) gives

$$k_2[\text{Br}][\text{H}_2] - k_{-2}[\text{HBr}][\text{H}] = k_3[\text{H}][\text{Br}_2] \qquad (2.5\text{-}21)$$

which yields the hydrogen atom concentration

$$[\text{H}] = \frac{k_2[\text{H}_2][\text{Br}]}{k_3[\text{Br}_2] + k_{-2}[\text{HBr}]} = \frac{k_2[\text{H}_2](K_1[\text{Br}_2])^{1/2}}{k_3[\text{Br}_2] + k_{-2}[\text{HBr}]} \qquad (2.5\text{-}22)$$

Therefore, the rate of production of HBr according to this reaction scheme is

$$\begin{aligned}
\text{R}_{\text{HBr}} &= \mathcal{R}_2 + \mathcal{R}_3 = 2\mathcal{R}_3 \\
&= 2k_3[\text{H}][\text{Br}_2] \\
&= 2k_3 \frac{k_2[\text{H}_2](K_1[\text{Br}_2])^{1/2}}{k_3[\text{Br}_2] + k_{-2}[\text{HBr}]}[\text{Br}_2] \\
&= 2k_2 K_1^{1/2} \frac{[\text{H}_2][\text{Br}_2]^{1/2}}{1 + k_{-2}[\text{HBr}]/k_3[\text{Br}_2]} \qquad (2.5\text{-}23)
\end{aligned}$$

within the accuracy of the quasi-steady-state approximations. This result agrees with the experimentally determined rate expression, Eq. (2.5-12). The inhibition term noted there turns out to be the ratio of the rates of the two H-consuming steps.

The early stages of the reaction require more detailed treatment, since Eqs. (2.5-18 and 2.5-19) do not hold during the initial buildup of H and Br. Measurements during this induction period might be used to determine $k_1$ in the presence of various colliders M. The method used by Bodenstein and Lutkemeyer (1924) was to determine the coefficient $k_{-1}$ from measurements of $R_{HBr}$ at high concentrations of Br produced by photochemical dissociation of $Br_2$; then $k_1$ was calculated as $k_{-1}K_1$.

The quasi-steady-state method comes from Bodenstein (1913), Christiansen (1919), Herzfeld (1919) and Polanyi (1920). Fuller accounts of research on the hydrogen-bromine reaction are given by Benson (1960) and Campbell and Fristrom (1958).

## Example 2.3. Homogeneous Catalysis by an Enzyme

Consider the reaction scheme of Michaelis and Menten (1913) for enzyme-catalyzed conversion of a substrate (reactant) S to a product P:

$$E + S \overset{1}{\rightleftharpoons} ES \tag{2.5-24}$$

$$ES \overset{2}{\to} E + P \tag{2.5-25}$$

Here E is the enzyme and ES is an intermediate complex. A mass balance on E gives

$$[E] + [ES] = [E]_0 \tag{2.5-26}$$

where $[E]_0$ is the initial enzyme concentration.

A quasi-steady-state assumption for [ES] (and thus for [E]) gives

$$k_1[E][S] - k_{-1}[ES] - k_2[ES] \approx 0 \tag{2.5-27}$$

or

$$k_1([E]_0 - [ES])[S] = (k_{-1} + k_2)[ES] \tag{2.5-28}$$

Hence, the concentration of the intermediate complex is

$$[ES] = \frac{k_1[E]_0[S]}{k_{-1} + k_2 + k_1[S]} \tag{2.5-29}$$

and the rate of the overall reaction $S \to P$ is

$$R_P = k_2[ES]$$
$$= \frac{k_1 k_2[E]_0[S]}{k_{-1} + k_2 + k_1[S]}$$
$$= \frac{k_2[E]_0[S]}{K + [S]} \tag{2.5-30}$$

as shown by Briggs and Haldane (1925). Here $K = (k_{-1} + k_2)/k_1$.

The rate $R_P$ is linear in the total enzyme concentration $[E]_0$ but non-linear in the substrate concentration. $R_P$ increases proportionally with $[S]$ when $[S]/K$ is small but approaches an upper limit $k_2[E]_0$ when $[S]/K$ is large. From Eq. (2.5-29) we see that the concentration of the intermediate approaches $[E]_0$ in the latter case. The constant $K$ is the value of $[S]$ at which $R_P$ attains one-half of its upper limit.

Michaelis and Menten (1913) treated the special case where $k_2 \ll k_{-1}$. Under this assumption, $K$ reduces to $1/K_1$ and the first reaction is effectively at equilibrium. Their overall rate expression corresponds to the final form in Eq. (2.5-30), though their constant $K$ has a different meaning. Thus, it is not possible to test the accuracy of the Michaelis-Menten equilibrium assumption by reaction rate experiments in the quasi-steady-state region. Rather, one would need additional measurements very early in the reaction to allow calculation of the rate coefficients $k_1$, $k_{-1}$, and $k_2$.

## Example 2.4. Heterogeneous Catalysis

When dealing with heterogeneous processes, it is important to recognize the physical steps as well as the chemical ones. Therefore, we begin by listing the steps for reactions in porous catalytic particles as presented by Hougen and Watson (1943, 1947). The following discussion is quoted from their book.

"In order that a reactant in the main fluid phase may be converted catalytically to a product in the main fluid phase, it is necessary that the reactant be transferred from its position in the fluid to the catalytic interface, be activatedly adsorbed on the surface, and undergo reaction to form the adsorbed product. The product must then be desorbed and transferred from the interface to a position in the fluid phase. The rate at which each of these steps occurs influences the distribution of concentrations in the system and plays a part in determining the over-all rate. Because of the differences in the mechanisms involved, it is convenient to classify these steps as follows when dealing with catalysts in the form of porous particles:

1. The mass transfer of reactants and products to and from the gross exterior surface of the catalyst particle and the main body of the fluid.
2. The diffusional and flow transfer of reactants and products in and out of the pore structure of the catalyst particle when reaction takes place at interior interfaces.
3. The activated adsorption of reactants and the activated desorption of products at the catalytic interface.
4. The surface reaction of adsorbed reactants to form chemically adsorbed products.

"In order to calculate the rate at which a catalytic reaction proceeds it is necessary to develop quantitative expressions for the rates of each of the individual steps which contribute to the mechanism. This requires consideration of the fundamental principles of activated adsorption, surface reactions, mass and heat transfer, and diffusion in porous solids."

Steps 1 and 2 of the above scheme are physical steps, to be treated in later chapters. Steps 3 and 4 are chemical and will be treated here. To do so, we consider the fluid state to be given at a point alongside the catalytic interface, where $\mathcal{R}$ is to be calculated. With this specification, a direct analog of the previous example is the conversion of a gaseous reactant A to a gaseous product P on the catalyst surface according to the following reaction scheme:

$$A + l \overset{1}{\rightleftharpoons} Al \qquad (2.5\text{-}31)$$

$$Al \overset{2}{\rightarrow} P + l \qquad (2.5\text{-}32)$$

Here $Al$ is a surface intermediate formed by adsorption of A onto a vacant surface site $l$. The total concentration of active sites on the catalytic surface, in moles per unit amount of catalyst, is

$$[l] + [Al] = [l]_0 \qquad (2.5\text{-}33)$$

Division of this equation by $[l]_0$ gives

$$\theta_l + \theta_{Al} = 1 \qquad (2.5\text{-}34)$$

in which $\theta_l$ and $\theta_{Al}$ are the fractions of vacant sites and sites occupied by A, respectively. A stationarity condition for $Al$ (and thus for $l$) gives

$$k_1 p_A \theta_l - k_{-1}\theta_{Al} - k_2\theta_{Al} = 0 \qquad (2.5\text{-}35)$$

or

$$k_1 p_A \left(1 - \theta_{Al}\right) = \left(k_{-1} + k_2\right)\theta_{Al} \qquad (2.5\text{-}36)$$

Hence, the fractional coverage of the catalytic surface by the intermediate is

$$\theta_{Al} = \frac{k_1 p_A}{k_{-1} + k_2 + k_1 p_A} \qquad (2.5\text{-}37)$$

and the rate of the overall reaction A $\rightarrow$ P is

$$\begin{aligned} R_P &= k_2\theta_{Al} \\ &= \frac{k_2 k_1 p_A}{k_{-1} + k_2 + k_1 p_A} \\ &= \frac{k_2 p_A}{\left(\frac{k_{-1}+k_2}{k_1}\right) + p_A} \qquad (2.5\text{-}38) \end{aligned}$$

These results are analogous to those of the previous example, except that here we have included the possibility of a true steady state in the catalytic particle. That possibility would also exist in the previous example if the enzyme were immobilized on a stationary porous particle in a flow reactor. The physical transport steps 1 and 2 noted above would then be necessary as for other supported catalysts.

Earlier investigations of heterogeneous catalysis were performed by Langmuir (1916, 1922) and by Hinshelwood (1926, 1940), without investigating (as Hougen and Watson did) steps 1 and 2 of the foregoing list. It has become customary to denote models of catalysis on a Langmuir surface as *Langmuir-Hinshelwood-Hougen-Watson models*, in recognition of the pioneering contributions of these four investigators.

Hougen and Watson (1947) derived general surface rate equations for catalysis on a Langmuir surface. However, they considered their general equations, implicit in $\mathcal{R}$, "so cumbersome as to be of little value." To get simpler equations, tractable on the mechanical calculators of that era, they assumed a single rate-controlling step in each of their subsequent investigations. Bradshaw and Davidson (1970) challenged this assumption and showed that the data of Franckaerts and Froment (1964) on ethanol dehydration were better fitted by treating adsorption, desorption, and surface reaction as mutually rate-controlling steps in the general surface rate formulation given but set aside by Hougen and Watson (1947, p. 918).

### Example 2.5. Bifunctional Catalysis

Sinfelt, Hurwitz, and Rohrer (1960) proposed essentially the following reaction scheme for the isomerization of $n$-pentane ($nC_5$) to $i$-pentane ($iC_5$) on a Pt/alumina catalyst:

$$nC_5(g) \overset{K_1}{\rightleftharpoons} nC_5^=(g) + H_2 \quad \text{(on Pt)} \tag{2.5-39}$$

$$l + nC_5^=(g) \overset{K_2}{\rightleftharpoons} nC_5^= l \qquad \text{(on alumina)} \tag{2.5-40}$$

$$nC_5^= l \overset{k_3}{\rightleftharpoons} iC_5^= l \qquad \text{(on alumina)} \tag{2.5-41}$$

$$iC_5^= l \overset{K_4}{\rightleftharpoons} iC_5^=(g) + l \qquad \text{(on alumina)} \tag{2.5-42}$$

$$H_2 + iC_5^=(g) \overset{K_5}{\rightleftharpoons} iC_5(g) \qquad \text{(on Pt)} \tag{2.5-43}$$

The $n$-pentane is dehydrogenated on the platinum to $n$-pentene, which migrates to the acidic sites $l$ on the alumina, where it is adsorbed and isomerizes to $i$-pentene. Then the $i$-pentene is desorbed and migrates to the platinum, where it is hydrogenated to $i$-pentane. Step 3 is treated as elementary and rate-controlling; the other four steps are treated as quasi-equilibrium reactions.

A quasi-stationary treatment of the intermediates gives equality of the rates of all five steps and of the overall reaction

$$nC_5 \rightleftharpoons iC_5 \tag{2.5-44}$$

The rate of step 3 is

$$\mathcal{R}_3 = k_3 \theta_{nC_5^= l} - k_{-3} \theta_{iC_5^= l} \tag{2.5-45}$$

in which the $\theta$'s are fractional coverages of the acidic sites:

$$\theta_{nC_5^= l} = K_2 \theta_l \, p_{nC_5^=} \quad \text{and} \quad \theta_{iC_5^= l} = (1/K_4) \theta_l \, p_{iC_5^=} \tag{2.5-46}$$

The sum of the coverages of the acidic sites is then (with $K_{-4} = 1/K_4$)

$$\begin{aligned}
1 &= \theta_l + \theta_{nC_5^= l} + \theta_{iC_5^= l} \\
&= \theta_l \left[ 1 + K_2 \, p_{nC_5^=} + K_{-4} \, p_{iC_5^=} \right]
\end{aligned} \tag{2.5-47}$$

From the three previous equations, the rate expression in terms of the olefin partial pressures is found to be

$$\mathcal{R}_3 = \frac{k_3 K_2 \, p_{nC_5^=} - k_{-3} K_{-4} \, p_{iC_5^=}}{1 + K_2 \, p_{nC_5^=} + K_{-4} \, p_{iC_5^=}} \tag{2.5-48}$$

with competitive adsorption as expressed by Eq. (2.5-47). Insertion of the quasi-equilibrium expressions for steps 1 and 5 finally gives

$$\begin{aligned}
\mathcal{R}_3 &= \frac{k_3 K_2 K_1 (p_{nC_5}/p_{H_2}) - k_{-3} K_{-4} K_{-5} (p_{iC_5}/p_{H_2})}{1 + K_2 K_1 (p_{nC_5}/p_{H_2}) + K_{-4} K_{-5} (p_{iC_5}/p_{H_2})} \\
&= \frac{k_3 K_2 K_1 (p_{nC_5} - p_{iC_5}/K)}{p_{H_2} + K_2 K_1 \, p_{nC_5} + K_{-4} K_{-5} \, p_{iC_5}}
\end{aligned} \tag{2.5-49}$$

in which $K = K_1 K_2 K_3 K_4 K_5$ is the equilibrium constant for the overall reaction in Eq. (2.5-44).

This example illustrates the use of quasi-equilibria and stationarity conditions to reduce a reaction scheme to a shorter form with fewer independent parameters. The original ten parameters of the five-reaction scheme are combined into four independent ones ($k_3, K, K_1 K_2$, and $K_{-4} K_{-5}$) in the final rate expression.

## Example 2.6. Composite Events; Stoichiometric Numbers

When a multistep sequence has a single controlling step, quasi-equilibria will prevail (in the approximate sense of Eqs. 2.5-2a,b) in the other steps. If the net production rates $R_i$ of the intermediate species are zero or negligible, one can then replace the sequence by a composite event to which the methods of Section 2.4 can be applied. Composite events are often used in applied kinetics, where a more detailed description may be unavailable or unduly complicated.

An early example of this concept is the reaction

$$2NO + Br_2 \rightarrow 2NOBr \tag{2.5-50}$$

which exhibits third-order kinetics as shown by Trautz and Dalal (1918):

$$\mathcal{R} = {}^1\!/_2 R_{NOBr} = k[NO]^2[Br_2] \qquad (2.5\text{-}51)$$

The rate was judged to be too large for a termolecular collision process; furthermore, the activation energy $-R\,d\ln k/d(1/T)$ was found to be negative. Therefore, these workers preferred the two-step mechanism

$$NO + Br_2 \overset{quasi-equil.}{\rightleftharpoons} NOBr_2 \qquad (2.5\text{-}52)$$
$$NOBr_2 + NO \to 2NOBr \qquad (2.5\text{-}53)$$

as a molecular explanation of the process. They noted that this mechanism yields Eqs. (2.5-50, 2.5-51) if the concentration of $NOBr_2$ remains very small, and that the negative overall activation energy can be attributed to the expected exothermicity of the quasi-equilibrium step. From a computational standpoint, however, their data are most directly described by the composite event (2.5-50) with the rate expression (2.5-51).

To construct a composite event from a detailed sequence, one can proceed as in Example 2.5 and represent the concentrations of intermediates in the controlling elementary step by quasi-equilibrium expressions (see Eq. 2.5-2b). However, it is simpler to construct the composite event in the form of Eq. (2.4-1); then the results of Section 2.4 can be applied directly. Thus, the intermediate $nC_5^=l$ in the controlling step is producible from $l + nC_5(g) - H_2(g)$ via the first two quasi-equilibrium reactions, whereas the intermediate $iC_5^=l$ is producible from $l + iC_5(g) - H_2(g)$ via the last two quasi-equilibrium reactions. With these substitutions into Eq. (2.5-41), we get

$$l + nC_5(g) - H_2(g) \overset{k_3'}{\rightleftharpoons} l + iC_5(g) - H_2(g) \qquad (2.5\text{-}54)$$

as the desired composite event, corresponding to a single occurrence of reaction 3 *and having the same transition state*. Taking the coefficients $\epsilon_{3i}$ and $\epsilon_{-3i}$ for Eq. (2.4-1) from this composite event as if it were elementary, we obtain the rate expression

$$\mathcal{R}_3 = k_3'\theta_l \left( p_{nC_5} - p_{iC_5}/K \right) p_{H_2}^{-1} \qquad (2.5\text{-}55)$$

by use of Eq. (2.4-5). Here $K$ is the equilibrium constant of the composite event, and thus of the overall reaction as written in Eq. (2.5-44). The stoichiometric coefficients $\nu_{3i}$ follow by use of Eq. (2.4-2). Corresponding substitutions for the intermediates in Eq. (2.5-47) lead to an expression for $\theta_l$ that is consistent with Eq. (2.5-49).

A composite event can be substituted for any reaction sequence that has a single rate-controlling step, provided that each intermediate species is quasi-equilibrated with its particular set of terminal species. To obtain the composite event, start with Eq. (2.4-1) for the controlling step, including its forward and reverse directions. Then replace each intermediate on the left by corresponding amounts of reactant species, and each intermediate on the right by corresponding amounts of product species, according to the given quasi-equilibrium reactions. The resulting modification of Eq. (2.4-1) gives the corresponding composite event, and the coefficients therein are the composite forward and reverse reaction orders. The stoichiometric coefficients are obtained from the net productions per event, as in Eq. (2.4-2).

The reaction orders in a composite event can include negative values, as illustrated by the last exponent in Eq. (2.5-55). Fractional orders are also possible, as long as they correspond to integer orders in the rate-controlling step and thus to whole numbers of each kind of elementary particle in the activated complex.

The *stoichiometric number* $\sigma_j$, introduced by Horiuti (1962) for reactions with a single rate-controlling step, is the number of times that step occurs in an overall reaction $j$ *as written*. This number is computable with reasonable precision from simultaneous observations of the forward and reverse reactions, and has been investigated for many reaction systems. Note, however, that $\sigma_j$ is necessarily unity if one constructs the overall reaction $j$ as a composite event in the manner just described. Thus, a reported stoichiometric number is simply the ratio of the investigator's chosen stoichiometric coefficients to those of the actual composite event, and is not needed further once the composite event has been constructed.

## 2.6 EVIDENCE FOR REACTION STEPS

The search for realistic reaction schemes for chemical processes is an active field of research with a rapidly growing arsenal of methods. In this section we summarize several kinds of evidence that have proved useful in such investigations.

Identification of the participating chemical species is most important. The completeness of reaction models is often limited by the difficulty of this step. Indirect evidence is commonly used to infer the existence of trace intermediates, but direct observations are becoming more frequent with modern spectroscopic methods.

Information about reaction model forms is sometimes available from observations of reactor hysteresis, oscillations or multiplicity (Feinberg 1977, 1980; Cutlip and Kenney 1978; Slin'ko and Slin'ko 1978; Cutlip 1979; Heinemann, Overholser, and Reddian, 1979; Aluko and Chang 1984, 1986; Cutlip et al., 1984; Mukesh et al., 1984; Graham and Lynch 1987; Harold and Luss 1987). New model forms have been discovered in this way by several investigators, showing the importance of keeping an open mind in modeling investigations.

Some reactions may be representable as quasi-equilibria over the range of the experiments. For such a step, Eq. (2.5-2b) gives a linear relation among the variables $\ln b_i$ at the reaction site. Thus, one can test for quasi-equilibria among the measurable species by determining which concentrations can be acceptably described by Eq. (2.5-2b). In case of doubt as to the appropriate reactions, one can determine a suitable set by performing this test with adjustable parameters $\nu_{ji}$ and functions $\ln K_j(T)$. The resulting coefficients, after adjustment to element-balanced integer values, define a set of net reactions that were too fast to measure in the given experiments. Such fast processes may render some slow steps redundant by bypassing them.

Kinetic steps are best identified by measuring the initial products formed from individual species (including postulated intermediates) or from simple mixtures. Isotopically labeled species have proved useful in such experiments. Initial products of homogeneous processes are observable in batch reactors at sufficiently short times or in flow reactors at points sufficiently near the inlet. The most advanced systems for initial product detection are molecular beam reactors (Herschbach 1976; Levine and Bernstein 1987) in which specific collisions are observed. Each of these techniques restricts the number of contributing reactions in a given experiment, so that their stoichiometry and rates can often be inferred.

The rank, NK, of the stoichiometric matrix $\nu$ can be estimated by fitting Eq. (2.2-3) to measurements of the species production rates. This value is the minimum number of reactions needed to model the stoichiometry; however, additional reactions may be needed to model the kinetics. This approach can be used also for complex mixtures, where the production rates may be available only as lumped values for groups of related species.

The site of each reaction should be identified as well as possible. This is commonly done by experiments with various amounts of surface or catalyst in the reactor, or with various surface compositions and pretreatments. For the chain reaction of hydrogen with chlorine, Pease (1934) included experiments with added oxygen as a gas-phase chain terminator to elucidate the locations of the normal initiation and termination reactions. Experience indicates, however, that adding a new species to a reacting mixture frequently raises more questions than it answers.

Whenever possible, the kinetics should be measured under conditions that minimize physical resistances to the heat and material transport required by the reaction. For heterogeneous catalysis, use of small particles and high flow velocities is strongly recommended since the work of Yoshida, Ramaswami, and Hougen (1962). For electrode reactions, the rotating-disk method of Levich (1962) gives controllable, spatially uniform states on the reaction surface. For homogeneous reactions, mechanical mixing often suffices to allow measurement of the chemical kinetics; however, diffusion becomes rate-controlling in the fastest ionic reactions (Eigen and Kustin 1960; Noyes 1961) and in the later stages of free-radical polymerization processes (Trommsdorf, Köhle, and Lagally 1948).

The experiments used to identify the reaction steps can also determine their initial rates as functions of the concentrations of reactants, catalysts, inhibitors and promoters. The reverse reaction rates should also be measured when feasible. The rates of elementary or composite events (in the sense of Example 2.6) are expected to be power functions of the concentrations under ideal-solution conditions; then the exponents $\epsilon_{ji}$ and $\epsilon_{-ji}$ can be found as the slopes of plots of $\ln \mathcal{R}_j$ versus $\ln b_i$. This method works if it is possible to establish $\mathcal{R}_j$ as a function of each $b_i$ individually while holding the temperature and the other $b$-values constant. For heterogeneous reactions this usually isn't possible, but measurements of reaction rates as functions of total pressure or concentration give useful evidence for testing reaction models, as shown by Yang and Hougen (1950); this method is widely used in the literature.

For a composite reaction, the values $\nu_{ji}$, $\epsilon_{ji}$, and $\epsilon_{ji}$ should conform reasonably to Eq. (2.5-56), and to Eq. (2.4-2) after division of the stoichiometric coefficients by a suitable constant. The divisor thus found is the stoichiometric number $\sigma_j$ for the reaction as originally written.

If any of the identified rates cannot be reasonably fitted with a power function of the corresponding concentrations, or if the powers are incompatible with Eq. (2.5-56), then the assumptions of the analysis should be reviewed. Such behavior might arise from one or more neglected chemical steps, or from nonideal solution behavior.

Various types of plots are available for testing special kinetic hypotheses. Some of these are used in problems at the end of this chapter. The reader is urged to start with simple plots to get a "feel" for the data and judge what kinds of rate expressions might be suitable. Graphical schemes are useful for preliminary selection of models, but the statistical discrimination methods of Chapters 6 and 7 are recommended for the later stages.

Addition of reaction steps generally allows a closer fit of the data, whether the added details are true or not. Thus we need criteria to decide when we have an adequate model, and when we need a better model or more data. Statistics gives objective approaches to these difficult questions, as we will demonstrate in Chapters 6 and 7, but alertness to mechanistic clues is most helpful in discovering good forms of reaction models.

Approximate reaction networks have become customary for modeling reactions in which the species are too numerous for a full accounting or chemical analysis. Lumped components or continuous distributions commonly take the place of single components in process models for refinery streams (Wei and Kuo 1969; Weekman 1969; Krambeck 1984; Astarita 1989; Chou and Ho 1989; Froment and Bischoff 1990). Polymerization processes are described in terms of moments of the distributions of molecular weight or other properties (Zeman and Amundson 1965; Ray 1972, 1983; Ray and Laurence 1977). Lumped components, or even hypothetical ones, are also prevalent in models of catalyst deactivation (Szepe and Levenspiel 1968; Butt 1984; Pacheco and Petersen 1984; Schipper et al. 1984;Froment and Bischoff 1990).

## PROBLEMS

### 2.A Computation of Production Rates from Reaction Rates

Calculate the local column vector $\mathbf{R}$ of species production rates for the reaction scheme in Section 2.1, given the vector

$$\boldsymbol{\mathcal{R}} = \begin{bmatrix} 0.7 \\ 0.2 \\ 0.1 \end{bmatrix} \quad \text{mole m}^{-3}\ \text{s}^{-1}$$

of local reaction rates.

### 2.B Computation of Reaction Rates from Production Rates

Calculate the full column vector $\boldsymbol{\mathcal{R}}$ of reaction rates in the scheme of Section 2.1, for an experiment that gave production rates $R_{C_2H_4} = -0.4$ and $R_{H_2O} = 0.5$ mole m$^{-3}$ s$^{-1}$. The feed stream was free of $C_2H_4O$, and the residence time in the reactor was short enough that the secondary reaction of combustion of $C_2H_4O$ could be neglected.

### 2.C Reaction Rate Near Equilibrium

Show that Eqs. (2.4-5) and (2.3-14) lead to the following expressions for the derivatives in Eq. (2.5-1):

$$\frac{\partial \mathcal{R}_j}{\partial b_m}\Big|_{\boldsymbol{b}^*, T^*} = -(\nu_{jm}/b_m^*)\mathcal{R}_{jf}(\boldsymbol{b}^*, T^*) \qquad (2.\text{C-1})$$

$$\frac{\partial \mathcal{R}_j}{\partial T}\Big|_{\boldsymbol{b}^*, T^*} = (\Delta\widetilde{\mathcal{E}}_j^\circ / RT^{*2})\mathcal{R}_{jf}(\boldsymbol{b}^*, T^*) \qquad (2.\text{C-2})$$

Here $\mathcal{R}_{jf}(\boldsymbol{b}^*, T^*) = \mathcal{R}_{jr}(\boldsymbol{b}^*, T^*)$ is the magnitude of each right-hand term in Eq. (2.4-5) at the particular equilibrium point $(\boldsymbol{b}^*, T^*)$.

### 2.D The $H_2$ + $Br_2$ Reaction

Extend the analysis in Example 2.2 to include reversibility of all the reactions. Reduce your expression for $R_{HBr}$ by use of a common denominator. See if the equilibrium expression obtained by setting $R_{HBr} = 0$ in your solution is consistent with the ideal-solution version of Eq. (2.3-7) for the overall reaction.

### 2.E Multiple Equilibria in Chemisorption on Pt

Chemical and physical evidence indicate the following species to be present in the hydrogenation of propylene ($C_3H_6$) on supported platinum (Pt):

1. $C_3H_6$      2. $H_2$      3. $C_3H_8$      4. M      5. HM
6. $C_3H_5M$      7. $C_3H_6M$      8. $C_3H_7M$      9. $C_3H_6M_2$      10. $C_3H_5M_3$

Here M denotes a surface atom of Pt; thus, species 4–10 are interfacial.
(*a*) Write a set of element-balanced reactions, each giving one molecule of one surface species 5, 6, 7, 8, 9, or 10 from positive or negative amounts of species 1, 2, and/or 4.

(*b*) Treating the reactions in (*a*) as equilibria, solve for each surface concentration $[A_i]$ in terms of $p_1$, $p_2$, and $[M]$. Here $[A_i]$ is expressed in molecules per surface Pt atom. (The fraction of the Pt surface atoms held by species $i$ is $m_i[A_i]$, where $m_i$ is the number of M atoms in the molecule $A_i$).
(*c*) Write out the equation

$$[M] + [HM] + [C_3H_5M] + \ldots = 1$$

for the sum of the surface concentrations $m_i A_i$ and insert the results found in (*b*) to get a cubic equation for $[M]$ at given partial pressures. Give an analytic expression for $[M]$, neglecting the triadsorbed species $C_3H_5M_3$.

## 2.F Estimation of Chemisorption Equilibrium from Calorimetry

The differential enthalpy of adsorption of hydrogen on supported Pt is about $-14$ kcal/mole of $H_2$ at high coverage and 308 K, according to calorimetric measurements by N. Cardona-Martinez [PhD thesis, University of Wisconsin–Madison (1989)], reported in Goddard et al. (1989). The standard entropy of adsorption, $\Delta S°$, is estimated as $-32$ cal mol$^{-1}$ K$^{-1}$ by assuming equal $S°$ values for $[M]$ and $[HM]$ and evaluating the standard-state entropy of $H_2$ at (308 K, 1.0 atm). Put these values into the units of your hydrogen chemisorption step in Problem 2.E or 2.G and calculate a numerical upper bound on $[M]$ for experiments in that chemical system at $T = 308$ K and $p_2 \geq 0.5$ atm, treating your equilibrium constant as independent of surface concentrations.

## 2.G Alternative Model of Chemisorption Equilibria

(*a*) Write a new set of equilibrium reactions for the system of Problem 2.E such that each reaction forms one molecule of species 4, 6, 7, 8, 9, or 10 from positive or negative amounts of species 1, 2, and/or 5.
(*b*) Express the M-balance as a cubic equation for $[HM]$, eliminating all other surface concentrations by use of the equilibrium expressions for the reactions obtained in (*a*).
(*c*) Noting that $[M] << [HM]$ according to the data of Problem 2.F, discuss the merits of this chemisorption model relative to that in Problem 2.E.

## 2.H Expansions for Standard-State Energy Functions

Verify the integrations in Eqs. (2.3-16) and (2.3-17).

## REFERENCES and FURTHER READING

Aluko, M., and H.-C. Chang, Multi-scale analysis of exotic dynamics in surface catalyzed reactions–II. Quantitative parameter space analysis of an extended Langmuir-Hinshelwood reaction scheme, *Chem. Eng. Sci.*, **39**, 51–64 (1984).

Aluko, M., and H.-C. Chang, Dynamic modelling of a heterogeneously catalyzed system with stiff Hopf bifurcations, *Chem. Eng. Sci.*, **41**, 317–331 (1986).

Aris, R., Prolegomena to the rational analysis of systems of chemical reactions II. Some addenda, *Arch. Rat. Mech. Anal.*, **22**, 356 (1968).

Aris, R., *Elementary Chemical Reactor Analysis*, Prentice-Hall, Englewood Cliffs, NJ (1969).

Aris, R., Reactions in continuous mixtures, *AIChE J.*, **35**, 539–548 (1989).

Aris, R., and G. R. Gavalas, On the theory of reactions in continuous mixtures, *Proc. Roy. Soc.*, **A260**, 351–393 (1966).

Aris, R., and R. S. H. Mah, Independence of chemical reactions, *Ind. Eng. Chem. Fundam.*, **2**, 90–94 (1963).

Arrhenius, S., Über die Reaktionsgeschwindigkeit bei der Inversion von Rohrzucker durch Säuren, *Z. physik. Chem.*, **4**, 226–248 (1889).

Ashmore, P. G., *Catalysis and Inhibition of Chemical Reactions*, Butterworths, London (1963).

Astarita, G., Lumping nonlinear kinetics: apparent overall order of reaction, *AIChE J.*, **35**, 529–538 (1989).

Avery, N. R., and N. Sheppard, The use of thermal desorption and electron energy loss spectroscopy for the determination of structures of unsaturated hydrocarbons chemisorbed on metal single-crystal surfaces I. Alk-1-enes on Pt(111), *Proc. Roy. Soc. London A*, **405**, 1–25 (1986).

Baltanas, M. A., and G. F. Froment, Computer generation of reaction networks and calculation of product distributions in the hydroisomerization and hydrocracking of paraffins on Pt-containing bifunctional catalysts, *Comput. Chem. Eng.*, **9**, 71–81 (1985).

Benson, S. W., *The Foundations of Chemical Kinetics*, McGraw-Hill, New York (1960).

Benson, S. W., *Thermochemical Kinetics*, 2nd edition, Wiley, New York (1976).

Blakemore, J. E., and W. H. Corcoran, Validity of the steady-state approximation applied to the pyrolysis of *n*-butane, *Ind. Eng. Chem. Process Des. Devel.*, **8**, 206–209 (1969).

Bodenstein, M., Eine Theorie der photochemischen Reaktionsgeschwindigkeiten. *Z. physik. Chem.*, **85**, 329–397 (1913).

Bodenstein, M., and S. C. Lind, Geschwindigkeit der Bildung des Bromwasserstoffs aus seiner Elementen, *Z. physik. Chem.*, **57**, 168–192 (1907).

Bodenstein, M., and H. Lutkemeyer, Die photochemische Bildung von Bromwasserstoff und die Bildungsgeschwindigkeit der Brommolekel aus den Atomen, *Z. physik. Chem.*, **114**, 208–236 (1924).

Boudart, M., Kinetics on ideal and real surfaces, *AIChE J.*, **2**, 62–64 (1956).

Boudart, M., D. E. Mears and M. A. Vannice, Kinetics of heterogeneous catalytic reactions, *Ind. Chim. Belg.*, **32**, 281–284 (1967).

Boudart, M., *Kinetics of Chemical Processes*, Prentice-Hall, Englewood Cliffs, NJ (1968).

Boudart, M., and G. Djéga-Mariadassou, *Kinetics of Heterogeneous Catalytic Reactions*, Princeton University Press, Princeton, NJ (1984).

Bowen, J. R., A. Acrivos and A. K. Oppenheim, Singular perturbation refinement to quasi-steady state approximation in chemical kinetics, *Chem. Eng. Sci.*, **18**, 177–188 (1963).

Box, G. E. P., Fitting empirical data, *Ann. N.Y. Acad. Sci.*, **86**, 792–816 (1960).

Bradshaw, R. W., and B. Davidson, A new approach to the analysis of heterogeneous reaction rate data, *Chem. Eng. Sci.*, **24**, 1519–1527 (1970).

Briggs, G. E., and J. B. S. Haldane, A note on the kinetics of enzyme action, *Biochem. J.*, **19**, 338–339 (1925).

Broadbelt, L. J., and J. Pfaendiner, Lexicography of kinetic modeling of complex reaction networks, *AIChE J.*, **51**, 2112–2121 (2005).

Brønsted, J. N., Zur Theorie der chemischen Reaktionsgeschwindigkeit, *Z. physik. Chem.*, **102**, 169–207 (1922).

Butt, J. B., Catalyst deactivation and regeneration, in *Catalysis Science and Technology*, J. R. Anderson and M. Boudart, eds., **6**, Springer, New York (1984), 1–63.

Campbell, E. H., and R. H. Fristrom, Reaction kinetics, thermodynamics, and transport in the hydrogen-bromine system – A survey of properties for flame studies, *Chem. Revs.*, **58**, 173–234 (1958).

Carberry, J. J., *Chemical and Catalytic Reaction Engineering*, McGraw-Hill, New York (1976).

Chang, H.-C., and M. Aluko, Multi-scale analysis of exotic dynamics in surface catalyzed reactions. I. Justification and preliminary model discriminations, *Chem. Eng. Sci.*, **39**, 36–50 (1984).

Chou, M. Y., and T. C. Ho, Lumping coupled nonlinear reactions in continuous mixtures, *AIChE J.*,**35**, 533–538 (1989).

Christiansen, J. A., On the reaction between hydrogen and bromine, *Kgl. Danske Videnskab. Selskab. Mat.-fys. Medd.*, **1**(14), 1–19 (1919).

Conway, B. E., Some chemical factors in the kinetics of processes at electrodes, *Progress in Reaction Kinetics*, **4**, G. Porter, ed., Pergamon, New York (1967), 399–483.

Curtiss, C. F., and J. O. Hirschfelder, Integration of stiff equations. *Proc. Natl. Acad. Sci. U. S. A.*, **38**, 235–243 (1952).

Cutlip, M. B., Concentration forcing of catalytic surface rate processes. I: Isothermal carbon monoxide oxidation over supported platinum, *AIChE J.*, **25**, 502–508 (1979).

Cutlip, M. B., and C. N. Kenney, Limit cycle phenomena during catalytic oxidation reactions over a supported platinum catalyst, *ACS Symp. Ser.*, **65**, 475–486 (1978).

Cutlip, M. B., C. N. Kenney, W. Morton, D. Mukesh and S. C. Capsakis, Transient and oscillatory phenomena in catalytic reactors, *I. Chem. E. Symp. Ser.*, **87**, 135–150 (1984).

Davies, C. W., Salt effects in solution kinetics, in *Progress in Reaction Kinetics*, G. Porter, ed., **1**, Pergamon, Oxford (1961), 161–186 .

Debye, P., Reaction rates in ionic solutions (with Discussion), *Trans. Electrochem. Soc.*, **82**, 265–272 (1942).

de Groot, S. R., and P. Mazur, *Non-Equilibrium Thermodynamics*, North-Holland, Amsterdam (1962).

Delahay, P., *Double Layer and Electrode Kinetics*, Wiley, New York (1965).

Deuflhard, P., G. Bader, and U. Nowak, LARKIN — a software package for the numerical simulation of large systems arising in chemical reaction kinetics, in *Modelling of Chemical Reaction Systems*, K. H. Ebert, P. Deuflhard, and W. Jäger, eds., Springer, Berlin (1981), 38–55.

Dumesic, J. A., B. A. Milligan, L. A. Greppi, V. R. Balse, K. T. Sarnowski, C. E. Beall, T. Kataoka, D. F. Rudd, and A. A. Trevino, A kinetic modelling approach to the design of catalysts: formulation of a catalyst design advisory program, *Ind. Eng. Chem. Res.*, **26**, 1399–1407 (1987).

Eckert, C. A., and M. Boudart, On the use of fugacities in gas kinetics, *Chem. Eng. Sci.*, **18**, 144–146 (1963).

Eckert, C. A., and R. A. Grieger, Mechanistic evidence for the Diels-Alder reaction from high-pressure kinetics, *J. Am. Chem. Soc.*, **52**, 7149–7153 (1970).

Eigen, M., and K. Kustin, The influence of steric factors in fast protolytic reactions as studied with HF, $H_2S$ and substituted phenols, *J. Am. Chem. Soc.*, **82**, 5952–5953 (1960).

Eigen, M., and K. Kustin, The study of very rapid reactions in solution by relaxation spectrometry, *ICSU Rev.*, **5**, 97–115 (1963).

Evans, M. G., and M. Polanyi, Some applications of the transition state method to the calculation of reaction velocities, especially in solution, *Trans. Faraday Soc.* **31**, 867–894, (1935).

Eyring, H., The activated complex in chemical reactions, *J. Chem. Phys.*, **3**, 107–115 (1935).

Eyring, H., J. Walter, and G. E. Kimball, *Quantum Chemistry.* Wiley, New York (1944).

Farrow, L. A., and D. Edelson, The steady-state approximation: fact or fiction? *Int. J. Chem. Kinet.*, **VI**, 787–800 (1974).

Feinberg, M., Mathematical aspects of mass action kinetics, in *Chemical Reactor Theory: A Review,* L. Lapidus and N. Amundson, eds., Prentice-Hall, Englewood Cliffs, NJ (1977), 1–78.

Feinberg, M., Chemical oscillations, multiple equilibria, and reaction network structure, in *Dynamics and Modelling of Reactive Systems,* W. E. Stewart, W. H. Ray, and C. C. Conley, eds., Academic Press, New York (1980), 59–130.

Fjeld, M., O. A. Asbjørnsen and K. J. Astrøm, Reaction invariants and their importance in the analysis of eigenvectors, state observability and controllability of the continuous stirred tank reactor, *Chem. Eng. Sci.*, **29**, 1917–1926 (1974).

Franckaerts, J., and G. F. Froment, Kinetic study of the dehydration of ethanol, *Chem. Eng. Sci.*, **19**, 807–818 (1964).

Frank-Kamenetskii, D. A., *Diffusion and Heat Transfer in Chemical Kinetics,* 2nd edition, translated from Russian, Plenum, New York (1969).

Fredrickson, A. G., and H. Tsuchiya, Microbial kinetics and dynamics, in *Chemical Reactor Theory: A Review,* L. Lapidus and N. R. Amundson, eds., Prentice-Hall, Englewood Cliffs, NJ (1977), 405–483.

Froment, G. F., The kinetics of complex catalytic reactions, *Chem. Eng. Sci.*, **42**, 1073–1087 (1987).

Froment, G. F., and K. B. Bischoff, *Chemical Reactor Analysis and Design,* 2nd edition, Wiley, New York (1990).

Gates, B. C., J. R. Katzer, and G. C. A. Schuitt, *Chemistry of Catalytic Processes,* McGraw-Hill, New York (1979).

Georgakis, C., and R. Aris, Diffusion, reaction, and the pseudo-steady-state hypothesis, *Math. Biosci.*, **25**, 237–258 (1975).

Gibbs, J. W., On the equilibrium of heterogeneous substances, *Trans. of the Connecticut Acad.* **III**, 108–248 (1875–1876) and 343–524 (1877–1878); *Collected Works,* **I**, 55–349, Yale University Press, New Haven, CT (1948).

Goddard, S. A., M. D. Amiridis, J. E. Rekoske, N. Cardona-Martinez, and J. A. Dumesic, Kinetic simulation of heterogeneous catalytic processes: ethane hydrogenolysis over Group VIII metals, *J. Catal.*, **117**, 155–169 (1989).

Golden, D. M., Measurement and estimation of rate constants, in *Dynamics and Modelling of Reactive Systems*, W. E. Stewart, W. H. Ray and C. C. Conley, eds., Academic Press, New York (1980), 315–331.

Graham, W. R. C., and D. T. Lynch, CO oxidation on Pt: model discrimination using experimental bifurcation behavior, *AIChE J.*, **33**, 792–800 (1987).

Greeley, J., J. K. Nørskov, and M. Mavrikakis, Electronic structure and catalysis on metal surfaces, *Annu. Rev. Phys. Chem.*, **53**, 319–348 (2002).

Halsey, G. D., Jr., Catalysis on non-uniform surfaces, *J. Chem. Phys.*, **17**, 758–761 (1949).

Happel, J., and P. H. Sellers, Analysis of the possible mechanisms for a catalytic reaction system, *Adv. Catal.*, **32**, 274–323 (1983).

Harold, M. P., and D. Luss, Use of multiplicity features for kinetic modeling: CO oxidation on $Pt/Al_2O_3$, *Ind. Eng. Chem. Res.*, **26**, 2099–2106 (1987).

Hase, W. L., S. L. Mondro, R. J. Duchovic and D. M. Hirst, Thermal rate constant for $H + CH_3 \rightarrow CH_4$ recombination. 3. Comparison of experiment and canonical variational transition state theory, *J. Am. Chem. Soc.*, **109**, 2916–2922 (1987).

Heinemann, R. F., K. A. Overholser, and G. W. Reddien, Multiplicity and stability of premixed laminar flames: an application of bifurcation theory, *Chem. Eng. Sci.*, **34**, 833–840 (1979).

Henri, V., Théorie générale de l'action de quelques diastases, *C. R. Acad. Sci. Paris*, **135**, 916–919 (1902).

Herschbach, D. R., *Chemical Kinetics*, University Park Press, Baltimore, MD (1976).

Herzfeld, K. F., Zur Theorie der Reaktionsgeschwindigkeiten in Gasen, *Ann. Physik.*, **59**, 635–667 (1919).

Hill, C. G., *An Introduction to Chemical Engineering Kinetics and Reactor Design*, Wiley, New York (1977).

Hinshelwood, C. N., *Kinetics of Chemical Change*, Clarendon Press, Oxford (1926, 1940).

Ho, T. C., and R. Aris, On apparent second-order kinetics, *AIChE J.*, **33**, 1050–1051 (1987).

Horiuti, J., On the statistical mechanical treatment of the absolute rate of chemical reaction, *Bull. Chem. Soc. Japan*, **13**, 210–216 (1938).

Horiuti, J., Significance and experimental determination of stoichiometric number, *J. Catal.*, **1**, 199–207 (1962).

Hougen, O. A., and K. M. Watson, Solid catalysts and reaction rates. General principles, *Ind. Eng. Chem.*, **35**, 529–541 (1943).

Hougen, O. A., and K. M. Watson, *Chemical Process Principles, Part Three*, Wiley, New York (1947).

Hutchison, P., and D. Luss, Lumping of mixtures with many parallel first-order reactions, *Chem. Eng. J.*, **1**, 129–136 (1970).

Johnston, H. S., *Gas Phase Reaction Rate Theory*, Ronald Press, New York (1966).

Kabel, R. L., and L. N. Johansen, Reaction kinetics and adsorption equilibria in the vapor-phase dehydration of ethanol, *AIChE J.*, **8**, 621–628 (1962).

Kee, R. J., F. M. Rupley, and J. A. Miller, Chemkin-II: A Fortran chemical kinetics package for the analysis of gas-phase chemical kinetics, *SANDIA REPORT SAND89-8009B, UC-706*, (reprinted January 1993).

Kittrell, J. R., W. G. Hunter and R. Mezaki, The use of diagnostic parameters for kinetic model building, *AIChE J.,* **12**, 1014–1017 (1966).

Kittrell, J. R., and J. Erjavec, Response surface methods in heterogeneous kinetic modelling, *Ind. Eng. Chem., Process Des. Dev.,* **7**, 321–327 (1968).

Krambeck, F. J., The mathematical structure of chemical kinetics in homogeneous single-phase systems, *Arch. Rational Mech. Anal.,* **38**, 317–347 (1970).

Krambeck, F. J., Accessible composition domains for monomolecular systems, *Chem. Eng. Sci.,* **39**, 1181–1184 (1984).

Kreevoy, M. M., and D. G. Truhlar, Transition state theory, in *Investigation of Rates and Mechanisms of Reactions,* 4th edition, C. F. Bernasconi, ed., **1**, 13–95, Wiley, New York (1986).

Laidler, K. J., *Chemical Kinetics,* 3rd edition, Harper & Row, New York (1987).

Laidler, K. J., and P. S. Bunting, *The Chemical Kinetics of Enzyme Action,* 2nd edition, Oxford University Press, London (1973).

Langmuir, I., The dissociation of hydrogen into atoms. III. The mechanism of the reaction, *J. Am. Chem. Soc.,* **38**, 1145–1156 (1916).

Langmuir, I., The constitution and fundamental properties of solids and liquids, *J. Am. Chem. Soc.,* **38**, 2221–2295 (1916).

Langmuir, I., The mechanism of the catalytic action of platinum in the reactions $2CO + O_2 = 2CO_2$ and $2H_2 + O_2 = 2H_2O$, *Trans. Faraday Soc.,* **17**, 621–654 (1922).

Lee, W. M., Nonstationary-state calculation of thermal polymerization of styrene, *J. Polym. Sci. B,* **6**, 603–607 (1968).

Levich, V. G., *Physicochemical Hydrodynamics,* 2nd edition, Prentice-Hall, Englewood Cliffs, NJ (1962), Chapter 2.

Levine, R. D., and R. B. Bernstein, *Molecular Reaction Dynamics and Chemical Reactivity,* Oxford University Press, New York (1987).

Lindemann, F. A., Discussion on "The radiation theory of chemical action." *Trans. Faraday Soc.,* **17**, 598–599 (1922).

Mahan, B. H., Activated complex theory of bimolecular reactions, *J. Chem. Educ.,* **51**, 709–711 (1974).

Marcus, R. A., Generalization of activated-complex theory. III. Vibrational adiabaticity, separation of variables, and a connection with analytical mechanics, *J. Chem. Phys.,* **43**, 1598–1605 (1965).

McCabe, R. W., Kinetics of ammonia decomposition on nickel, *J. Catal.,* **79**, 445–450 (1983).

Michaelis, L., and M. L. Menten, Die Kinetik der Invertinwirkung, *Biochem. Z.,* **49**, 333–369 (1913).

Moore, J. W., and R. G. Pearson, *Kinetics and Mechanism,* 3rd edition, Wiley, New York (1981).

Mukesh, D., W. Morton, C. N. Kenney, and M. R. Cutlip, Island models and the catalytic oxidation of carbon monoxide and carbon monoxide-olefin mixtures, *Surf. Sci.,* **138**, 237–257 (1984).

Newman, J. S., *Electrochemical Systems,* 2nd edition, Prentice-Hall, Englewood Cliffs, NJ (1991).

Noyes, R. M., Effects of diffusion rates on chemical kinetics, in *Progress in Reaction Kinetics,* G. Porter, ed., **1**, Pergamon, Oxford (1961), 129–160.

Pacheco, M. A., and E. E. Petersen, On the development of a catalyst fouling model, *J. Catal.,* **88**, 400–408 (1984).

Pease, R. N., Kinetics of the thermal hydrogen-chlorine reaction, *J. Am. Chem. Soc.*, **56**, 2388–2391 (1934).

Pelzer, H., and E. Wigner, Über die Geschwindigkeitskonstante von Austauschreaktionen, *Z. Physik. Chem. B*, **15**, 445–471 (1932).

Petzold, L. R., A description of DASSL: A differential/algebraic system solver, in *Scientific Computing*, eds. R. S. Stepleman et al., North-Holland, Amsterdam, (1983).

Poling, B. E., J. M. Prausnitz and J. P. O'Connell, *The Properties of Gases and Liquids*, 5th edition, McGraw-Hill, New York (2000).

Polanyi, M., Reaktionsisochore und Reaktionsgeschwindigkeit vom Standpunkte der Physik, *Z. Elektrochem.*, **26**, 50–54 (1920).

Pritchard, H. O., R. G. Sowden and A. F. Trotman-Dickenson, Studies in energy transfer II. The isomerization of cyclopropane—a quasi-unimolecular reaction, *Proc. Roy. Soc. A*, **217**, 563–571 (1953).

Ray, W. H., On the mathematical modeling of polymerization reactors, *J. Macromol. Sci.–Rev. Macromol. Chem.*, **C8**, 1–56 (1972).

Ray, W. H., Current problems in polymerization reaction engineering, *ACS Symp. Ser.*, **226**, 101–133 (1983).

Ray, W. H., and R. L. Laurence, Polymerization reaction engineering, in *Chemical Reactor Theory: A Review*, L. Lapidus and N. R. Amundson, eds., Prentice-Hall, Englewood Cliffs, NJ (1977), 532–582.

Razón, L. F., and R. A. Schmitz, Multiplicities and instabilities in chemically reacting systems — a review, *Chem. Eng. Sci.*, **42**, 1005–1047 (1987).

Rideal, E. K., A note on a simple molecular mechanism for heterogeneous catalytic reactions, *Proc. Camb. Phil. Soc.*, **35**, 130–132 (1939).

Robinson, P. J., and K. A. Holbrook, *Unimolecular Reactions*, Wiley, New York (1972).

Schipper, P. H., K. R. Graziani, B. C. Choi and M. P. Ramage, The extension of Mobil's kinetic reforming model to include catalyst deactivation, *I. Chem. E. Symp. Ser.*, **87**, 33–44 (1984).

Schneider, D. R., and G. V. Reklaitis, On material balances for chemically reacting systems, *Chem. Eng. Sci.*, **30**, 243–247 (1975).

Schubert, E., and H. Hofmann, Reaction engineering 1. Stoichiometry of complex reactions, *Int. Chem. Eng.*, **16**, 132–136 (1976).

Seinfeld, J. H., Dynamics of aerosols, in *Dynamics and Modelling of Reactive Systems*, W. E. Stewart, W. H. Ray, and C. C. Conley, eds., Academic Press, New York (1980).

Shinnar, R., and C. A. Feng, Structure of complex catalytic reactions: thermodynamic constraints in kinetic modeling and catalyst evaluation, *Ind. Eng. Chem. Fundam.*, **24**, 153–170 (1985).

Sinfelt, J. H., H. Hurwitz, and J. C. Rohrer, Kinetics of $n$-pentane isomerization over Pt-$Al_2O_3$ catalyst, *J. Phys. Chem.*, **64**, 892–894 (1960).

Sinfelt, J. H., H. Hurwitz, and R. A. Shulman, Kinetics of methylcyclohexane dehydrogenation over Pt-$Al_2O_3$, *J. Phys. Chem.*, **64**, 1559–1562 (1960).

Slin'ko, M. G., and M. M. Slin'ko, Self-oscillations of heterogeneous catalytic reaction rates, *Catal. Rev.*, **17**, 119–153 (1978).

Smith, J. M., *Chemical Engineering Kinetics*, 3rd edition, McGraw-Hill, New York (1981).

Smith, W. R., and R. W. Missen, *Chemical Reaction Equilibrium Analysis*, Wiley, New York (1982).

Smoluchowski, M. V., Versuch einer mathematischen Theorie der Koagulationskinetik kolloider Lösungen.*Z. physik. Chem.*, **92**, 129–168 (1917).

Smoot, L. D., and P. J. Smith, Modeling pulverized-coal reaction processes, in *Pulverized-Coal Combustion and Gasification*, L. D. Smoot and D. T. Pratt, eds., Plenum, New York (1979).

Sohn, H. Y., and M. E. Wadsworth, (eds.), *Rate processes of extractive metallurgy*, Plenum Press, New York (1979).

Somorjai, G. A., *Chemistry in Two Dimensions: Surfaces.* Cornell University Press, Ithaca, NY (1981).

Sørensen, J. P., *Simulation, Regression and Control of Chemical Reactors by Collocation Techniques*, Doctoral Thesis, Danmarks tekniske Højskole, Lyngby (1982).

Sørensen, J. P., and W. E. Stewart, Structural analysis of multicomponent reactor models: I. Systematic editing of kinetic and thermodynamic values, *AIChE J.*, **26**, 98–104 (1980).

Sørensen J. P., and W. E. Stewart, Structural analysis of multicomponent reactor models: II. Formulation of mass balances and thermodynamic constraints, *AIChE J.*, **26**, 104–111 (1980).

Stewart, W. E., and J. P. Sørensen, Computer-aided modelling of reaction networks, in *Foundations of Computer-Aided Process Design*, Vol. II, R. S. H. Mah and W. D. Seider, eds., Engineering Foundation, New York (1981), 335–366.

Szekely, J., J. W. Evans and H. Y. Sohn, *Gas-Solid Reactions*, Academic Press, New York (1976).

Szepe, S., and O. Levenspiel, Catalyst deactivation, *4th Eur. Symp. React. Eng.*, 265–276 (1968).

Taylor, T. I., Hydrogen isotopes in the study of hydrogenation and exchange, in *Catalysis*, **V**, 257–403, P. H. Emmett, ed., Reinhold, New York (1957).

Thomas, J. M., and W. J. Thomas, *Introduction to the Principles of Heterogeneous Catalysis*, Academic Press, New York (1967).

Trautz, M., and V. P. Dalal, Die Geschwindigkeit der Nitrosylbromidbildung $2NO_2 + Br_2 = 2NOBr$, *Z. Anorg. Chem.*, **102**, 149–172 (1918).

Trommsdorf, E., H. Köhle and P. Lagally, Zur Polymerisation des Methacrylsäuremethylesters, *Makro. Chemie*, **1**, 169–198 (1948).

Truhlar, D. G., A. D. Isaacson and B. C. Garrett, Generalized transition state theory, in *Theory of Chemical Reaction Dynamics*, **4**, M. Baer, ed., CRC Press, Boca Raton, Florida (1985), 65–137.

Tsuchida, E., and S. Mimashi, Analysis of oligomerization kinetics of styrene on the electronic digital computer, *J. Polym. Sci. A*, **3**, 1401–1434 (1965).

van t'Hoff, J. H., *Études de dynamique chimique*, Muller, Amsterdam (1884).

Voevodsky, V. V., and V. N. Kondratiev, Determination of rate constants for elementary steps in branched chain reactions, in *Progress in Reaction Kinetics*, G. Porter, ed., **1**, Pergamon, Oxford (1961), 43–65.

Waissbluth, M., and R. A. Grieger, Activation volumes of fast protein reactions: the binding of bromophenol blue to $\beta$-lactoglobulin B. *Arch. Biochem. Biophysics*, **159**, 639–645 (1973).

Weekman, V. W., Kinetics and dynamics of catalytic cracking selectivity in fixed-bed reactors, *Ind. Eng. Chem. Process Des. Devel.*, **8**, 385 (1969).

Wei, J., and C. D. Prater, The structure and analysis of complex reaction systems, *Adv. Catal.*, **13**, 203–392 (1962).

Wei, J., and J. C. W. Kuo, A lumping analysis in monomolecular reaction systems, *Ind. Eng. Chem. Fundam.*, **8**, 114–123 (1969).

Westerterp, K. R., W. P. M. van Swaaij, and A. A. C. M. Beenakers, *Chemical Reactor Design and Operation*, Wiley, New York (1984).

Westmoreland, P. R., J. B. Howard, J. P. Longwell, and A. M. Dean, Prediction of rate constants for combustion and pyrolysis reactions by bimolecular QRRK, *AIChE J.*, **32**, 1971–1979 (1986).

Whitwell, J. C., and S. R. Dartt, Independent reactions in the presence of isomers, *AIChE J.*, **19**, 1114–1120 (1973).

Wilkinson, F., *Chemical Kinetics and Reaction Mechanisms*, Van Nostrand Reinhold, New York (1980).

Woodward, R. B., and R. Hoffman, *The Conservation of Orbital Symmetry*, Verlag Chemie, Weinheim/Bergstrasse (1970).

Wynne-Jones, W. F. K., and H. Eyring, Reactions in condensed phases, *J. Chem. Phys.*, **3**, 492–502 (1935).

Yang, K. H., and O. A. Hougen, Determination of mechanism of catalyzed gaseous reactions, *Chem. Eng. Prog.*, **46**, 146–157 (1950).

Yoshida, F., D. Ramaswami, and O. A. Hougen, Temperatures and partial pressures at the surfaces of catalyst particles, *AIChE J.*, **8**, 5–11 (1962).

Zeman, R. J., and N. R. Amundson, Continuous polymerization models I, II, *Chem. Eng. Sci.*, **30**, 331, 637 (1965).

# Chapter 3
# Chemical Reactor Models

Reactor models, like the reaction models they include, are important tools for process investigation, design, and discovery. A well-constructed reactor model allows exploration of process alternatives with far less effort, time, and expense than traditional scale-up methods using pilot-plant and semiworks experiments. This approach is gaining wide acceptance among engineers, along with recognition of its greater reliance on the investigators' experimental, mathematical, and statistical skills and ingenuity, which we aim to nurture with this book.

In this chapter we give models for several reactor types, emphasizing their use to get process information. To have predictive value, the calculations should include not only the reaction rates, extents, and heat effects, but also the fluid states at any heterogeneous reaction sites as analyzed for fixed-bed reactors beginning in Section 3.3. In this chapter we also introduce the reader to Athena Visual Studio, the software package that Michael Caracotsios created for easy coding and analysis of differential and algebraic models. Broader coverage of chemical reactor modeling is provided in Froment and Bischoff (1990) and in the expert compilation *Chemical Reactor Theory: A Review*, edited by Leon Lapidus and Neal Amundson (1977) and dedicated to the memory of Richard W. Wilhelm.

## 3.1 MACROSCOPIC CONSERVATION EQUATIONS

Macroscopic mass, energy, and momentum balances provide the simplest starting point for reactor modeling. These equations give little spatial detail, but provide a first approximation to the performance of chemical reactors. This section builds on Chapter 22 of Bird, Stewart, and Lightfoot (2002). A table of notation is given at the end of the current chapter.

### 3.1.1 Material Balances

Consider a gas or liquid system of volume $V$, containing NS chemical species and having an inlet plane "1" and an outlet plane "2." The system boundaries may also include a mass transfer surface "0." Let $m_{i,\text{tot}}$ be the total mass [g] of species $i$ in the system at time $t$. A mass balance on chemical species $i$ for the system can then be written as

$$\frac{d}{dt}m_{i,\text{tot}} = w_{i1} - w_{i2} + w_{i0} + r_{i,\text{tot}} \qquad i = 1, 2, \ldots, \text{NS} \qquad (3.1\text{-}1)$$

Here $w_{i0}$ is the rate of mass transfer $[g/s]$ of species $i$ into the system, and $r_{i,\text{tot}}$ is the total rate of mass production of species $i$ in the system:

$$r_{i,\text{tot}} = \int_V r_i \, dV + \int_S r_i^s \, dS \qquad (3.1\text{-}2)$$

The integrals over $V$ and $S$ are the total rates of production of species $i$ in the system by *homogeneous* and *heterogeneous* chemical reactions, respectively. The rate $r_{i,\text{tot}}$ is in $[g/s]$, while $r_i$ and $r_i^s$ are in $[g/cm^3 s]$ and $[g/cm^2 s]$, respectively.

We normally use a full set of NS material balances, even though the number of independent balances is limited to the rank NK of the stoichiometric matrix $\nu$ for the given reaction scheme as shown in Section 2.1, and in Aris (1969). Exceptions must be made, however, when one or more constraints are imposed, such as quasi-equilibrium for some reactions or pseudo-steady state (better called *quasi-conservation*) for some chemical species; then each active constraint will replace a mass balance. By these procedures, we avoid catastrophic cancellations that might occur in subtractions performed to reduce the number of species variables from NS to NK.

Summation of Eq. (3.1-1) over all the species gives the total mass balance,

$$\frac{d}{dt} m_{\text{tot}} = w_1 - w_2 + w_0 \qquad (3.1\text{-}3)$$

This balance does not contain a reaction term, since mass is conserved very accurately in chemical reactions.

Corresponding material balances can be written in molar units. Let $M_{i,\text{tot}} = m_{i,\text{tot}}/M_i$ be the total number of moles of species $i$ in the system, and let $W$ denote a molar flow expressed in $[moles/s]$. Then the molar balance on species $i$ is

$$\frac{d}{dt} M_{i,\text{tot}} = W_{i1} - W_{i2} + W_{i0} + R_{i,\text{tot}} \qquad i = 1, 2, \ldots, \text{NS} \qquad (3.1\text{-}4)$$

in which $W_{i1} = w_{i1}/M_i$, $W_{i2} = w_{i2}/M_i$, and $W_{i0} = w_{i0}/M_i$. Again, the production term can be expressed as the sum of the homogeneous and heterogeneous contributions,

$$R_{i,\text{tot}} = \int_V R_i \, dV + \int_S R_i^s \, dS \qquad i = 1, 2, \ldots, \text{NS} \qquad (3.1\text{-}5)$$

this time in units of $[moles/s]$. Summation of the species molar balances gives the total molar balance,

$$\frac{d}{dt} M_{\text{tot}} = W_1 - W_2 + W_0 + \sum_{i=1}^{\text{NS}} R_{i,\text{tot}} \qquad (3.1\text{-}6)$$

The summation term is essential, since moles are not generally conserved in chemical reactions. The term $W_0 = \sum_i W_{i0}$ is the total molar transfer rate $[moles/s]$ into the system.

### 3.1.2 Total Energy Balance

The total energy balance for the system takes the form

$$
\begin{aligned}
\frac{d}{dt}\left(U + K + \varPhi\right)_{\text{tot}} = {} & \sum_{i=1}^{\text{NS}} W_{i1}\left(\overline{H}_i + M_i\widehat{K} + M_i\widehat{\varPhi}\right)_1 \\
& - \sum_{i=1}^{\text{NS}} W_{i2}\left(\overline{H}_i + M_i\widehat{K} + M_i\widehat{\varPhi}\right)_2 \\
& + \sum_{i=1}^{\text{NS}} W_{i0}\left(\overline{H}_i + M_i\widehat{K} + M_i\widehat{\varPhi}\right)_0 + Q + W_m,
\end{aligned}
\tag{3.1-7}
$$

in which the summands are averages over their respective surfaces. Thus, the rate of change of the total energy (internal, kinetic, and potential) of the system is equal to the net inflow of enthalpy, kinetic energy, and potential energy plus the rate of heat input $Q$ and the mechanical power input $W_m$.

The overlined quantities $\overline{H}_i$ are partial molar enthalpies, described in Bird, Stewart, and Lightfoot (2002), Section 19.3. If heats of mixing are neglected, as is usual for gas mixtures at moderate pressures, the partial molar enthalpies $\overline{H}_i$ reduce to pure-component molar enthalpies $\widetilde{H}_i$. The internal energy $U$ and the enthalpies must all be based on consistent datum states for the chemical elements.

### 3.1.3 Momentum Balance

A momentum balance on the macroscopic system gives

$$
\begin{aligned}
\frac{d}{dt}\boldsymbol{P}_{\text{tot}} = {} & \left(\frac{<\rho v_n \boldsymbol{v}>}{<\rho v_n>}w + p\boldsymbol{S}\right)_1 \\
& - \left(\frac{<\rho v_n \boldsymbol{v}>}{<\rho v_n>}w + p\boldsymbol{S}\right)_2 \\
& + \left(\frac{<\rho v_n \boldsymbol{v}>}{<\rho v_n>}w\right)_0 + \boldsymbol{F} + m_{tot}\boldsymbol{g}
\end{aligned}
\tag{3.1-8}
$$

in which $\boldsymbol{P}_{\text{tot}}$ is the total momentum vector for the system, and the terms in angle brackets ($<>$) are averages over those boundary surfaces. The boundary normal stress $[\boldsymbol{n} \bullet \boldsymbol{\tau}]$ is neglected on $\boldsymbol{S}_1$ and $\boldsymbol{S}_2$. The vector $\boldsymbol{F}$ is usually obtained by difference, as the resultant of the forces exerted on the system by all its boundaries *except* $\boldsymbol{S}_1$ and $\boldsymbol{S}_2$.

From Newton's second law, as well as Eq. (3.1-8), we know that force has the dimensions of mass times acceleration; thus, Eq. (3.1-8) and other fluid-mechanical formulas are most easily expressed by choosing mass (M), length (L), and time (t) as fundamental dimensions, and expressing force (F) in units of $ML/t^2$. By following this well-established custom, we avoid the need for the factor $g_c$ (standard gravitational acceleration), which appears if one treats both mass and force as fundamental dimensions. The Standard International unit of force is named, appropriately, the Newton $(1 \text{ Newton} = 1(kg)(meter)/s^2$.

### 3.1.4 Mechanical Energy Balance

The mechanical energy balance is not a fundamental principle; rather, it is a corollary (Bird 1957; Bird, Stewart, and Lightfoot 2002) of the equation of motion. For constant fluid density, the macroscopic mechanical energy balance takes the form

$$
\begin{aligned}
\frac{d}{dt} (K + \Phi)_{tot} = {} & w_1 \left[ \widehat{K} + \widehat{\Phi} + \frac{p}{\rho} \right]_1 \\
& - w_2 \left[ \widehat{K} + \widehat{\Phi} + \frac{p}{\rho} \right]_2 \\
& + w_0 \left[ \widehat{K} + \widehat{\Phi} + \frac{p}{\rho} \right]_0 + W_m - E_v
\end{aligned}
\tag{3.1-9}
$$

and for steady flow with the density $\rho$ represented as a function of $p$ alone, it becomes

$$
\begin{aligned}
0 = {} & w_1 \left[ \widehat{K} + \widehat{\Phi} + \int_{p_a}^{p} \frac{dp}{\rho} \right]_1 \\
& - w_2 \left[ \widehat{K} + \widehat{\Phi} + \int_{p_a}^{p} \frac{dp}{\rho} \right]_2 \\
& + w_0 \left[ \widehat{K} + \widehat{\Phi} + \int_{p_a}^{p} \frac{dp}{\rho} \right]_0 + W_m - E_v
\end{aligned}
\tag{3.1-10}
$$

The term $E_v$ is the rate of viscous dissipation of mechanical energy; its estimation is described in Bird, Stewart, and Lightfoot (2002). $p_a$ is a reference pressure, in units of $[M/Lt^2]$; the resulting SI unit is the Pascal.

Equations (3.1-4) and (3.1-7) are particularly important in reactor applications. In the following examples, we show how they simplify for some common reactor systems.

### Example 3.1. Balance Equations for Batch Reactors

Consider a well-stirred batch reactor of constant fluid volume $V$ in which the reactions occurring are homogeneous. As the system for our macroscopic balances, we choose the fluid in the reactor volume $V$; then the inflows $W_{i1}$ and outflows $W_{i2}$ vanish and no mass-transfer surface "0" is required. The species inventories $M_{i,tot}$ are expressed in terms of fluid concentrations as $Vc_i$. Then Eq. (3.1-4) takes the form

$$
V \frac{dc_i}{dt} = R_{i,tot} = VR_i \qquad i = 1, \dots, NS
\tag{3.1-11}
$$

It is common practice to define *extents of production* of the various species per unit volume of a batch reactor as

$$\xi_i = c_i - c_{iI} \qquad i = 1, \ldots, NS \qquad (3.1\text{-}12)$$

These extents are measured from the initial concentrations $c_{iI}$, so they vanish at the start of the operation. With these substitutions, Eq. (3.1-11) takes the form

$$\frac{d\xi_i}{dt} = R_i \qquad i = 1, \ldots, NS \qquad (3.1\text{-}13)$$

with initial condition

$$\xi_i = 0 \text{ at } t = 0 \qquad i = 1, \ldots, NS \qquad (3.1\text{-}14)$$

to be used for reactor simulation or to analyze data from reactor experiments.

For this system, the total energy balance in Eq. (3.1-7) gives

$$\frac{dU_{\text{tot}}}{dt} = Q \qquad (3.1\text{-}15)$$

when kinetic and potential energy changes and mechanical power input $W_m$ are neglected. Setting $U_{\text{tot}} = H_{\text{tot}} - pV$ in this equation, we get

$$(m\widehat{C}_p)_{\text{tot}}\frac{dT}{dt} + \sum_{i=1}^{NS} \overline{H}_i R_i = Q \qquad (3.1\text{-}16)$$

for this constant-volume system. For reactors heated or cooled by a heat exchange jacket, the heat input rate $Q$ is commonly represented as

$$Q = U_E A_E (T_e - T) \qquad (3.1\text{-}17)$$

Here $U_E$ is the overall heat transfer coefficient, based on the heat exchange area $A_E$, external fluid temperature $T_e$, and reactor fluid temperature $T(t)$. Another way of accomplishing the heat exchange, and the stirring as well, is to circulate the reactor fluid through a heat exchanger, which then forms part of the reactor volume. This method is useful for highly exothermic reactions.

Combining the two expressions for $Q$, we get the differential equation

$$(m\widehat{C}_p)_{\text{tot}}\frac{dT}{dt} = U_E A_E (T_e - T) - \sum_{i=1}^{NS} \overline{H}_i R_i \qquad (3.1\text{-}18)$$

for the temperature of the well-mixed reactor fluid. The summation term with negative sign affixed is the heat production from chemical reactions.

**Example 3.2. Balance Equations for CSTR Reactors**

Consider the steady operation of a perfectly-stirred tank reactor (CSTR) of constant volume $V$, in which only homogeneous reactions occur. Again, the system for the balances is chosen as the fluid contents of the reactor. Eq. (3.1-4) then gives

$$W_{i2} - W_{i1} = R_{i,\text{tot}} = VR_i \qquad i = 1, \ldots, \text{NS} \qquad (3.1\text{-}19)$$

The net outflow $W_{i2} - W_{i1}$ can be written as $F\xi_i$, where $F$ is the total molar feed rate and $\xi_i$ is a corresponding *extent of production* of species $i$ in moles per mole of feed. Then Eq. (3.1-19) becomes

$$F\xi_i = VR_i. \qquad i = 1, \ldots, \text{NS} \qquad (3.1\text{-}20)$$

The energy balance in Eq. (3.1-7), when combined with Eq. (3.1-17), gives the following expressions for the required heat input rate and external fluid temperature for steady operation

$$Q = \sum_{i=1}^{\text{NS}} \left(W_i \overline{H}_i\right)_2 - \sum_{i=1}^{\text{NS}} \left(W_i \overline{H}_i\right)_1 \qquad (3.1\text{-}21a)$$

$$= W_2 \tilde{H}_2 - W_1 \tilde{H}_1 = U_E A_E (T_e - T) \qquad (3.1\text{-}21b)$$

when kinetic and potential energy changes and work of stirring are neglected.

For transient CSTR operation, Eqs. (3.1-4), (3.1-7), and (3.1-17) give

$$V \frac{dc_i}{dt} = W_{i1} - W_{i2} + VR_i \qquad i = 1, \ldots, \text{NS} \qquad (3.1\text{-}22)$$

and

$$\frac{d}{dt} U_{tot} = \sum_{i=1}^{\text{NS}} \left(W_i \overline{H}_i\right)_1 - \sum_{i=1}^{\text{NS}} \left(W_i \overline{H}_i\right)_2 + Q \qquad (3.1\text{-}23a)$$

$$= W_1 \tilde{H}_1 - W_2 \tilde{H}_2 + U_E A_E (T_e - T) \qquad (3.1\text{-}23b)$$

In the literature, these equations are commonly simplified by assuming a constant volumetric input flow rate $q$ and constant density and pressure throughout the operation; then Eqs. (3.1-20) and (3.1-23b) reduce to

$$\frac{dc_i}{dt} = (q/V)(c_{i1} - c_{i2}) + R_i \qquad i = 1, \ldots, \text{NS} \qquad (3.1\text{-}24)$$

$$(m\widehat{C}_p)_{tot} \frac{dT}{dt} + \sum_{i=1}^{\text{NS}} \overline{H}_i R_i = w_1(\widehat{H}_1 - \widehat{H}_2) + U_E A_E (T_e - T) \qquad (3.1\text{-}25)$$

## Example 3.3. Balances for Plug-Flow Homogeneous Reactors (PFR)

Consider the steady operation of a tubular reactor in which only homogeneous reactions occur. The state of the fluid now depends on the downstream distance $z$. Our chosen system for the macroscopic balances consists of the reactor contents between $z$ and $z + \Delta z$. Equation (3.1-4) then gives

$$
\begin{aligned}
0 &= W_i|_z - W_i|_{z+\Delta z} + \Delta R_{i,\text{tot}} \\
&= -\Delta W_i + < R_i > \Delta V \qquad i = 1, \ldots, \text{NS}
\end{aligned}
\tag{3.1-26}
$$

in which $\Delta V = (\pi D^2/4)\Delta z$, and $<>$ denotes an average over the volume $\Delta V$. Division by $\Delta V$ gives, in the limit as $\Delta V \to 0$,

$$
\frac{dW_i}{dV} = < R_i > \qquad i = 1, \ldots, \text{NS}
\tag{3.1-27}
$$

as the set of differential mass-balance equations. Alternatively, this set of equations can be written in terms of the total molar feed rate $F$ to the reactor and a set of extents of production

$$
\xi_i = \frac{W_i - W_{i1}}{F} \qquad i = 1, \ldots, \text{NS}
\tag{3.1-28}
$$

expressed in moles per mole of feed. These extents satisfy the differential equation system

$$
F\frac{d\xi_i}{dV} = < R_i > \qquad i = 1, \ldots, \text{NS}
\tag{3.1-29}
$$

with the inlet condition

$$
\xi_i = 0 \text{ at } V = 0 \qquad i = 1, \ldots, \text{NS}
\tag{3.1-30}
$$

A corresponding steady-state treatment of Eq. (3.1-7) gives, for gas-phase operation at moderate pressure (so that $d\widetilde{H}_i = \widetilde{C}_{pi}dT$),

$$
\sum_{i=1}^{\text{NS}} W_i \widetilde{C}_{pi}\frac{dT}{dV} = -\sum_{i=1}^{\text{NS}} R_i \overline{H}_i + \frac{dA_E}{dV}U_E(T_e - T)
\tag{3.1-31}
$$

as the differential equation for the temperature of the reacting fluid. Here $dA_E/dV$ is the incremental heat exchange area per unit reactor volume.

## Example 3.4. Balances for Plug-Flow Fixed-Bed Reactors

Consider the steady operation of a fixed-bed catalytic reactor in which only heterogeneous reactions occur. The practical measure of reaction space is now the catalyst mass rather than the reactor volume, which can vary according to (i) the density to which the catalyst bed is packed and (ii) the volume fraction of inert particles used by the experimenter to dilute the catalyst, thus reducing temperature excursions in the reactor. Let $m_c(z)$ be the catalyst mass in an interval $(0, z)$ of the reactor, and let $\widehat{R}_i \Delta m_c$ be the molar production rate of species $i$ in a catalyst mass increment $\Delta m_c$. Then Eq. (3.1-4) for that increment takes the form

$$\Delta W_i = \widehat{R}_i \Delta m_c \qquad i = 1, \ldots, \text{NS} \qquad (3.1\text{-}32)$$

Taking the limit as $\Delta m_c \to 0$, we get the differential equations

$$\frac{dW_i}{dm_c} = \widehat{R}_i \qquad i = 1, \ldots, \text{NS} \qquad (3.1\text{-}33)$$

for the species molar flow rates and

$$F \frac{d\xi_i}{dm_c} = \widehat{R}_i \qquad i = 1, \ldots, \text{NS} \qquad (3.1\text{-}34)$$

for the species extents of production $\xi_i = [W_i - W_{i1}]/F$ in moles of $i$ per mole of reactor feed. These extents satisfy the inlet conditions

$$\xi_i = 0 \text{ at } m_c = 0 \qquad i = 1, \ldots, \text{NS} \qquad (3.1\text{-}35)$$

For steady state, Eq. (3.1-7) gives

$$0 = \sum_{i=1}^{\text{NS}} [W_{i1}\widetilde{H}_{i1} - W_{i2}\widetilde{H}_{i2} + W_{i0}\widetilde{H}_{i0}] + Q \qquad (3.1\text{-}36)$$

when kinetic and potential energy changes, heats of mixing, and mechanical power input $W_m$ are neglected. Equating the resulting heat flow $Q$ to that of Eq. (3.1-17) would give an expression for the temperature profile along the reactor, were it not for the temperature difference required to transfer heat of reaction between the catalyst particles and the fluid. This temperature difference, and the corresponding differences between the compositions in the fluid and at the particle exteriors, require careful treatment to model a reactor properly for process scale-up. In the following sections, we analyze these phenomena to get better models of fixed-bed operations.

## 3.2 HEAT AND MASS TRANSFER IN FIXED BEDS

We define local transfer coefficients for a binary fluid in a packed bed by the relations

$$q_0 = h_{\mathrm{loc}}(T_0 - T_b) \tag{3.2-1}$$

$$J_{A0}^\star = k_{x,\mathrm{loc}}(x_{A0} - x_{Ab}) \tag{3.2-2}$$

for the conduction and diffusion fluxes into the fluid from a representative interfacial area of the two-phase system. These relations correspond to Eqs. (22.1-7,8) of Bird, Stewart, and Lightfoot (2002), with the reference temperature $T^\circ$ for enthalpy equated to $T_0$.

Extensive data on forced convection of gases[1] and liquids[2] through packed beds have been critically analyzed to obtain the following correlation:[3]

$$j_H = j_D = 2.19\mathrm{Re}^{-2/3} + 0.78\mathrm{Re}^{-0.381} \tag{3.2-3}$$

Here the Chilton-Colburn $j$-factors and the Reynolds number Re are defined by

$$j_H = \frac{h_{\mathrm{loc}}}{\widehat{C}_p G_0}\left(\frac{\widehat{C}_p \mu}{k}\right)^{2/3} \tag{3.2-4}$$

$$j_D = \frac{k_{x,\mathrm{loc}}}{c G_0}\left(\frac{\mu}{\rho \mathcal{D}_{AB}}\right)^{2/3} \tag{3.2-5}$$

$$\mathrm{Re} = \frac{D_p G_0}{(1-\varepsilon)\mu\psi} = \frac{6G_0}{a\mu\psi} \tag{3.2-6}$$

and $a$ is the external surface area of particles per unit bed volume. In these equations, $G_0 = w/S$ is the streamwise mass flux averaged over the bed cross-section, and the physical properties are all evaluated at a common reference state, for which we choose the arithmetic mean of $(T_0, x_{A0})$ and $(T_b, x_{Ab})$. $\psi$ is a particle-shape factor, with a defined value of 1 for spheres and a fitted value[3] of 0.92 for cylindrical pellets of diameter equal to their length.

For small Re, Equation (3.2-3) yields the asymptote

$$j_H = j_D = 2.19\mathrm{Re}^{-2/3} \tag{3.2-7}$$

[1] B. W. Gamson, G. Thodos, and O. A. Hougen, *Trans. AIChE*, **39**, 1–35 (1943); C. R. Wilke and O. A. Hougen, *Trans. AIChE*, **41**, 445–451 (1945).
[2] L. K. McCune and R. H. Wilhelm, *Ind. Eng. Chem.*, **41**, 1124-1134 (1949); J. E. Williamson, K. Bazaire, and C. J. Geankoplis, *Ind. Eng. Chem. Fund*, **2**, 126–129 (1963); E. J. Wilson and C. J. Geankoplis, *Ind. Eng. Chem. Fund.*, **5**, 9–14 (1966).
[3] R. B. Bird, W. E. Stewart, and E . N. Lightfoot, *Transport Phenomena*, 2nd edition, Wiley, New York (2002), 441–442.

consistent with boundary layer theory[4] for creeping flow with RePr$\gg$ 1. This asymptote represents the creeping-flow mass-transfer data for liquids very well.

The exponent $2/3$ in Eqs. (3.2-4,5) is an asymptote given by boundary layer theory for steady laminar flows[4] and for steadily driven turbulent flows.[5] This dependence is consistent with the cited data for values above 0.6 of the Prandtl number $\mathrm{Pr} = \widehat{C}_p\mu/k$ and of the Schmidt number $\mathrm{Sc} = \mu/\rho\mathcal{D}_{AB}$.

In the following section we apply these results to multicomponent simulations of fixed-bed reactors.

## 3.3 INTERFACIAL STATES IN FIXED-BED REACTORS

Realistic analysis of fixed-bed reactor experiments requires calculation of interfacial states. Laboratory reactors are typically much shorter than full-scale units and operate with smaller axial velocities, producing significant departures of the interfacial states from the measurable values in the mainstream fluid and consequent difficulties in establishing catalytic reaction models. Interfacial temperatures and partial pressures were calculated with $j_H$ and $j_D$, and used in estimating reaction model parameters, in a landmark paper by Yoshida, Ramaswami, and Hougen (1962). Here we give an updated analysis of interfacial states in fixed-bed reactor operations for improved treatment of catalytic reaction data.

For a single chemical reaction in a catalyst particle at steady state, mass balances give the expressions

$$N_{i0} = \frac{\widehat{\mathcal{R}}_j}{\hat{a}}\nu_{ji} \qquad i = 1, \mathrm{NS} \tag{3.3-1}$$

for the interfacial molar fluxes into the surrounding fluid. Here $\widehat{\mathcal{R}}_j$ is the rate (i.e., the frequency per unit catalyst mass) of molar reaction event $j$, $\hat{a}$ is the particle's external area per unit catalyst mass, and the stoichiometric coefficients $\nu_{ji}$ are defined as in Eq. (2.1-1). An energy balance then gives the interfacial energy flux as

$$q_0 = -\sum_{i=1}^{\mathrm{NS}} N_{i0}\widetilde{H}_i(T_0) \tag{3.3-2}$$

an accurate result for gases at moderate pressures.

---

[4] W. E. Stewart, *AIChE J.*, **9**, 528–535 (1963); R. Pfeffer, *Ind. Eng. Chem. Fund.*, **3**, 380-383 (1964): J. P. Sørensen and W. E. Stewart, *Chem. Eng. Sci.* **29**, 833-837 (1974).

[5] W. E. Stewart, *AIChE J.*, **33**, 2008-2016 (1987); corrigenda **34**, 1030 (1988).

Now consider the calculation of the particle-fluid temperature difference. Mass conservation on and within the catalyst particles gives a vanishing mass-average velocity $v$ at their exterior surfaces when pressure-driven flow through the particles is neglected. The heat transfer coefficient $h_{loc}$ (hereafter abbreviated as $h$) can then be predicted by Eqs. (3.2-3,4) if we neglect the usually unimportant Soret (thermal diffusion) and Dufour (concentration-driven heat flux) effects, as Yoshida, Ramiswami, and Hougen did. To improve on this prediction, we correct $h$ for the interfacial molar fluxes using the correction factor

$$\theta_T = \frac{\phi_T}{\exp\phi_T - 1} \tag{3.3-3}$$

adapted from the film model given in Section 22.8 of Bird, Stewart, and Lightfoot (2002). Here $\theta_T$ is the ratio of the heat transfer coefficient $h^\bullet$ in the presence of mass transfer to the coefficient $h$ in its absence [predicted by Eqs. (3.2-3,4)], and

$$\phi_T = \sum_{i=1}^{NS} N_{i0}\widetilde{C}_{pi}(T_0)/h \tag{3.3-4}$$

is an extension of the dimensionless rate factor $\phi_T$ of Bird, Stewart, and Lightfoot (2002) to transfer of multiple species. Then the predicted exterior temperature of the catalyst particle considered here is

$$T_0 = T_b + q_0/h^\bullet = T_b + q_0/(h\theta_T) \tag{3.3-5}$$

The preceding five equations may be solved by successive substitution, beginning with the assumption $T_0 = T_b$ to initiate a rapidly convergent iteration process.

Next, we predict the interfacial gas composition on the chosen particle, using a detailed multicomponent diffusion law. For this calculation we use the array

$$[\boldsymbol{x}] = \begin{pmatrix} x_1 \\ \vdots \\ x_{NS-1} \end{pmatrix} \tag{3.3-6}$$

of species mole fractions and the array

$$[\boldsymbol{J}^\star] = \begin{pmatrix} J_1^\star \\ \vdots \\ J_{NS-1}^\star \end{pmatrix} \tag{3.3-7}$$

of species molar fluxes with respect to the molar average velocity $\boldsymbol{v}^\star$. This choice of variables yields the species conservation equations

$$c\left[\frac{\partial}{\partial t}[\boldsymbol{x}] + (\boldsymbol{v}^\star \bullet \nabla[\boldsymbol{x}])\right] = -(\nabla \bullet [\boldsymbol{J}^\star]) \tag{3.3-8}$$

and the flux equations

$$c\nabla[x] = -[A][J^*] \qquad (3.3\text{-}9)$$

analyzed by Stewart and Prober[1] for multicomponent isothermal systems. A diffusivity matrix $[D]$, the inverse of $[A]$, was used by Toor[2] in his independent analysis of multicomponent mass-transfer problems.

Though the molar average velocity is not often used in fluid mechanics, it works well in boundary-layer developments. The tangential components of the velocities $v$ (mass average) and $v^*$ (molar average) agree closely in boundary layers because the tangential diffusion fluxes there are unimportant; this fact was used by Stewart (1963) to solve the boundary-layer diffusion equation for variable-density three-dimensional flows. The present choice of variables gives an accurate representation of diffusion normal to the interface, and thus of the interphase mass-transfer rates and driving forces.

Kinetic theory (Maxwell 1860, 1868; Stefan 1871, 1872; Curtiss and Hirschfelder 1949) gives the elements of $[A]$ as follows in the moderate-density gas region:

$$A_{ij} = \frac{x_i}{\mathcal{D}_{i\text{NS}}} - \frac{x_i}{\mathcal{D}_{ij}} \quad i,j = 1,\dots,\text{NS}-1, \quad j \neq i \qquad (3.3\text{-}10a)$$

$$A_{ii} = \frac{x_i}{\mathcal{D}_{i\text{NS}}} + \sum_{\substack{j=1 \\ j \neq i}}^{\text{NS}} \frac{x_i}{\mathcal{D}_{ij}} \quad i = 1,\dots,\text{NS}-1 \qquad (3.3\text{-}10b)$$

In this first approximation of kinetic theory, the divisors $\mathcal{D}_{ij}$ and $\mathcal{D}_{i\text{NS}}$ are the *binary* gas-phase diffusivities of the corresponding pairs of species at $c$ and $T$.

In the following development, Eqs. (3.3-8) and 3.3-9) are linearized by evaluating each coefficient in them at the reference state $(T_0+T_b)/2$, $([x_0]+[x_b])/2$. With this assumption, the matrix $A$ is reducible to diagonal form by the transformation

$$[P]^{-1}[A][P] = \begin{bmatrix} \bar{A}_1 & & \\ & \ddots & \\ & & \bar{A}_{\text{NS}-1} \end{bmatrix} \qquad (3.3\text{-}11)$$

as proved by Stewart and Prober (1964) for isothermal systems. Here $[P]$ is the matrix of column eigenvectors of $[A]$, and $\bar{A}_1,\dots\bar{A}_{\text{NS}-1}$ are the corresponding eigenvalues. These eigenvalues are positive real numbers at any locally stable thermodynamic state of the mixture (Stewart and Prober 1964), including states where one or more mole fractions vanish, and are invariant

---

[1] W. E. Stewart and R. Prober, *Ind. Eng. Chem. Fund.*, **3**, 224–235 (1964).

[2] H. L. Toor, *AIChE J.*, **10**, 460–465 (1964).

to similarity transformations of $[A]$ into other systems of state variables and fluxes (Cullinan 1965). The matrices $[P]^{-1}$, $[P]$ and the eigenvalues $\bar{A}_i$ are computable efficiently by the subroutines DSYTRD, DORGTR, and DSTEQR, described in the *LAPACK Users' Guide* by Anderson et al., 1992, and available from www.netlib.org.

Equation (3.3-11) indicates that the following transformations of state and flux arrays will be useful:

$$[\bar{x}] = [P]^{-1}[x] = \begin{bmatrix} \bar{x}_1 \\ \vdots \\ \bar{x}_{NS-1} \end{bmatrix} ; \qquad \begin{bmatrix} x_1 \\ \vdots \\ x_{NS-1} \end{bmatrix} = [P][\bar{x}] \qquad (3.3\text{-}12,13)$$

$$[\bar{J}^\star] = [P]^{-1}[J^\star] = \begin{bmatrix} \bar{J}_1^\star \\ \vdots \\ \bar{J}_{NS-1}^\star \end{bmatrix} ; \qquad \begin{bmatrix} J_1^\star \\ \vdots \\ J_{NS-1}^\star \end{bmatrix} = [P][\bar{J}^\star] \qquad (3.3\text{-}14,15)$$

Hereafter, in the manner of Stewart and Prober (1964), a bar is placed above each array computed by premultiplication with $[P]^{-1}$. A bar is also placed above each element of any arrays that are thus computed.

Premultiplication of Eq. (3.3-7) by $[P]^{-1}$ gives uncoupled flux equations

$$c\nabla\bar{x}_i = -\bar{\lambda}_i\bar{J}_i^\star \text{ for } i = 1,\dots,NS-1 \qquad (3.3\text{-}16)$$

having the form of Fick's first diffusion law for $NS-1$ corresponding binary systems. Equation (3.3-6) correspondingly transforms to the uncoupled equation set

$$c\left[\frac{\partial\bar{x}_i}{\partial t} + (\boldsymbol{v}^\star \bullet \bar{x}_i)\right] = -\left(\nabla \bullet \bar{J}_i^\star\right) \text{ for } i = 1,\dots,NS-1 \qquad (3.3\text{-}17)$$

Thus, the transformed state variables $\bar{x}_i$ and transformed fluxes $\bar{J}_i^\star$ for each $i$ satisfy the flux equations and conservation equations of a binary system (Stewart and Prober 1964) with the same $\boldsymbol{v}^\star$ function (laminar or turbulent) as the multicomponent system and with a material diffusivity $\bar{D}_i = c/\bar{\lambda}_i$, giving a Schmidt number $\mu/\rho\bar{D}_i = (\mu/\rho)(\bar{\lambda}_i/c)$.

In a binary flow system with interfacial fluid composition $x_{A0}$ and bulk fluid composition $x_{Ab}$, the local mass-transfer coefficient $k_x^\bullet$ corrected for the interfacial total molar flux $cv_0^\star = [N_{A0} + N_{B0}]$ satisfies

$$N_{A0} - x_{A0}[N_{A0} + N_{B0}] = k_x^\bullet(x_{A0} - x_{Ab}) \qquad (3.3\text{-}18)$$

Corresponding relations hold for multicomponent mass transfer with transformed compositions $\bar{x}_{i0}$ and $\bar{x}_{ib}$:

$$\bar{N}_{i0} - \bar{x}_{i0}\sum_{j=1}^{NS} N_{j0} = \bar{k}_{xi}^\bullet(\bar{x}_{i0} - \bar{x}_{ib}) \qquad i = 1,\dots,NS-1 \qquad (3.3\text{-}19)$$

Equation (3.3-19) corresponds to a matrix equation

$$\left[ \bar{N}_0 - \bar{x}_0 \sum_{j=1}^{NS} N_{j0} \right] = [\bar{k}_x^{\bullet}][\bar{x}_0 - \bar{x}_b] \tag{3.3-20}$$

in which $\bar{k}_x^{\bullet}$ is the diagonal matrix of elements $\bar{k}_{x1}^{\bullet}, \ldots, \bar{k}_{xNS-1}^{\bullet}$, $\bar{N}_0$ is the array of transformed interfacial fluxes $\bar{N}_{10}, \ldots, \bar{N}_{NS-1,0}$, and $[\bar{x}_0]$ and $[\bar{x}_b]$ are the arrays of transformed interfacial and bulk mole fractions. Application of Eqs. (3.3-12) through (3.3-15) to this result gives

$$\left[ N_0 - x_0 \sum_{j=1}^{NS} N_{j0} \right] = [\boldsymbol{P}][\bar{k}_x^{\bullet}][\boldsymbol{P}]^{-1}[x_0 - x_b] \tag{3.3-21}$$

for the multicomponent molar fluxes when the terminal compositions $[x_0]$ and $[x_b]$ are given.

A fixed-bed reactor experiment can provide the arrays $[N_0]$ and $[x_b]$, but the interfacial composition $[x_0]$ needs to be calculated, as it is not measurable and may differ considerably from the bulk composition $[x_b]$, as shown by Yoshida, Ramaswami, and Hougen (1962). The following equation from Stewart and Prober (1964) gives the molar composition $[x_0]$ of the gas at the outer surface of a representative catalyst particle:

$$[x_0] = [x_b] + [\boldsymbol{P}] \left[ 1 / \left( k_x^{\bullet} + \sum_{j=1}^{NS} N_{j0} \right) \right] [\boldsymbol{P}]^{-1} \left[ N_0 - x_b \sum_{j=1}^{NS} N_{j0} \right] \tag{3.3-22}$$

Here the array between $[\boldsymbol{P}]$ and $[\boldsymbol{P}]^{-1}$ is a diagonal matrix, with its $i$th diagonal element given by the reciprocal of $\bar{k}_{xi}^{\bullet} + \sum_{j=1}^{NS} N_{j0}$.

The transformed mass transfer coefficients $\bar{k}_{xi}^{\bullet}$ may be predicted as follows from the results of Section 3.2 and the $i$th eigenvalue of the matrix $\boldsymbol{A}$ introduced above:

$$\bar{k}_{xi}^{\bullet} = \theta_{xi} \bar{k}_{xi} = \theta_{xi} j_D c G_0 \left( \frac{\mu \bar{A}_i}{\rho c} \right)^{2/3} \qquad i = 1, \ldots, NS - 1 \tag{3.3-23}$$

Here $\theta_{xi}$ is the interfacial-velocity correction factor for $\bar{k}_{xi}$. We recommend the film model

$$\theta_{xi} \approx \frac{\phi_{xi}}{(\exp \phi_{xi} - 1)}; \qquad \phi_{xi} = \sum_{j=1}^{NS} N_{j0}/k_{xi} \tag{3.3-24,25}$$

for this factor, in place of Table 1 of Stewart (1963), to include an estimate of the boundary-layer thickness factor $\delta/\delta^0$ that was mentioned but not provided there.

For analysis of fixed-bed reactor data, Eqs. (3.3-5) and (3.3-23) should be used to calculate interfacial states at representative points along the reactor. Use of these corrected interfacial states, and the intraparticle states of Section 3.4, will allow improved estimates of reaction model parameters by the methods of Chapters 6 and 7.

## Example 3.5. Calorimetric Measurement of Reaction Rate

The reactor design for the propylene hydrogenation experiments of Shabaker (1965) is described here, as analyzed further by Lu (1988).

Shabaker (1965) designed a multibed calorimetric reactor and used it for differential rate measurements on hydrogenation of propylene over Sinclair-Baker Standard RD-150 catalyst. The calorimetric method, with a 10-junction bismuth-antimony thermopile outside each catalyst bed, allowed precise rate measurements at propylene mole fractions down to 0.02 [versus 0.15 for the gas chromatographic method of Rogers (1961)] and gave continuous information on the catalyst activity. Data were taken simultaneously on fine-ground catalyst (40-50 Tyler mesh size) and full-size particles (extrudate of 0.058 inch diameter and 0.06 to 0.2 inch length); the data from the fine-ground catalyst are considered here.

Interpolation in the screen size table of *Perry's Chemical Engineers' Handbook* (Perry, Green, and Mahoney 1984) gives the screen openings as 0.0147 mm and 0.0111 mm, respectively, for the 40 and 50 Tyler mesh sizes. Treating the ground particles as spheres and using the pellet density $\rho_p$ of 1.22 g/cm$^3$, we calculate the specific surface of the 40-50 mesh fraction as $\hat{a} = [6/D_{\max} + 6/D_{\min}]/2\rho_p = 3890$ cm$^2$/g.

The reactor shell was about 8 feet tall and 8 inches in outer diameter (O.D.). Provision was made for simultaneous reaction rate measurements on five small catalyst charges mounted in series inside a 3/8 inch I.D. thin-wall stainless steel tube centered in the shell; the data from the upstream bed are considered here.

Tiny amounts of catalyst, on the order of 50 mg, were inserted into the reactor tube in steel catalyst boats 1 inch long. The ends of the boats had female threads, allowing them to be sealed roughly with screws while weighing and transporting new charges to the reactor. The fines were supported in thin beds 1/4 inch in diameter on 200-mesh screens soldered into the boat walls. The boat for whole pellets had an axial cylindrical hole 0.127 inch in diameter and 1/2 inch long. The pellets were stacked end to end in a basket made from two 0.029 inch diameter spring steel wires formed into U-bends 1/2 inch long, held in perpendicular planes by a drop of solder at their intersection. The diameters of the hole and the wires were chosen to give a close-fitting basket for pellets of 0.058 inch diameter while avoiding mechanical damage to the catalyst.

The annular space between the reactor tube and shell contained the thermopiles and was thermally insulated with Santocell-powdered silica gel.

Thermal conductivities about one-fifth that of still air were attained by evacuating the shell and insulation to a total pressure of 10 microns Hg.

Very effective heat-exchange sections were provided between the catalyst boats with 15 inch lengths of steel pinion gear rod. These sections were designed to provide good thermal contact between the gas and the tube wall so that the gas temperature rise caused by reaction in the catalyst bed could be measured by the temperature change along the wall. Minimal clearance of about 0.003 inch was left between the gear teeth and the tube wall, so that the gas stream was effectively separated into 16 trapezoidal channels 0.048 inch deep between the gear teeth. The length of these channels sufficed for full development of the temperature rise across each bed.

The gear teeth were stripped for a 9/16-inch length from the ends of each gear rod. The first and last 3/16 inch of each rod were threaded to fit the 1/4 inch female threads in the ends of the catalyst boats. Thus, the reactor core assembly was joined securely for accurate positioning in the reactor tube. The reactor was designed so that the boats and gear sections could be inserted or removed individually when a fitting in the process line below the reactor was opened.

During a run, the gas flowed down between the gear teeth to the center of the rod, and then through an axial hole in the rod to enter the boat. Screens of 200-mesh were soldered across the ends of the center holes in the rods to prevent catalyst fines from being blown upward out of their boat. These screens also served to center the basket in the boat containing the full-size pellets.

Teflon O-rings were placed between each boat and the shoulder of the adjacent gear rod to seal the small clearance between the boat circumference and the tube wall. The need for these seals was dramatically demonstrated by a runaway reaction observed on fines at $1°C$ caused by insufficient heat transfer from the particles to the gas stream when the seals were omitted. Since the O-rings became worn after repeated removal of the core assembly from the reactor, they were replaced frequently.

The maximum hydrogenation rate observed in Shabaker's experiments was $\widehat{\mathcal{R}}_1 = 0.52$ mols/hr/g at a bulk gas temperature $T_b$ of 307 K, pressure $p$ of 3.05 atm, and propylene mole fraction $x_b$ of 0.0404 in a catalyst bed having a catalyst charge of 33.9 mg and a wall diameter of 1/4 inch.


## 3.4 MATERIAL TRANSPORT IN POROUS CATALYSTS

Two approaches, identified by Jackson (1977) in his excellent monograph, have become popular for modeling material transport in porous media. The dusty-gas approach, introduced by James Clerk Maxwell (1860, 1868) and generalized by E. A. Mason and collaborators, models a porous solid as a swarm of very massive particles. The smooth-field approach of Stewart and collaborators (Johnson and Stewart 1965; Feng and Stewart 1973;

Feng, Kostrov, and Stewart 1974) applies the dusty-gas equations to a thoroughly cross-linked network of cylindrical pore segments with distributed sizes and orientations. Example 7.2 and the latter two articles give strong experimental evidence for the smooth-field approach for nonreacting systems. The smooth-field approach is used exclusively in this section, though a different approach has been given by Jackson (1977) to describe reactions in micropores.

### 3.4.1 Material Transport in a Cylindrical Pore Segment

Consider the flow and diffusion of NS chemical species in the $\ell$ direction of a cylindrical pore segment within a solid. We apply the flux equations of Mason and Evans (1969) to such a segment and add a surface diffusion term, giving

$$N_{i\ell} = N_{i\ell}^{(v)} + N_{i\ell}^{(g)} + N_{i\ell}^{(s)} \qquad i = 1, \dots, \text{NS} \qquad (3.4\text{-}1)$$

The right-hand terms are the fluxes of species $i$ by viscous flow, gaseous diffusion, and surface diffusion.

The viscous flow contribution is

$$N_{i\ell}^{(v)} = -c_i \frac{r^2}{8\mu} \frac{dp}{d\ell} \qquad i = 1, \dots, \text{NS} \qquad (3.4\text{-}2)$$

for laminar flow in a pore of radius $r$. The gaseous diffusion term, $N_{i\ell}^{(g)}$, is obtained by solving the following system of equations:

$$\frac{dc_i}{d\ell} = \frac{-N_{i\ell}^{(g)}}{\mathcal{D}_{iK}(r)} + \sum_{j=1}^{\text{NS}} \frac{N_{j\ell}^{(g)} x_i - N_{i\ell}^{(g)} x_j}{\mathcal{D}_{ij}} \qquad i = 1, \dots, \text{NS} \qquad (3.4\text{-}3)$$

Here $\mathcal{D}_{iK}(r)$ is a Knudsen diffusivity, which we take as the limiting value for free-molecule flow in a long tube of radius $r$ Ångstroms,

$$\mathcal{D}_{iK}(r) = 9.7 \times 10^{-5} r \sqrt{T/M_i} \qquad i = 1, \dots, \text{NS} \qquad (3.4\text{-}4)$$

and the $\mathcal{D}_{ij}$ are binary gas-phase diffusivities at the local molar density and temperature. The surface diffusion term is written as

$$N_{i\ell}^{(s)} = -\mathcal{D}_{is}(r) \frac{dc_i}{d\ell} \qquad i = 1, \dots, \text{NS} \qquad (3.4\text{-}5)$$

with the flux and diffusivity based on the pore cross section $\pi r^2$. This expression is limited to low surface coverages; at high coverages the adsorbed species would interact, and a matrix of surface diffusivities $\mathcal{D}_{ijs}$ would be required.

The column vector of species fluxes along a pore segment is then

$$[N_\ell] = -[c]\frac{r^2}{8\mu}\frac{dp}{d\ell} - [F(r)]^{-1}[\frac{dc}{d\ell}] - [D_s(r)][\frac{dc}{d\ell}] \qquad (3.4\text{-}6)$$

Here $[N_\ell]$ and $[c]$ are NS-element vectors with elements $N_{i\ell}$ and $c_i$, $[D_s(r)]$ is a diagonal matrix of surface diffusivities $D_{is}(r)$, and $[F(r)]$ is the matrix formed by the coefficients in Eq. (3.4-3):

$$F_{ij}(r) = -\frac{x_i}{D_{ij}} \qquad \begin{matrix} i = 1,\dots,\text{NS} \\ (i \neq j) \end{matrix} \qquad (3.4\text{-}7a)$$

$$F_{ii}(r) = \frac{1}{D_{iK}(r)} + \sum_{\substack{h=1 \\ h \neq i}}^{\text{NS}} \frac{x_h}{D_{ih}} \qquad (3.4\text{-}7b)$$

Equation (3.4-6) gives an explicit solution of the flux equations of Mason and Evans (1969) for a capillary tube, plus a vector of surface diffusion terms. These equations involve the tacit assumption that each gaseous molecule is small relative to the pore radius $r$.

### 3.4.2 Material Transport in a Pore Network

Now consider the total material flux in a porous solid. We postulate a thoroughly cross-linked main pore network with singly attached dead-end pore branches.

Because of the cross-linking, the fluid states in neighboring pores of the main network are similar. The gradients along each of these pores can then be approximated as

$$\frac{dp}{d\ell} = (\boldsymbol{\delta}_\ell \bullet \nabla p) \qquad (3.4\text{-}8a)$$

$$\frac{dc_i}{d\ell} = (\boldsymbol{\delta}_\ell \bullet \nabla c_i) \qquad (3.4\text{-}8a)$$

Here $\boldsymbol{\delta}_\ell$ is a unit vector along the given pore segment, and $p$ and $c_i$ are smoothed profiles for the main pore network.

Let $f(r, \Omega)$ denote the void fraction of main-network pores per unit intervals of pore radius $r$ and orientation (solid angle) $\Omega$. Then the smoothed molar flux of species $i$ in the medium is

$$\boldsymbol{N}_i = \int_r \int_\Omega \boldsymbol{\delta}_\ell N_{i\ell} f(r, \Omega) d\Omega dr \qquad (3.4\text{-}9)$$

With Eqs. (3.4-6) and (3.4-8), this gives

$$[\boldsymbol{N}] = -\int_r \int_\Omega [c]\frac{r^2}{8\mu}\boldsymbol{\delta}_\ell\boldsymbol{\delta}_\ell \bullet \nabla p f(r, \Omega) d\Omega dr$$

$$- \int_r \int_\Omega [F(r)]^{-1}\boldsymbol{\delta}_\ell\boldsymbol{\delta}_\ell \bullet \nabla[c] f(r, \Omega) d\Omega dr \qquad (3.4\text{-}10)$$

$$- \int_r \int_\Omega [D_s(r)]\boldsymbol{\delta}_\ell\boldsymbol{\delta}_\ell \bullet \nabla[c] f(r, \Omega) d\Omega dr$$

This result can be expressed in terms of measurable quantities, under the assumption that $p$ and $[c]$ are independent of $\Omega$ and $r$ in the main network:

$$[\boldsymbol{N}] = -\frac{1}{\mu}[c]\boldsymbol{B}\bullet\nabla p - \int_{r=0}^{\infty}[\boldsymbol{F}(r)]^{-1}\boldsymbol{\kappa}(r)\bullet\nabla[c]d\epsilon(r) - [\boldsymbol{\mathcal{D}}_s]\bullet\nabla[c] \quad (3.4\text{-}11)$$

Here $\epsilon(r)$ is the measured void fraction in pores of radii less than $r$. The permeability $\boldsymbol{B}$ and the tortuosity $\boldsymbol{\kappa}(r)$ are symmetric second-order tensors according to the preceding two equations; they can be predicted as integrals over $r$ and $\Omega$ but are better obtained by fitting Eq. (3.4-11) to material flux measurements for the given porous solid.

If $\nabla p$ and $\nabla[c]$ are directed along a laboratory axis $z$, Eq. (3.4-11) gives

$$[N_z] = -\frac{1}{\mu}[c]\beta_{zz}\frac{dp}{dz} - \int_{r=0}^{\infty}[\boldsymbol{F}(r)]^{-1}\frac{d[c]}{dz}\kappa_{zz}(r)d\epsilon(r) - [\boldsymbol{\mathcal{D}}_{s,zz}]\frac{d[c]}{dz} \quad (3.4\text{-}12)$$

This corresponds to the usual arrangement for testing cylindrical catalyst pellets. If the pore system is isotropic (randomly oriented) then $\boldsymbol{B}$, $\boldsymbol{\kappa}$, and $\boldsymbol{\mathcal{D}}_s$ are also isotropic, with magnitudes $B$, $\kappa$, and $\mathcal{D}_s$.

### 3.4.3 Working Models of Flow and Diffusion in Isotropic Media

If the main pore network is isotropic and has a uniform pore size $r_1$, then Eq. (3.4-12) takes the form

$$[\boldsymbol{N}] = -[c]\frac{\beta}{\mu}\nabla p - \epsilon\kappa[\boldsymbol{F}(r_1)]^{-1}\nabla[c] - [\boldsymbol{\mathcal{D}}_s]\nabla[c] \quad (3.4\text{-}13)$$

which has $\beta$, $\epsilon\kappa$, $r_1$, and $\mathcal{D}_{is}$ of each species as adjustable parameters. This result includes the constant-pressure equations of Evans et al. (1961), Scott and Dullien (1962), and Rothfeld (1963), the variable-pressure equations of Mason and Evans (1969), and of Gunn and King (1969), and the surface diffusion flux as given, for instance, by Hwang and Kammermeyer (1966).

If the pore network is isotropic and $\kappa$ is independent of $r$, then Eq. (3.4-12) gives

$$[\boldsymbol{N}] = -[c]\frac{\beta}{\mu}\nabla p - \kappa\int_{r=0}^{\infty}[\boldsymbol{F}(r)]^{-1}d\epsilon(r)\nabla[c] - [\boldsymbol{\mathcal{D}}_s]\nabla[c] \quad (3.4\text{-}14)$$

This result includes Eq. 10 of Johnson and Stewart (1965) and Eq. 2 of Satterfield and Cadle (1968) as special cases. Equation (3.4-11) predicts $\kappa = 1/3$ for isotropic pore systems with no dead-end branches; this value is accurate within a factor of 2 for many pelleted catalysts (Johnson and Stewart 1965; Brown et al. 1969; Horak and Schneider 1971).

If the integral in Eq. (3.4-12) is approximated by an $m$-term summation, then for isotropic media we obtain

$$[\boldsymbol{N}] = -[c]\frac{\beta}{\mu}\nabla p - \sum_{k=1}^{m}W_k[\boldsymbol{F}(r_k)]^{-1}\nabla[c] - [\boldsymbol{\mathcal{D}}_s]\nabla[c] \quad (3.4\text{-}15)$$

The constants $W_k$ and $r_k$ may be regarded as lumped contributions to a tortuosity-corrected void fraction distribution. A one- or two-term lumping is generally adequate. The form with $m = 1$ corresponds to Eq. (3.4-14); the form with $m = 2$ is accurate for wider pore size distributions.

### 3.4.4 Discussion

Equation (3.4-15) with $m = 2$ is the most reliable of the working models. It is recommended for accurate work.

Equation (3.4-12) is useful for making estimates in the absence of diffusion data and can be used to advantage in planning new catalyst formulations. The model of Wakao and Smith (1962) can also be used in this way, but it does not predict the effects of pressure as accurately (Brown et al. 1969).

The dusty-gas model of Mason et al. (1967) yields the gaseous flux terms of Eq. (3.4-13) as a first kinetic-theory approximation, regardless of the dust size distribution. However, when written for a network of cylindrical pores, this model is consistent with Eq. (3.4-6), which predicts a dependence of $[N_\ell]$ on the pore size distribution. The latter interpretation seems more realistic and is supported by the excellent results obtained with Eq. (3.4-15); see Example 7.2.

The models studied here are for systems with pores wider than the gaseous molecules. This assumption is implicit in Eqs. (3.4-2) to (3.4-4), which fail in the presence of molecules too large to pass through the pores.

### 3.4.5 Transport and Reaction in Porous Catalysts

Given the transport fluxes for all species inside the catalyst particle, as modeled in Section 3.4.3, a steady-state mass balance considering the simultaneous transport and chemical reaction gives

$$0 = -\nabla \cdot N_i + R_i \qquad i = 1, \dots, \text{NS} \qquad (3.4\text{-}16)$$

in which the production rates are evaluated from the reaction rates and the stoichiometry, as in Eq. (2.1-7). The reaction and production rate expressions may take many forms, as discussed in Section 2.5. Langmuir-Hinshelwood-Hougen-Watson rate expressions are often good choices for describing reaction and production rates inside catalyst particles. The interior of the catalyst particle is often close to isothermal, and an energy balance inside the particle is often not required.

In the packed-bed reactor, the molar concentrations and temperature at the exterior of the catalyst particle are coupled to the respective fluid-phase concentrations and temperature through the interfacial fluxes given in Eqs. (3.3-1) and (3.3-2). Overall mass and heat transfer coefficients are often used to describe these interfacial fluxes, similar in structure to Eqs. (3.2-1) and (3.2-2). Complete solution of the packed-bed reactor model then requires solution of the reaction-diffusion problem inside the particle simultaneously with the fluid-phase mass and energy balances, Eqs. (3.1-33) and (3.1-36). Froment and Bischoff (1990), and Rawlings and Ekerdt (2002) provide further detailed discussion of solving packed-bed reactor models and provide some numerical example calculations.

## 3.5 GAS PROPERTIES AT LOW PRESSURES

Gas transport properties are required to apply the theory given in Sections 3.3 and 3.4. Viscosities of pure nonpolar gases at low pressures are predicted from the Chapman-Enskog kinetic theory with a Lennard-Jones 12-6 potential. The collision integrals for viscosity and thermal conductivity with this potential are computed from the accurate curve-fits given by Neufeld et al. (1972).

Lennard-Jones parameters for $H_2$ at temperatures above 300 K are taken from Hirschfelder, Curtiss, and Bird (1954) as $\sigma = 2.915$Å, $\epsilon/\kappa = 38.0$K. Values for normal fluids are calculated from Correlation iii of Tee et al. (1966)

$$\sigma(p_c/T_c)^{1/3} = 2.3442 \exp(-0.1303\omega) \tag{3.5-1}$$

$$\frac{\epsilon}{\kappa T_c} = -0.8109 \exp(-0.6228\omega) \tag{3.5-2}$$

Here $p_c$ is the fluid's critical pressure in atm, $T_c$ is the fluid's critical temperature in Kelvins, and $\omega$ is the fluid's acentric factor as defined in Poling et al. (2001). With these parameters, the kinetic theory reproduces the viscosities of the 14 investigated normal fluids with a RMS deviation of 2.13 percent.

Thermal conductivities of pure nonpolar gases at low pressures are predicted by the method of Chung et al. as presented in Eqs. (10-3.14,15) of Poling et al. (2001), who report that this method "except for polar compounds, yields values quite close to those reported experimentally." Viscosities of nonpolar gas mixtures at low pressures are calculated by the method of Wilke (1950) from the pure-component values at the same temperature and pressure,

$$\mu_m = \frac{\sum_\alpha x_\alpha \mu_\alpha}{\sum_\beta x_\beta \Phi_{\alpha\beta}} \tag{3.5-3}$$

in which

$$\Phi_{\alpha\beta} = \sqrt{\frac{M_\beta}{8(M_\beta + M_\alpha)}} \left[ 1 + \sqrt{\frac{\mu_\alpha}{\mu_\beta}} \left( \frac{M_\beta}{M_\alpha} \right)^{1/4} \right]^2 \tag{3.5-4}$$

Thermal conductivities, $k_m$, of nonpolar gas mixtures at low pressures are calculated from a formula like Eq. (3.5-3), but with $\mu$ replaced everywhere by $k$ as in the book *Transport Phenomena* by Bird, Stewart, and Lightfoot

(2002). This formula differs from that of Mason and Saxena (1958) in that their empirical coefficient 1.065 is replaced by 1 (with approval of Dr. Mason) to give the known result for mixtures of identical species.

Binary diffusivities $\mathcal{D}_{\alpha\beta}$ at low pressures are predicted from the Chapman-Enskog kinetic theory with binary Lennard-Jones parameters predicted as follows from the pure-component values:

$$\epsilon_{\alpha\beta} = \sqrt{\epsilon_{\alpha\alpha}\epsilon_{\beta\beta}}; \qquad \sigma_{\alpha\beta} = (\sigma_{\alpha\alpha} + \sigma_{\beta\beta})/2 \qquad (3.5\text{-}5,6)$$

## 3.6 NOTATION

| | |
|---|---|
| $A_E$ | heat exchange area, $L^2$ |
| $\hat{a}$ | particle area per unit mass, $L^2/M$ |
| $\widehat{C}_p$ | specific heat at constant pressure, $L^2/t^2T$ |
| $\widehat{C}_v$ | specific heat at constant volume, $L^2/t^2T$ |
| $\widetilde{C}_p$ | molar heat capacity at constant pressure, $ML^2/t^2T/\text{mole}$ |
| $\widetilde{C}_{pi0}$ | molar heat capacity of species $i$ at interface, $ML^2/t^2T/\text{mole}$ |
| $E_v$ | viscous dissipation term in mechanical energy balance, $ML^2/t^3$ |
| $\boldsymbol{F}$ | force exerted on fluid by its boundaries (except "1" and "2"), $ML/t^2$ |
| $\boldsymbol{g}$ | gravitational acceleration, $L/t^2$ |
| $\widehat{H}$ | specific enthalpy, $L^2/t^2$ |
| $\widetilde{H}$ | molar enthalpy, $ML^2/t^2/\text{mole}$, |
| $\overline{H}_i$ | partial molar enthalpy of species $i$, $ML^2/t^2/\text{mole}$ |
| $h$ | heat transfer coefficient uncorrected for $\phi_T$, $M/t^3T$ |
| $h^\bullet$ | $= \theta_T h$, heat transfer coefficient corrected for $\phi_T$, $M/t^3T$ |
| $\boldsymbol{J}_i^\star$ | molar flux of species $i$ relative to molar average velocity, $\text{moles}/L^2t$ |
| $K_{\text{tot}}$ | total kinetic energy in system, $ML^2/t^2$ |
| $\widehat{K}$ | kinetic energy per unit mass of stream, $L^2/t^2$ |
| $k$ | thermal conductivity, $Ml/t^2T$ |
| $\bar{k}_{xi}$ | mass transfer coefficient uncorrected for $\phi_{xi}$, $\text{moles}/L^2t$ |
| $\bar{k}_{xi}^\bullet$ | $= \theta_{xi}\bar{k}_{xi}$, mass transfer coefficient corrected for $\phi_{xi}$, $\text{moles}/L^2t$ |
| $M_i$ | molecular weight of species $i$, $M/\text{mole}$ |
| $\mathcal{M}_{\text{tot}}$ | total moles in system |
| $\mathcal{M}_{i,\text{tot}}$ | moles of species $i$ in system |
| $m_{\text{tot}}$ | total mass in system |
| $m_{i,\text{tot}}$ | mass of species $i$ in system |
| $N_{i0}$ | molar flux of species $i$ into system, relative to the interface, $\text{moles}/L^2t$ |
| $\boldsymbol{P}$ | system momentum, $ML/t$ |
| $Q$ | rate of heat input to system, $ML^2/t^3$ |
| $p$ | pressure, $M/Lt^2$ |
| $q_0$ | conductive energy flux from particles into fluid, $M/t^3$ |
| $\text{R}_i$ | local rate of production of species $i$ by homogeneous reactions, $\text{moles}/L^3t$ |
| $\text{R}_i^s$ | local rate of production of species $i$ per unit area by heterogeneous reactions, $\text{moles}/L^2t$ |

$R_{i,\text{tot}}$    total rate of production of species $i$ in system, moles/$t$

$\mathcal{R}_i$    local rate of homogeneous reaction $i$, moles/$L^3t$

$\widehat{\mathcal{R}}_j$    local rate of heterogeneous reaction $j$ per unit mass of catalyst, moles/$Mt$

$S$    area of surface or bed cross section, $L^2$

$\boldsymbol{S}$    flow cross section vector, $L^2$

$T_0$    local exterior temperature of catalyst particles

$T_b$    bulk fluid temperature in chosen cross section

$t$    time, $t$

$U_E$    overall heat transfer coefficient based on area $A_E$, $M/t^3T$

$U_{\text{tot}}$    total internal energy in system, $ML^2/t^2$

$V$    system volume, $L^3$

$\boldsymbol{v}$    mass-average velocity, $L/t$

$\boldsymbol{v}^\star$    molar average velocity, $L/t$

$W_m$    mechanical power input to system, $ML^2/t^3$

$W$    total molar flow rate, moles/$t$

$W_{ik}$    molar flow rate of species $i$ through surface $S_k$, moles/$t$

$W_{i0}$    molar rate of addition of species $i$ to system through mass transfer surface, moles/$t$

$W_0$    total molar input to system through mass transfer surface, moles/$t$

$\boldsymbol{w}$    mass flow rate, $M/t$

$w_{ik}$    mass flow rate of species $i$ through surface $S_k$, $M/t$

$w_{i0}$    mass rate of addition of species $i$ to system through mass transfer surface, $M/t$

$w_0$    total rate of mass input to system through mass transfer surface, $M/t$

$x_i$    mole fraction of species $i$

$x_{i0}$    value of $x_i$ in interfacial fluid

$x_{ib}$    bulk value of $x_i$ in gas stream

$\mu$    viscosity, $M/Lt$

$\nu_{ji}$    stoichiometric coefficient for species $i$ in reaction $j$

$\theta_T$    $= h^\bullet/h$, mass-transfer correction factor for $h$

$\theta_{xi}$    $= \bar{k}_{xi}^\bullet/\bar{k}_{xi}$, mass-transfer correction factor for $\bar{k}_{xi}$

$x_i$    mole fraction of species $i$

$\xi_i$    extent of production of species $i$

$\rho$    density, $M/L^3$

$\boldsymbol{\tau}$    viscous stress tensor, $ML^{-1}t^{-2}$

$\Phi_{\text{tot}}$    system potential energy, $ML^2/t^2$

$\widehat{\Phi}$    potential energy per unit mass of given stream, $L^2/t^2$

$\phi_{xi}$    mass-transfer rate factor for $k_{xi}$, dimensionless

$\phi_T$    mass-transfer rate factor for $h$, dimensionless

*Superscripts and Overlines*

$\bullet$    value corrected for mass-transfer rates

*Subscripts*

$i, j$      chemical species

$\alpha, \beta$      chemical species

$m$      mixture value

0      value in fluid next to catalyst exterior surface

$1, 2$      value at inlet plane ("1") or outlet plane ("2")

*Mathematical Operations*

$<>$      average over indicated region

## REFERENCES and FURTHER READING

Anderson, E., Z. Bai, C. Bischof, J. Demmel, J. J. Dongarra, J. Du Croz, A. Greenbaum, S. Hammarling, A. McKenney, S. Ostrouchov, and D. Sorensen, *LAPACK Users' Guide*, Society for Industrial and Applied Mathematics, Philadelphia, PA, (1992).

Aris, R., *Elementary Reactor Analysis*, Prentice-Hall, Englewood Cliffs, NJ (1969).

Bird, R. B., The equations of change and the macroscopic mass, momentum and energy balances, *Chem. Eng. Sci.*, **6**, 123–131 (1957).

Bird, R. B., W. E. Stewart, and E. N. Lightfoot, *Transport Phenomena*, Wiley, New York (2002).

Brown, L. F., H. W. Haynes, and W. H. Manogue, *J. Catal.*, **14**, 3, 220 (1969).

Cullinan, H. T., Analysis of the flux equations of multicomponent diffusion, *Ind. Eng. Chem. Fundam.*, **4**, 133–139 (1965).

Curtiss, C. F., and J. O. Hirschfelder, *J. Chem. Phys.*, **17**, 550-555 (1949).

Evans III, R. B., E. A. Mason, and G. M. Watson, *J. Chem. Phys.*, **35**, 2076 (1961).

Feng, C. F., V. V. Kostrov, and W. E. Stewart, Multicomponent diffusion of gases in porous solids: models and experiments, *Ind. Eng. Chem. Fundam.*, **13**, 5–9 (1974).

Feng, C. F., and W. E. Stewart, Practical models for isothermal diffusion and flow in porous media, *Ind. Eng. Chem. Fundam.*, **12**, 143–146 (1973).

Froment, G. F., and K. B. Bischoff, *Chemical Reactor Analysis and Design*, 2nd edition, Wiley, New York (1990).

Gamson, B. W., G. Thodos, and O. A. Hougen, *Trans. AIChE*, **39**, 1–35 (1943).

Gunn, R. D., and C. J. King, *AIChE J.*, **15**, 4, 507 (1969).

Hirschfelder, J. O., C. F. Curtiss, and R. B. Bird, *Molecular Theory of Gases and Liquids*, Wiley, New York (1954).

Horak, J., and P. Schneider, Comparison of some models of porous media for gas diffusion, *Chem. Eng. J.*, **2**, 26 (1971).

Hwang, S. T., and K. Kammermeyer, *Can. J. Chem. Eng.*, **44**, 2, 82 (1966).

Jackson, R., *Transport in Porous Catalysts*, Elsevier, New York (1977).

Johnson, M. F. L., and W. E. Stewart, Pore structure and gaseous diffusion in solid catalysts, *J. Catalysis*, **4**, 248–252 (1965).

Lapidus, L., and N. R. Amundson, eds., *Chemical Reactor Theory: A Review*, Prentice-Hall, Englewood Cliffs, NJ (1977).

Lu, Y. T., Reaction modeling of propylene hydrogenation over alumina-supported platinum, M.S. Thesis, University of Wisconsin–Madison (1988).

Mason, E. A., and R. B. Evans III, *J. Chem. Educ.*, **46**, 359 (1969).

Mason, E. A., A. P. Malinauskas, and R. B. Evans III, *J. Chem. Phys.*, **46**, 3199 (1967).

Mason, E. A., and S. C. Saxena, *Phys. Fluids*, **1**, 5, 361 (1958).

Maxwell, J. C., *Phil. Mag.*, **XIX**, 19–32 (1860); **XX**, 21–32, 33–36 (1868).

McCune, L. K., and R. H. Wilhelm, *Ind. Eng. Chem.*, **41**, 1124-1134 (1949).

Mears, D. E., Tests for transport limitations in experimental catalytic reactors, *IEC Process Des. Dev*, **10**, 541–547 (1971).

Neufeld, P. D., A. R. Jansen and R. A. Aziz, *J. Chem. Phys.*, **57**, 1100-4102 (1972).

Perry, R. H., D. W. Green, and J. O. Maloney, *Perry's Chemical Engineers' Handbook*, 5th edition, McGraw-Hill, New York, (1984).

Pfeffer, R., *Ind. Eng. Chem. Fund.*, **3**, 380-383 (1964).

Poling, B. E., J. M. Prausnitz, and J. P. O'Connell, *The Properties of Gases and Liquids*, 5th edition, McGraw-Hill, New York (2000).

Rase, H. F., *Fixed-Bed Reactor Design and Diagnostics: Gas-Phase Reactions*, Butterworths, Boston (1990).

Rawlings, J. B., and J. G. Ekerdt, *Chemical Reactor Analysis and Design Fundamentals*, Nob Hill Publishing, Madison, Wisconsin (2002).

Rogers, G. B., Kinetics of the catalytic hydrogenation of propylene, PhD thesis, University of Wisconsin–Madison (1961).

Rothfeld, L. R., *AIChE J.*, **9**, 1, 19 (1963).

Satterfield, C.N., and P. J. Cadle, *Ind. Eng. Chem. Fundam.*, **7**, 202 (1968).

Scott, D. S., F. A. L. Dullien, *AIChE J.*, **8**, 3, 293 (1962).

Shabaker, R. H., Kinetics and effectiveness factors for the hydrogenation of propylene on a platinum-alumina catalyst, Ph. D. thesis, University of Wisconsin–Madison (1965).

Sørensen, J. P., E. W. Guertin and W. E. Stewart, Computational models for cylindrical catalyst particles, *AIChE J.*, **19**, 969–975, 1286 (1973); **21**, 206 (1975).

Sørensen, J. P., and W. E. Stewart, Computation of forced convection in slow flow through ducts and packed beds — IV. Convective boundary layers in cubic arrays of spheres, *Chem. Eng. Sci.*, **29**, 833-837 (1974).

Sørensen, J. P., and W. E. Stewart, Collocation analysis of multicomponent diffusion and reactions in porous catalysts, *Chem. Eng. Sci.*, **37**, 1103–1114 (1982); **38**, 1373 (1983).

Stefan, J., *Sitzungber. Kais. Akad. Wiss. Wien*, **LXIII**(2), 63-124 (1871); **LXV**(2), 323–363 (1872).

Stephenson, J. L., and W. E. Stewart, Optical measurements of porosity and fluid motion in packed beds, *Chem. Eng. Sci.*, **41**, 2161–2170 (1986).

Stewart, W. E., *AIChE J.*, **9**, 528–535 (1963).

Stewart, W. E. *AIChE J.*, **33**, 2008–2016 (1986).

Stewart, W. E. *AIChE J.*, **34**, 1030 (1988).

Stewart, W. E., D. F. Marr, T. T. Nam, and A. M. Gola-Galimidi, Transport modeling of packed-tube reactors — I. Framework for a data-based approach, *Chem. Eng. Sci.*, **46**, 2905–2911 (1991).

Stewart, W. E., and R. Prober, *Ind. Eng. Chem. Fund.*, **3**, 224–235 (1964).

Stewart, W. E., and J. P. Sørensen, Transient reactor analysis by orthogonal collocation, *Fifth European Symposium on Chemical Reaction Engineering*, pages B8-75 to B8-88, C2-9, and C2-9 (1972).

Stewart, W. E., and J. P. Sørensen, Computer-aided modelling of reaction networks. In *Foundations of Computer-aided Process Design*, Vol. **II**, R. S. H. Mah and W. D. Seider, eds., Engineering Foundation, New York, (1981), 335–366.

Stewart, W. E., J. P. Sørensen, and B. C. Teeter, Pulse-response measurement of thermal properties of small catalyst pellets, *Ind. Eng. Fundam.*, **17**, 221–224 (1976); **18**, 438 (1979).

Stewart, W. E., and J. V. Villadsen, Graphical calculation of multiple steady states and effectiveness factors for porous catalysts, *AIChE J.*,**15**, 28–34, 951 (1969).

Taylor, G. I., Dispersion of soluble matter in solvent flowing slowly through a tube, *Proc. Roy. Soc.*, **A219**, 186–203 (1953).

Tee, L. S., S. Gotoh, and W. E. Stewart, Molecular parameters of normal fluids: I. The Lennard-Jones potential, *Ind. Eng. Chem. Fundam.*, **5**, 356–362 (1966).

Toor, H. L., *AIChE J.*, **10**, 460–465 (1964).

Uppal, A., W. H. Ray, and A. B. Poore, On the dynamic behavior of continuous stirred tank reactors, *Chem. Eng. Sci.*, **29**, 967–985 (1974).

Uppal, A., W. H. Ray, and A. B. Poore, The classification of the dynamic behavior of continuous stirred tank reactors – influence of reactor residence time, *Chem. Eng. Sci.*, **31**, 205–214 (1976).

Varma, A., and R. Aris, Stirred pots and empty tubes, *Chemical Reactor Theory: A Review*, L. Lapidus and N. R. Amundson, eds., Prentice-Hall, Englewood Cliffs, NJ (1977) Chapter 2.

Villadsen, J. V., and W. E. Stewart, Solution of boundary-value problems by orthogonal collocation, *Chem. Eng. Sci.*, **22**, 1483-1501 (1967); **23**, 1515 (1968).

Wakao, N., and J. M. Smith, *Chem. Eng. Sci.*, **17**, 11, 825 (1962).

Wang, J. C., and W. E. Stewart, New descriptions of dispersion in flow through tubes: convolution and collocation methods, *AIChE J.*, **29**, 493–498 (1983).

Wang, J. C., and W. E. Stewart, Multicomponent reactive dispersion in tubes: collocation vs. radial averaging.*AIChE J.*, **35**, 490–499, 1582 (1989).

Weidman, D. L., and W. E. Stewart, Catalyst particle modelling in fixed-bed reactors, *Chem. Eng. Sci.*, **45**, 2155-2160 (1990).

Wilke, C. R., *J. Chem. Phys.*, **18**, 517-519 (1950).

Wilke, C. R., and O. A. Hougen, *Trans. AIChE*, **41**, 445–451 (1945).

Williamson, J. E., K. Bazaire, and C. J. Geankoplis, *Ind. Eng. Chem. Fund*, **2**, 126–129 (1963).

Wilson, E. J., and C. J. Geankoplis, *Ind. Eng. Chem. Fund.*, **5**, 9–14 (1966).

Yoshida, F., D. Ramaswami, and O. A. Hougen, Temperatures and partial pressures at the surfaces of catalyst particles, *AIChE J.*,**8**, 5–11 (1962).

# Chapter 4

# Introduction to
# Probability and Statistics

In Chapters 2 and 3 we studied the formulation of models from chemical and physical principles and the solution of models by numerical methods. In this chapter we begin the study of statistical methods and their role in model development.

Statistics is the science of efficient planning and use of observations with the aid of probability theory. It began with investigations in astronomy; later it came into wide use throughout science and engineering. The following areas of statistics are considered in this book:

1. *Inference:* Estimation of process parameters and states from data. See Chapters 5–7 and Appendix C.
2. *Criticism:* Analysis of residuals (departures of data from fitted models). This topic is included in Chapters 6 and 7 and Appendix C.
3. *Design of Experiments:* Construction of efficient test patterns for these activities. Factorial designs are well treated by Box, Hunter, and Hunter (1978). Sequential procedures for experimental design are presented in Chapters 6 and 7 and Appendix C, and in the references cited there.

This chapter outlines the complementary roles of these three areas in model development and gives some basic results of probability theory.

## 4.1 STRATEGY OF DATA-BASED INVESTIGATION

Scientific investigations involving data should seek a rational model of observed phenomena rather than a mere curve-fit of measured values. This viewpoint is also gaining favor in many industrial organizations as more powerful model-building and computing techniques become available. The approach is commonly known as the *scientific method* and proceeds by alternate formulation and testing of hypotheses, as illustrated in the following diagram adapted from Box and Tiao (1973).

The diagrammed investigation begins with a hypothesis (model), which might come from inspiration or from an analysis of existing information. To test this hypothesis, the investigator designs some experiments, performs

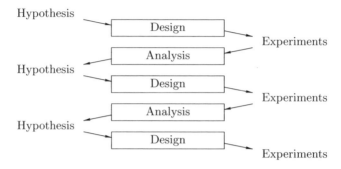

**Figure 4.1:** Iterative process of scientific investigation.

them or requests them, and analyzes the results by fitting the model to the data and examining the residuals. This analysis may indicate that the model is adequate for the intended use or may suggest a revised model, the testing of which may call for further experiments. The stopping criterion will depend on various circumstances, such as the urgency of the project, the risks associated with the current model, the availability of funds to continue the investigation, and the needs of other projects.

Many physical and chemical principles have been developed in this way, including the law of definite proportions, the periodic table of the elements, the structures of molecules, and the currently favored mechanisms of many chemical reactions. High-energy physics has progressed impressively through such iterations on many projects.

The analyses of the experiments are crucial, since they extract information about the candidate models and give clues for alterations that may be needed. Inference and criticism are both essential in these stages. The experiments should be designed for efficiency in gaining the desired information; some strategies for doing this are described in Chapters 6 and 7 and implemented in the package GREGPLUS of Appendix C.

## 4.2 BASIC CONCEPTS IN PROBABILITY THEORY

Probability theory deals with the expected frequencies of various events in random sampling. The set of events considered in the sampling is called the *sample space* of the given problem, and may be discrete (like "heads" and "tails" in coin tossing) or continuous (like the set of values on the real number line).

For a discrete sample space of mutually exclusive events $E_i$, one defines a corresponding set of event probabilities $p_i = p(E_i)$, which may be given or may be estimated from observations. The $p_i$ have the properties

$$\sum p_i = 1 \tag{4.2-1}$$

$$0 \leq p_i \leq 1 \quad \text{for all } i \tag{4.2-2}$$

$$p(E_i \text{ or } E_j) = p_i + p_j \quad \text{for } i \neq j \tag{4.2-3}$$

For a continuous sample space, one associates the probabilities not with points, but with differential regions of the space. Then the probability of obtaining a sample value in the interval $(x_0 - \Delta x/2, x_0 + \Delta x/2)$ is expressed as $p(x_0)\Delta x$, in which $p(x_0)$ is the value at $x_0$ of the *probability density function* $p(x)$. This function has the properties

$$\int_{-\infty}^{\infty} p(x)dx = 1 \tag{4.2-4}$$

$$p(x) \geq 0 \tag{4.2-5}$$

The probability of obtaining a sample value in the region $x < \xi$ is then

$$F(\xi) = \int_{-\infty}^{\xi} p(x)dx = \int_{-\infty}^{\xi} dF(x) \tag{4.2-6}$$

and is known as the *distribution function* corresponding to $p(x)$. We see that $F(\xi)$ is a nondecreasing function of $\xi$, since the density $p(x)$ is everywhere nonnegative.

For later discussions, it is useful to extend the concepts of probability density and distribution to sample spaces of discrete points $x_i$. Equations (4.2-4) to (4.2-6) hold directly here also, if we define the density function for this case as a set of spikes of zero width (impulse functions), with included probability $p(x_i)$ for each $x_i$. The resulting distribution function $F(\xi)$ is the combined probability of the sample points $x_i < \xi$.

Often the probability of an event depends on one or more related events or conditions. Such probabilities are called *conditional*. We will write $p(A|B)$ for the probability of event $A$ given $B$ (or the probability density of $A$ given $B$ if the sample space of $A$ is continuous).

Suppose that events $A$ and $B$ are discrete members of their respective sample spaces. Let $p(B)$ denote the probability of occurrence of $B$, and let $p(A, B)$ denote the probability of occurrence of both $A$ and $B$. These values are related by the formula

$$p(A, B) = p(A|B)p(B) \quad \text{if } p(B) > 0 \tag{4.2-7}$$

derived by Thomas Bayes (1763) in his classic paper on inverse probability.

Equation (4.2-7) also holds if one or both of $A$ and $B$ belongs to a continuous sample space. The function $p(A, B)$ is then a probability density with respect to each such argument. This interpretation of Eq. (4.2-7) is used extensively in the following chapters.

A *random variable*, $Z$, may be constructed on a sample space by specifying its values $\{Z_1, Z_2, \ldots\}$ on the sample points. Such variables are useful
in modeling random errors of observations. The mean value of $Z$ over the
sample space is then given by the relations

$$\mathrm{E}(Z) = \begin{cases} \sum_i Z_i p_i & \text{for a space of integers } i \\ \int_{-\infty}^{\infty} Z(x)p(x)dx & \text{for a space of real numbers } x \end{cases} \qquad (4.2\text{-}8)$$

or more concisely as an inner product over the sample space:

$$\mathrm{E}(Z) = \langle Z, p \rangle \qquad (4.2\text{-}9)$$

The notation $\mathrm{E}(Z)$ denotes the *expectation* of the random variable $Z$.
Another important measure of a random variable is its *variance*,

$$\mathrm{Var}(Z) \equiv \mathrm{E}\left[(Z - \mu)^2\right] \qquad (4.2\text{-}10)$$

Here $\mu$ denotes the mean value $\mathrm{E}(Z)$. Thus the variance measures the
expected scatter of observations of the random variable about its mean.

### Example 4.1. Tossing a Die

The possible outcomes of a fair toss of a die form a sample space $E_1, \ldots, E_6$
in which $E_i$ denotes the event that face $i$ comes up. Under ideal conditions,
all six outcomes are equally likely; application of Eq. (4.2-1) then gives
$p_i = 1/6$ for each sample point.

Choosing the score of the next toss as a random variable $Z$, we get the
set of values $\{Z_i\} = \{i\}$. The expectation of $Z$ is then

$$\mu = \mathrm{E}(Z) = \sum_{i=1}^{6} Z_i p_i = (1 + 2 + 3 + 4 + 5 + 6)/6 = \frac{7}{2}$$

and the variance of $Z$ is

$$\begin{aligned}
\mathrm{Var}(Z) &= \sum_{i=1}^{6}(Z_i - \mu)^2 p_i \\
&= \left[(-2.5)^2 + (-1.5)^2 + (-0.5)^2 + (0.5)^2 + (1.5)^2 + (2.5)^2\right]/6 \\
&= \frac{35}{12}
\end{aligned}$$

## Example 4.2. The Univariate Normal Distribution

The error of an observation $y_u$ in the $u$th event of a sequence may be modeled as a random variable $\varepsilon_u$. For a continuous random variable with range $(-\infty, \infty)$, a commonly used probability density model is the normal error curve

$$p(\varepsilon_u | \sigma) = \frac{1}{\sqrt{2\pi}\sigma} \exp\left(-\frac{\varepsilon_u^2}{2\sigma^2}\right) \tag{4.2-11}$$

the origin of which is discussed in Section 4.3. The parameter $\sigma$ is known as the *standard deviation*. The corresponding probability of finding $\varepsilon_u$ less than a chosen value $\xi$ is

$$F(\xi | \sigma) = \int_{-\infty}^{\xi} p(\varepsilon_u | \sigma) d\varepsilon_u = \frac{1}{\sqrt{2\pi}\sigma} \int_{-\infty}^{\xi} \exp\left(-\frac{\varepsilon_u^2}{2\sigma^2}\right) d\varepsilon_u \tag{4.2-12}$$

and is known as the *(cumulative) normal distribution*. These two functions are plotted in Figure 4.2 for a standardized random variable $x = \varepsilon_u/\sigma$.

The expectation of the random variable $\varepsilon_u$ is

$$E(\varepsilon_u) = \frac{1}{\sqrt{2\pi}\sigma} \int_{-\infty}^{\infty} \varepsilon_u \exp\left(-\frac{\varepsilon_u^2}{2\sigma^2}\right) d\varepsilon_u \tag{4.2-13}$$

Noting that the integrand is antisymmetric in $\varepsilon_u$, we see that

$$E(\varepsilon_u) = 0 \qquad \text{(Univariate normal distribution)} \tag{4.2-14}$$

Thus, the expectation of the error vanishes for Eq. (4.2-11).

The variance of the normally distributed random real variable $\varepsilon_u$ is

$$\text{Var}(\varepsilon_u) = \frac{1}{\sqrt{2\pi}\sigma} \int_{-\infty}^{\infty} \varepsilon_u^2 \exp\left(-\frac{\varepsilon_u^2}{2\sigma^2}\right) d\varepsilon_u \tag{4.2-15}$$

Evaluation of the integral (see Prob. 4.B) gives

$$\text{Var}(\varepsilon_u) = \sigma^2 \qquad \text{(Univariate normal distribution)} \tag{4.2-16}$$

Thus, the variance of the univariate normal error distribution is equal to the square of the standard deviation.

## 4.3 DISTRIBUTIONS OF SUMS OF RANDOM VARIABLES

Games of chance inspired several early investigations of the random sum $S_n$, representing the number of successes in $n$ Bernoulli trials:

$$S_n = X_1 + \ldots + X_n \tag{4.3-1}$$

Here the random variable $X_j$ equals 1 if trial $j$ gives a success and 0 if trial $j$ gives a failure; the probabilities of these two outcomes for a Bernoulli trial are $p$ and $q = 1 - p$, respectively.

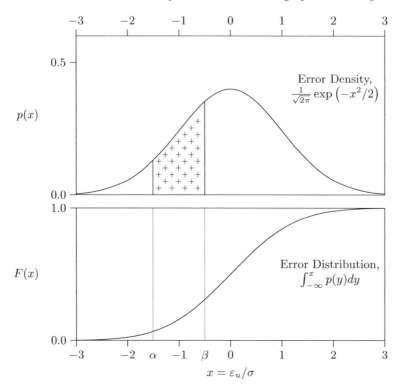

**Figure 4.2:** The univariate normal density and normal distribution. The hatched area is $F(\beta) - F(\alpha)$.

Early workers knew that the probability of exactly $k$ successes in $n$ such trials (with $k \leq n$) is

$$b(k|n, p) = \frac{n!}{k!\,(n-k)!}p^k q^{n-k} = \binom{n}{k}p^k q^{n-k} \qquad (4.3\text{-}2)$$

This expression is the $k$th term in the binomial expansion of $(p + q)^n$; the resulting distribution of $S_n$ is known as the *binomial distribution*. The probability that $S_n$ will fall in a preassigned range of nonnegative integers is then

$$\Pr\{\alpha \leq S_n \leq \beta\} = b(\alpha|n, p) + b(\alpha + 1|n, p) + \ldots + b(\beta|n, p) \qquad (4.3\text{-}3)$$

Calculations with these formulas proved tedious, so asymptotic approximations were sought. DeMoivre (1733) obtained the following analytic asymptotes [see Feller (1968), Chapter 7]:

$$b(k|n, p) \sim \frac{1}{\sqrt{2\pi npq}}\exp\left(-\frac{(k-np)^2}{2npq}\right) \qquad \text{as} \qquad \frac{(k-np)^3}{n^2} \to 0 \quad (4.3\text{-}4)$$

and

$$\Pr\{\alpha \le S_n \le \beta\} \sim \Phi\left(\frac{\beta - np + \frac{1}{2}}{\sqrt{npq}}\right) - \Phi\left(\frac{\alpha - np - \frac{1}{2}}{\sqrt{npq}}\right)$$

$$\text{as} \quad \frac{(\alpha - np)^3}{n^2} \to 0 \quad \text{and} \quad \frac{(\beta - np)^3}{n^2} \to 0$$

(4.3-5)

Here $\Phi(\bullet)$ is the normal distribution function

$$\Phi(x) \equiv \frac{1}{\sqrt{2\pi}} \int_{-\infty}^{x} \exp\left(-\frac{y^2}{2}\right) dy$$

(4.3-6)

so named because of its prevalence in sampling problems. Here $\sim$ indicates that the relative difference between the left-hand and right-hand members goes to zero in the indicated limit. Laplace (1810, 1812) generalized these results to describe distributions of errorin observations of real (noninteger) variables.

Equations (4.3-4) and (4.3-5) are the first of several important limit theorems that establish conditions for asymptotic convergence to normal distributions as the sample space grows large. Such results are known as *central limit theorems*, because the convergence is strongest when the random variable is near its central (expectation) value. The following two theorems of Lindeberg (1922) illustrate why normal distributions are so widely useful.

The simpler of the theorems goes as follows. Let $\{X_1, X_2, \ldots\}$ be a sequence of independently and identically distributed random variables, each having the same finite expectation $\mu$ and the same finite variance $\sigma^2$. Define

$$S_n = X_1 + \ldots + X_n$$

Then for every forward interval $[\alpha, \beta]$ on the real number line, the following limit holds as $n \to \infty$,

$$\Pr\left\{\alpha < \frac{S_n - n\mu}{\sigma n^{1/2}} < \beta\right\} \longrightarrow \Phi(\beta) - \Phi(\alpha)$$

(4.3-7)

in which $\Phi$ is the normal distribution function, defined in Eq. (4.3-6).

The more general of the theorems goes as follows. Let $X_1, X_2, \ldots$ be mutually independent, scalar random variables with distributions $F_1, F_2, \ldots$ such that

$$\mathrm{E}(X_k) = 0, \qquad \mathrm{Var}(X_k) = \sigma_k^2$$

(4.3-8,9)

and define

$$s_n^2 = \sigma_1^2 + \ldots + \sigma_n^2$$

(4.3-10)

Assume that with increasing $n$, $s_n^2$ grows unboundedly relative to each of the individual variances $s_k^2$, so that for every positive $t$,

$$s_n^{-2} \sum_{k=1}^{n} \int_{|y| > t s_n} y^2 dF_k(y) \to 0 \qquad \text{as} \quad n \to \infty$$

(4.3-11)

Then the distribution of the standardized sum

$$S_n^* = (X_1 + \ldots + X_n)/s_n \qquad (4.3\text{-}12)$$

tends as $n \to \infty$ to the normal distribution with zero mean and unit variance, given by Eq. (4.3-6) with $S_n^*$ as the random variable $x$.

These theorems are useful for assessing the uncertainties of sums obtained in public opinion polls or in replicated laboratory experiments. In this interpretation the values $k = 1, \ldots, n$ denote successive observations, and the distribution sampled is that of their sum $X_1 + \ldots + X_n$. Condition (4.3-11) of the second Lindeberg theorem shows this sum to be asymptotically normally distributed with variance $s_n^2$, provided that the variance $\sigma_k^2$ of each observation becomes negligible relative to $s_n^2$ when $n$ is large. This interpretation has been used in sampling theories of parameter estimation.

The second theorem of Lindeberg can be viewed in another way to obtain a model of random error in a single observation. The various contributions to the error can be modeled as random variables $X_k$ with individual distributions $F_k$ and included in the sums in descending order of their variances $\sigma_k^2$. The number, $n$, of such contributions can be rather large, and increases with the thoroughness of study of the method of observation. Lindeberg's theorem tells us that a sequence of models so constructed for the total error $S_n$ will converge to a normal distribution with increasing $n$, provided that (4.3-11) is satisfied. This condition will be satisfied if, with increasing $n$, the combined variance $s_n^2$ grows large relative to each of the individual variances $\sigma_k^2$. Careful observers and instrument makers, who strive to reduce the principal causes of error to the levels of the others, thus create conditions that favor normal distributions of the total error $\varepsilon_u$. This interpretation of the detailed Lindeberg theorem is implied whenever one treats future observations as normally distributed, as we will do in the following chapters. Laplace (1810) used his own central limit theorem in a similar manner to justify the normal error curve used by Gauss (1809) in his derivation of the method of least squares.

These central limit theorems are also relevant to physical models based on random processes. These theorems tell us that normal distributions can arise in many ways. Therefore, the occurrence of such a distribution tells very little about the mechanism of a process; it indicates only that the number of random events involved is large.

## 4.4 MULTIRESPONSE NORMAL ERROR DISTRIBUTIONS

A corresponding normal distribution is available for multiresponse data, that is, for interdependent observations of two or more measurable quantities. Such data are common in experiments with chemical mixtures, mechanical structures, and electric circuits as well as in population surveys and econometric studies. Modeling with multiresponse data is treated in Chapter 7 and in the software of Appendix C.

The error in a vector $\boldsymbol{y}_u = \{y_{iu}\}$ of $m$ observed responses in an event $u$ may be modeled as a random $m$-vector $\boldsymbol{\varepsilon}_u$ with elements $\varepsilon_{iu}$. By analogy with Eq. (4.2-11), let $E(\boldsymbol{\varepsilon}_u) = \boldsymbol{0}$. Then the variance $\sigma^2 = E(\varepsilon_u^2)$ used in Eq. (4.2-11) may be generalized to a symmetric matrix[1]

$$\boldsymbol{\Sigma} = E\left(\boldsymbol{\varepsilon}_u \boldsymbol{\varepsilon}_u^T\right) = \boldsymbol{\Sigma}^T \qquad (4.4\text{-}1)$$

known as the *covariance matrix*. Its elements are the expectations

$$\sigma_{ij} = E\left(\varepsilon_{iu}\varepsilon_{ju}\right) = \sigma_{ji} \qquad i,j = 1,\ldots,m \qquad (4.4\text{-}2)$$

and are called *covariances*; the diagonal elements $\sigma_{ii}$ are also called *variances*.

The multiresponse counterpart of $(1/\sigma^2)$ is the inverse covariance matrix $\boldsymbol{\Sigma}^{-1}$, which exists only if $\boldsymbol{\Sigma}$ has full rank. This condition is achievable by selecting a linearly independent set of responses, as described in Chapter 7. Then the exponential function in Eq. (4.2-11) may be generalized to

$$\exp\left(-\frac{1}{2}\boldsymbol{\varepsilon}_u^T \boldsymbol{\Sigma}^{-1}\boldsymbol{\varepsilon}_u\right)$$

This expression goes to zero with increasing magnitude of any error component $\varepsilon_{iu}$, since the argument is nonpositive whenever $\boldsymbol{\Sigma}^{-1}$ exists; see Problem 4.C.

Setting the error density function proportional to the exponential function just given, and adjusting to unit total probability, one obtains

$$p(\boldsymbol{\varepsilon}_u|\boldsymbol{\Sigma}) = (2\pi)^{-m/2}|\boldsymbol{\Sigma}|^{-1/2}\exp\left(-\frac{1}{2}\boldsymbol{\varepsilon}_u^T \boldsymbol{\Sigma}^{-1}\boldsymbol{\varepsilon}_u\right) \qquad (4.4\text{-}3)$$

as the normal error density function for a vector of $m$ linearly independent responses in a single event. Here $|\boldsymbol{\Sigma}|$ is the determinant of the covariance matrix $\boldsymbol{\Sigma}$. This density function is very important for analysis of multiresponse data; we will use it extensively in Chapter 7.

Figure 4.3 shows some contours of constant error density $p(\boldsymbol{\varepsilon}_u|\boldsymbol{\Sigma})$ for systems with two responses. In all four cases the responses are linearly independent, but in the last two they are statistically correlated because the off-diagonal elements of $\boldsymbol{\Sigma}$ are nonzero.

## 4.5 STATISTICAL INFERENCE AND CRITICISM

Inference and criticism are complementary facets of statistical analysis. Both are rooted in probability theory, but they address different questions about a model.

---

[1] The notation $\boldsymbol{\Sigma}$ is well established in the literature. It is distinguished from a summation sign by its bold font and lack of summation limits.

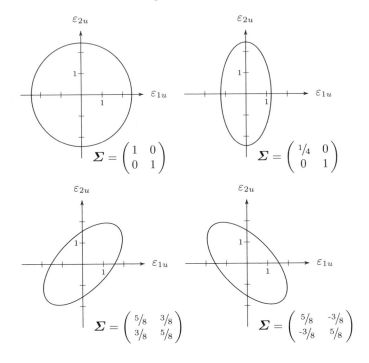

**Figure 4.3:** Contours of constant density $p(\boldsymbol{\varepsilon}_u|\boldsymbol{\Sigma})$ for the bivariate normal distribution. Each contour encloses a probability content of 0.95.

Inference deals with the estimation of the parameters $\boldsymbol{\theta}$ of a postulated model from given data $\boldsymbol{y}$. A natural way of doing this is to analyze the *posterior density function* $p(\boldsymbol{\theta}|\boldsymbol{y})$, constructed from the data and the postulated model according to Bayes' theorem as described in Chapters 5–7. This approach includes the well-known method of least squares, as well as more general methods to be described in Chapter 7 and Appendix C.

Criticism seeks to determine if a fitted model is faulty. This is done by examining the residuals (departures of the data from the fitted model) for any evidence of unusual or systematic errors. Sampling theory and diagnostic plots of the residuals are the natural tools for statistical criticism; their use is demonstrated in Chapters 6 and 7 and in Appendix C.

This view of the complementary roles of Bayesian inference and sampling theory is drawn from a landmark paper by Box (1980) and is followed consistently in this book. The controversy between Bayesians and sample theorists remains a lively one, but is becoming reconciled through the growing recognition that both approaches are useful and necessary.

## PROBLEMS

### 4.A Univariate Normal Density and Distribution

In Figure 4.2, let $\alpha, \beta$ be $(-1.5, -0.5)$. Look up $F(\alpha)$ and $F(\beta)$ in a table of the normal distribution and verify $F(\beta) - F(\alpha)$ by numerical integration of the univariate normal density function.

### 4.B Moments of the Normal Distribution

The $k$th moment of the univariate normal distribution is

$$\int_{-\infty}^{\infty} x^k p(x)\,dx \tag{4.B-1}$$

in the notation of Figure 4.2. Evaluate this integral for $k=0$, 1, and 2, using a table of integral formulas, and show that the result is consistent with Eqs. (4.2-4), (4.2-14), and (4.2-16).

### 4.C Properties of the Covariance Matrix $\Sigma$

Consider the real $m \times m$ covariance matrix $\Sigma$ and real random error vector $\varepsilon_u$ defined in Section 4.4. $\Sigma$ satisfies

$$\Sigma = \int_{\varepsilon_u} \varepsilon_u \varepsilon_u^T\, p(\varepsilon_u | \Sigma)\, d\varepsilon_{1u} \ldots d\varepsilon_{mu} \tag{4.C-1}$$

in which $\varepsilon_{1u}, \ldots, \varepsilon_{mu}$ are the elements of the $m$-vector $\varepsilon_u$.
(a) Use Eq. (4.C-1) to write out the quadratic form $x^T \Sigma x$, and show that this scalar product is nonnegative for every real $m$-vector $x$. (Hint: a volume integral is nonnegative if its integrand is nonnegative).
(b) A real, symmetric matrix $A$ is called *positive definite* if $x^T A x > 0$ for every conforming nonzero real vector $x$. Extend the result of (a) to show that the covariance matrix $\Sigma$ in Eq. (4.C-1) is positive definite if the scalar random variables $\varepsilon_{1u}, \ldots, \varepsilon_{mu}$ are linearly independent, that is, if there is no nonzero $m$-vector $x$ such that $x^T \varepsilon_u$ vanishes over the sample space of the random vector $\varepsilon_u$.

### 4.D Variance of a Function of Several Random Variables

Consider a function $F = \sum_i a_i Z_i$ of several random variables $Z_i$ with expectations $E(Z_i) = 0$ and variances $\text{Var}(Z_i) = \sigma_i^2$. Show that $F$ has expectation

$$E(F) = 0 \tag{4.D-1}$$

and variance

$$\text{Var}(F) = \sum_i \sum_j a_i a_j C_{ij} \tag{4.D-2}$$

in which the elements $C_{ij}$ are the covariances

$$C_{ij} = E[(Z_i - \mu_i)(Z_j - \mu_j)] \tag{4.D-3}$$

of the random variables. Show that Eq. (4.D-2) reduces to

$$\text{Var}(F) = \sum_i a_i^2 \sigma_i^2 \qquad (4.\text{D-}4)$$

if the covariance matrix $\boldsymbol{C} = \{C_{ij}\}$ is diagonal.

## REFERENCES and FURTHER READING

Bayes, T. R., An essay towards solving a problem in the doctrine of chances, *Phil. Trans. Roy. Soc. London*, **53**, 370–418 (1763). Reprinted in *Biometrika*, **45**, 293–315 (1958).

Box, G. E. P., and G. C. Tiao, *Bayesian Inference in Statistical Analysis*, Addison-Wesley, Reading, MA (1973). Reprinted by Wiley, New York (1992).

Box, G. E. P., Sampling and Bayes' inference in scientific modelling and robustness (with Discussion), *J. Roy. Statist. Soc. A*, **143**, 383–430 (1980).

Box, G. E. P., W. G. Hunter, and J. S. Hunter, *Statistics for Experimenters*, Wiley, New York (1978).

De Moivre, A., *Approximatio ad Summam Terminorum Binomii $\overline{a + b}|^n$ in Seriem expansi*, London (1733). Photographically reproduced in R. C. Archibald, A rare pamphlet of Moivre and some of his discoveries, *Isis*, **8**, 671–683 (1926).

Eisenhart, C., and M. Zelen, *Elements of Probability, Handbook of Physics*, 2nd edition, E. U. Condon and H. Odishaw (eds.), McGraw-Hill, New York (1967) Chapter 12.

Feller, W., *An Introduction to Probability Theory and Its Applications, Volume I*, Wiley, New York (1968).

Feller, W., *An Introduction to Probability Theory and Its Applications, Volume II*, 2nd edition, Wiley, New York (1971).

Gauss, C. F., *Theoria motus corporum coelestium in sectionibus conicis solem ambientium*, Perthas et Besser, Hamburg (1809); *Werke*, **7**, 240–254. Translated as *Theory of Motion of the Heavenly Bodies Moving about the Sun in Conic Sections* by C. H. Davis. Little, and Brown, Boston (1857); Dover, New York (1963).

Laplace, P. S., Supplément au Mémoire Sur les approximations des fonctions qui sont fonctions de trés-grands nombres, *Mémoires de l'Académie des sciences des Paris*, (1810) 565.

Laplace, P. S., *Théorie analytique des probabilités*, Courcier, Paris (1812).

Lindeberg, J. W., Eine neue Herleitung des Exponentialgesetzes in der Wahrscheinlichkeitsrechnung, *Math. Zeit.*, **15**, 211–225 (1922).

Stigler, S. M., *The History of Statistics*, Harvard University Press, Cambridge, MA (1986).

# Chapter 5
# Introduction to Bayesian Estimation

Bayes' theorem is fundamental in parameter estimation. Published over two centuries ago (Bayes 1763) from the last work of an English clergyman, this theorem is a powerful tool for data-based analysis of mathematical models.

From observations and any available prior information, Bayes' theorem infers a probability distribution for the parameters of a postulated model. This *posterior distribution* tells all that can be inferred about the parameters on the basis of the given information. From this function the most probable parameter values can be calculated, as well as various measures of the precision of the parameter estimation. The same can be done for any quantity predicted by the model.

Bayes demonstrated his theorem by inferring a posterior distribution for the parameter $p$ of Eq. (4.3-2) from the observed number $k$ of successes in $n$ Bernoulli trials. His distribution formula expresses the probability, given $k$ and $n$, that $p$ "lies somewhere between any two degrees of probability that can be named." The subtlety of the treatment delayed its impact until the middle of the twentieth century, though Gauss (1809) and Laplace (1810) used related methods. Stigler (1982, 1986) gives lucid discussions of Bayes' classic paper and its various interpretations by famous statisticians.

Bayesian procedures are important not only for estimating parameters and states, but also for decision making in various fields. Chapters 6 and 7 include applications to model discrimination and design of experiments; further applications appear in Appendix C. The theorem also gives useful guidance in economic planning and in games of chance (Meeden, 1981).

## 5.1 THE THEOREM

Suppose that we have a postulated model $p(\boldsymbol{y}|\boldsymbol{\theta})$ for predicting the probability distribution of future observation sets $\boldsymbol{y}$ for each permitted value $\boldsymbol{\theta}$ of a list of parameters. (An illustrative model is given in Example 5.1.) Suppose that a prior probability function $p(\boldsymbol{\theta})$ is available to describe our information, beliefs, or ignorance regarding $\boldsymbol{\theta}$ before any data are seen. Then, using Eq. (5.1-7), we can predict the joint probability of $\boldsymbol{y}$ and $\boldsymbol{\theta}$ either as

$$p(\boldsymbol{y}, \boldsymbol{\theta}) = p(\boldsymbol{y}|\boldsymbol{\theta})p(\boldsymbol{\theta}) \qquad (5.1\text{-}1)$$

or as

$$p(\boldsymbol{y}, \boldsymbol{\theta}) = p(\boldsymbol{\theta}|\boldsymbol{y})p(\boldsymbol{y}) \qquad (5.1\text{-}2)$$

Equating the two expressions for $p(\boldsymbol{y}, \boldsymbol{\theta})$, one obtains Bayes' theorem

$$p(\boldsymbol{\theta}|\boldsymbol{y}) = \frac{p(\boldsymbol{y}|\boldsymbol{\theta})p(\boldsymbol{\theta})}{p(\boldsymbol{y})} \qquad (5.1\text{-}3)$$

for making inferences about the parameter vector $\boldsymbol{\theta}$. The function $p(\boldsymbol{\theta}|\boldsymbol{y})$ gives the posterior probability content of each element of $\boldsymbol{\theta}$-space, conditional on the model and the data $\boldsymbol{y}$ at hand. This function can be interpreted generally as a *posterior density* (probability per unit volume in $\boldsymbol{\theta}$-space),[1] or as an array of *posterior probabilities* if $\boldsymbol{\theta}$ takes only discrete values. The *posterior distribution* of $\boldsymbol{\theta}$ is defined by integration of this function in the manner of Eq. (4.2-6), or by summation as indicated there if $\boldsymbol{\theta}$ is discrete-valued.

The quantity $p(\boldsymbol{y})$ is constant when the data and model are given. This can be seen by normalizing $p(\boldsymbol{\theta}|\boldsymbol{y})$ to unit probability content

$$p(\boldsymbol{\theta}|\boldsymbol{y}) = p(\boldsymbol{y}|\boldsymbol{\theta})p(\boldsymbol{\theta})/C \qquad (5.1\text{-}4)$$

and noting that the constant

$$C = \begin{cases} \int p(\boldsymbol{y}|\boldsymbol{\theta})p(\boldsymbol{\theta})\, d\boldsymbol{\theta} & \text{in general, or} \\ \sum_k p(\boldsymbol{y}|\boldsymbol{\theta}_k)p(\boldsymbol{\theta}_k) & \text{for discrete } \boldsymbol{\theta} \end{cases} \qquad (5.1\text{-}5)$$

corresponds to $p(\boldsymbol{y})$ of Eq. (5.1-3).

Since $p(\boldsymbol{\theta}|\boldsymbol{y})$ in Eq. (5.1-3) is conditional on $\boldsymbol{y}$, the quantity $p(\boldsymbol{y}|\boldsymbol{\theta})$ on the right-hand side is to be evaluated as a function of $\boldsymbol{\theta}$ at the given $\boldsymbol{y}$. This interpretation of $p(\boldsymbol{y}|\boldsymbol{\theta})$ is awkard, because the function $p(A|B)$ is normally conditional on $B$ (see Eq. (4.2-7)), but here it is conditional on $A$ instead. Fisher (1922) resolved this difficulty by introducing the likelihood function

$$\ell(\boldsymbol{\theta}|\boldsymbol{y}) \equiv p(\boldsymbol{y}|\boldsymbol{\theta}) \qquad (5.1\text{-}6)$$

to stand for $p(\boldsymbol{y}|\boldsymbol{\theta})$ when $\boldsymbol{y}$ is given. With this substitution, Bayes' theorem takes the form

$$p(\boldsymbol{\theta}|\boldsymbol{y}) \propto \ell(\boldsymbol{\theta}|\boldsymbol{y})p(\boldsymbol{\theta})$$
$$\text{(Posterior)} \propto \text{(Likelihood)} \times \text{(Prior)} \qquad (5.1\text{-}7)$$

whether the permitted values of $\boldsymbol{\theta}$ are continuous or discrete.

Equation (5.1-7) expresses completely the information provided about the parameter vector $\boldsymbol{\theta}$ of the postulated model. The prior expresses whatever information or belief is provided before seeing the data, and the likelihood function expresses the information provided by the data.

---

[1] The definition of a density function on a discrete sample space is discussed under Eq. (4.2-6).

Equation (5.1-7) allows very detailed estimation. Maximizing $p(\boldsymbol{\theta}|\boldsymbol{y})$ gives a modal (most probable) estimate $\widehat{\boldsymbol{\theta}}$ of the parameter vector. Integration of $p(\boldsymbol{\theta}|\boldsymbol{y})$ over any region of $\boldsymbol{\theta}$ gives the posterior probability content of that region. Thus, one can make very direct probability statements about the parameters of a postulated model using all the available information.

So far, the functional form of the prior is arbitrary. One might elect to use a function $p(\boldsymbol{\theta})$ based on personal knowledge or belief; such *informative priors* are illustrated in Section 5.2. For objective analysis, however, one needs an unprejudiced function $p(\boldsymbol{\theta})$; such functions are known as *noninformative priors* and are studied beginning in Section 5.3.

## Example 5.1. Likelihood Function for the Normal Mean

Consider a set $y_1, \ldots, y_n$ of measurements of a physical or chemical constant $\theta$. These data are to be modeled as

$$y_u = \theta + \varepsilon_u \tag{5.1-8}$$

with errors $\varepsilon_u$ normally distributed as in Example 4.2 and $\sigma$ given:

$$p(\varepsilon_u|\sigma) = \frac{1}{\sqrt{2\pi}\sigma} \exp\left(-\frac{\varepsilon_u^2}{2\sigma^2}\right) \tag{5.1-9}$$

If $\theta$ is given and $y_u$ is not yet measured, these two equations give the density function

$$p(y_u|\theta,\sigma) = \frac{1}{\sqrt{2\pi}\sigma} \exp\left[-\frac{(y_u - \theta)^2}{2\sigma^2}\right] \tag{5.1-10}$$

for prospective values of $y_u$ that might be observed in a single proposed experiment. This result leads in turn to the composite density function

$$
\begin{aligned}
p(\boldsymbol{y}|\theta,\sigma) &= \prod_{u=1}^{n} p(y_u|\theta,\sigma) \\
&= \left(\frac{1}{\sqrt{2\pi}\sigma}\right)^n \exp\left[-\frac{1}{2\sigma^2}\sum_{u=1}^{n}(y_u - \theta)^2\right]
\end{aligned}
\tag{5.1-11}
$$

for prospective values of the observation vector $\boldsymbol{y} = \{y_1, \ldots, y_n\}^T$ from $n$ independent experiments.

Once data are obtained, $\theta$ can be inferred by insertion of Eq. (5.1-11) and the data vector $\boldsymbol{y}$ into Eq. (5.1-4). However, at this stage it is clearer to regard the right-hand member of Eq. (5.1-11) as a likelihood function

$$
\begin{aligned}
\ell(\theta|\boldsymbol{y},\sigma) &= \left(\frac{1}{\sqrt{2\pi}\sigma}\right)^n \exp\left[-\frac{1}{2\sigma^2}\sum_{u=1}^{n}(\theta - y_u)^2\right] \\
&\propto \exp\left[-\frac{1}{2\sigma^2}\sum_{u=1}^{n}(\theta - y_u)^2\right]
\end{aligned}
\tag{5.1-12}
$$

in the manner of Eq. (5.1-6). The notation $\ell(\theta|\mathbf{y}, \sigma)$ places the given quantities on the right of the partition, as is usual for conditional functions.

Introduction of the mean value

$$\bar{y} = \frac{1}{n} \sum_{u=1}^{n} y_u \qquad\qquad (5.1\text{-}13)$$

into Eq. (5.1-12) gives

$$\ell(\theta|\mathbf{y}, \sigma) \propto \exp \left\{ -\frac{1}{2\sigma^2} \sum_{u=1}^{n} [(\theta - \bar{y}) + (\bar{y} - y_u)]^2 \right\}$$

$$\propto \exp \left\{ -\frac{1}{2\sigma^2} \left[ n(\theta - \bar{y})^2 + 2(\theta - \bar{y}) \sum_{u=1}^{n} (\bar{y} - y_u) + \sum_{u=1}^{n} (\bar{y} - y_u)^2 \right] \right\}$$

$$(5.1\text{-}14)$$

The sum of $(\bar{y} - y_u)$ in the last line is zero according to Eq. (5.1-13), and the last summation is independent of $\theta$. Hence, the equation reduces to

$$\ell(\theta|\mathbf{y}, \sigma) \propto \exp \left[ -\frac{n}{2\sigma^2} (\theta - \bar{y})^2 \right] \qquad\qquad (5.1\text{-}15)$$

This result is important and will be used in later examples.

## 5.2 BAYESIAN ESTIMATION WITH INFORMATIVE PRIORS

Two examples with informative priors are given here. The first deals with well-defined prior knowledge and a discrete parameter space; the second illustrates subjective priors and a continuous parameter space.

### Example 5.2. Estimation with Known Prior Probabilities

Four coins are provided for a game of chance. Three of the coins are fair, giving $p(\text{Heads}) = p(\text{Tails}) = 1/2$ for a fair toss, but the fourth coin has been altered so that both faces show "Heads," giving $p(\text{Heads}) = 1$ and $p(\text{Tails}) = 0$.

A coin is randomly chosen from the four and is tossed six times, giving the following data:

| $u$ | $y_u$ |
|-----|-------|
| 1 | Heads |
| 2 | Heads |
| 3 | Heads |
| 4 | Heads |
| 5 | Heads |
| 6 | Heads |

Let $\theta = 0$ denote the choice of a fair coin and $\theta = 1$ the choice of the altered one. Then the information in the first paragraph gives the prior probabilities

$$p(\theta=0) = 3/4 \quad \text{and} \quad p(\theta=1) = 1/4$$

for the two points in this discrete parameter space.

The likelihoods are calculated from Eq. (5.1-6):

| $u$ | $y_u$ | $\ell(\theta=0\|y_u)$ | $\ell(\theta=1\|y_u)$ |
|---|---|---|---|
| 1 | Heads | 1/2 | 1 |
| 2 | Heads | 1/2 | 1 |
| 3 | Heads | 1/2 | 1 |
| 4 | Heads | 1/2 | 1 |
| 5 | Heads | 1/2 | 1 |
| 6 | Heads | 1/2 | 1 |

Application of Bayes' theorem in the form (5.1-7) gives the posterior probabilities

$$p(\theta=0|y_1,\dots,y_6) \propto \frac{3}{4}\left(\frac{1}{2}\right)^6 = \frac{3}{256}$$

$$p(\theta=1|y_1,\dots,y_6) \propto \frac{1}{4}(1)^6 = \frac{64}{256}$$

Normalization to unit total probability then gives

$$p(\theta=0|\boldsymbol{y}) = \frac{3}{3+64} = \frac{3}{67}, \qquad p(\theta=1|\boldsymbol{y}) = \frac{64}{67}$$

Corresponding results after $n = 0,\dots,6$ experiments are tabulated below to show the progress of our learning from the accumulated data:

| $n$=no. of expts. | $p(\theta=0\|y_{\dots,n})$ | $p(\theta=1\|y_{\dots,n})$ |
|---|---|---|
| 0 | 3/4 | 1/4 |
| 1 | 3/5 | 2/5 |
| 2 | 3/7 | 4/7 |
| 3 | 3/11 | 8/11 |
| 4 | 3/19 | 16/19 |
| 5 | 3/35 | 32/35 |
| 6 | 3/67 | 64/67 |

After six experiments the posterior probabilities strongly favor $\theta = 1$, but $\theta = 0$ is not ruled out. The posterior probability $p(\theta = 0|y_{\dots,n})$ would rise to 1 if "Tails" appears at any time, since such an event requires the altered coin.

## Example 5.3. Estimation with Subjective Priors[1]

Investigators A and B are about to take measurements of a certain physical constant, $\theta$. Before taking any data, their beliefs regarding $\theta$ are expressed by the normal density functions

$$p_A(\theta) = \frac{1}{\sqrt{2\pi}(20)} \exp\left[-\frac{1}{2}\left(\frac{\theta - 900}{20}\right)^2\right]$$

$$p_B(\theta) = \frac{1}{\sqrt{2\pi}(80)} \exp\left[-\frac{1}{2}\left(\frac{\theta - 800}{80}\right)^2\right] \tag{5.2-1}$$

which are plotted in Figure 5.1.

The observations are to be modeled by Eq. (5.1-8), with normally distributed random errors having expectation zero and variance $\sigma^2 = 40^2$. Hence, the likelihood function after $n$ observations will be given by Eq. (5.1-15). Normalizing that equation to unit total area, as in Problem 5.B, we obtain

$$\ell(\theta|\boldsymbol{y}, \sigma) = \sqrt{\frac{n}{2\pi\sigma^2}} \exp\left[-\frac{n}{2}\left(\frac{\theta - \bar{y}}{\sigma}\right)^2\right] \tag{5.2-2}$$

as the "standardized likelihood function" after $n$ observations.

The resulting posterior densities for the two investigators are

$$p_i(\theta|\boldsymbol{y}, \sigma) \propto \ell(\theta|\boldsymbol{y}, \sigma)p_i(\theta)$$

$$\propto \exp\left[-\frac{n}{2}\left(\frac{\theta - \bar{y}}{\sigma}\right)^2 - \frac{1}{2}\left(\frac{\theta_i - \theta}{\sigma_i}\right)^2\right] \qquad i = A, B. \tag{5.2-3}$$

Each of these expressions can be written as a normal density function (see Problem 5.C), with mode $\widehat{\theta}_i$ and variance $\bar{\sigma}_i^2$ given by [2]

$$\widehat{\theta}_i = \left(\frac{\theta_i}{\sigma_i^2} + \frac{n\bar{y}}{\sigma^2}\right) \bigg/ \left(\frac{1}{\sigma_i^2} + \frac{n}{\sigma^2}\right) \qquad i = A, B \tag{5.2-4}$$

$$\frac{1}{\bar{\sigma}_i^2} = \frac{1}{\sigma_i^2} + \frac{n}{\sigma^2} \qquad i = A, B \tag{5.2-5}$$

The likelihood and posterior density functions are plotted in Figure 5.1 after one observation with $y_1 = 850$ and after 100 observations with mean $\bar{y} = 870$.

---

[1]  Adapted from Box and Tiao (1973, 1992) with permission.
[2]  The notation $\widehat{\theta}$ is standard for maximum-density estimates, but $\bar{\sigma}_i^2$ is specific to this example.

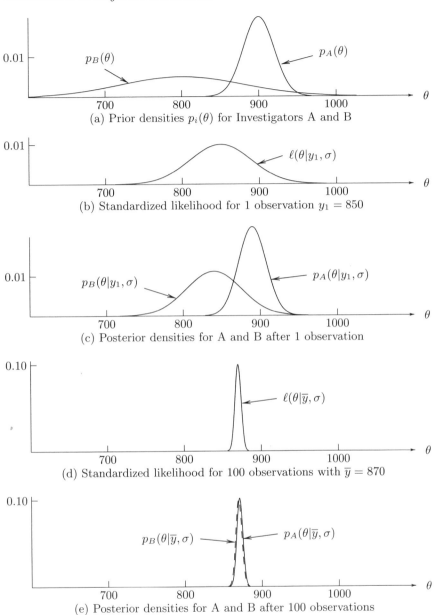

**Figure 5.1:** Priors, likelihoods, and posteriors for Example 5.3.

We see from plots (a), (c), and (e) in Figure 5.1 that the posterior density functions for the two investigators come together as data accumulate. This happens because the region of appreciable likelihood narrows down, so that each investigator's prior density is nearly constant across this important region of $\theta$. In situations of this kind, we say that the estimation is *dominated* by the likelihood.

For small data sets, the choice of the prior remains important. Investigators A and B could then reach agreement more readily by using a *noninformative* prior. Such priors are discussed in the following sections. Another type of prior, appropriate for discrimination among rival models, will appear in Chapters 6 and 7.

## 5.3 INTRODUCTION TO NONINFORMATIVE PRIORS

The construction of a noninformative prior is a nontrivial task, requiring analysis of likelihood functions for prospective data. The construction is simplest when the likelihood $\ell(\theta)$ for a single parameter is *data-translated* in some coordinate $\phi(\theta)$; then the noninformative prior density takes the form $p(\phi) = $ const. over the permitted range of $\phi$. This rule is a special form of a more general one derived by Jeffreys (1961) (see Section 5.4); we illustrate it here by two examples.

### Example 5.4. Prior for a Normal Mean

The likelihood function for $\theta$ in Examples 5.1 and 5.3 depends on the data only through the mean value $\bar{y}$, that is,

$$\ell(\theta|\boldsymbol{y}, \sigma) = \ell(\theta|\bar{y}, \sigma) \propto \exp\left[-\frac{n}{2\sigma^2}(\theta - \bar{y})^2\right] \qquad (5.3\text{-}1)$$

Any change in the data will translate the likelihood curve along the $\theta$ axis a distance exactly equal to the change in $\bar{y}$. This data-translation property is illustrated by the likelihood curves for various $\bar{y}$ values, plotted in Figure 5.2.

Since $\ell(\theta|\bar{y}, \sigma)$ is data-translated in $\theta$, the function $\phi(\theta)$ here is simply $\theta$. The noninformative prior is accordingly

$$p(\theta) = c \qquad (5.3\text{-}2)$$

The resulting posterior density function is thus proportional to the likelihood,

$$p(\theta|\boldsymbol{y}, \sigma) \propto \exp\left[-\frac{n}{2\sigma^2}(\theta - \bar{y})^2\right] \qquad (5.3\text{-}3)$$

and gives a normal distribution of $\theta$ with mode $\bar{y}$ and variance $\sigma^2/n$.

### Example 5.5. Prior for a Normal Standard Deviation

The standard deviation $\sigma$ of a normally distributed variable $y$ with known mean $\mu$ has the likelihood function (see Eq. (5.1-11))

$$\ell(\sigma|\boldsymbol{y}, \mu) \propto \sigma^{-n} \exp\left[-\frac{1}{2\sigma^2}\sum_{u=1}^{n}(y_u - \mu)^2\right]$$

$$\propto (\sigma/s)^{-n} \exp\left[-\frac{ns^2}{2\sigma^2}\right] \qquad (5.3\text{-}4)$$

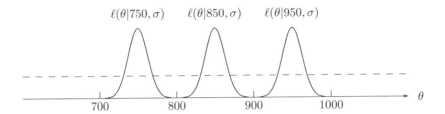

**Figure 5.2:** Noninformative prior (dashed line) for the normal mean $\theta$, and likelihood curves $\ell(\theta|\bar{\boldsymbol{y}}, \sigma)$ after 10 observations with $\sigma = 40$. Compare with Figure 5.1.

when a data vector $\boldsymbol{y} = \{y_1, \ldots, y_n\}^T$ is given. Here $s^2$ is the sample variance

$$s^2 = (1/n) \sum_{u=1}^{n} (y_u - \mu)^2 \qquad (5.3\text{-}5)$$

Equation (5.3-4) shows that $\ell(\sigma|\boldsymbol{y}, \mu)$ is exactly expressible as $\ell(\sigma|s)$. This form is handy for the following treatment.

Since $\ell(\sigma|s)\,d\sigma = \ell(\ln \sigma|s)\,d\ln \sigma = \ell(\ln \sigma|s)d\sigma/\sigma$, we can rewrite Eq. (5.3-4) as

$$\ell(\ln \sigma|s) \propto (\sigma/s)^{1-n} \exp[-(n/2)(s/\sigma)^2] \qquad (5.3\text{-}6)$$

Taking the natural logarithm of each member, we get

$$\begin{aligned}
\ln \ell(\ln \sigma|s) &= \text{const.} + (1-n)(\ln \sigma - \ln s) - (n/2)(s/\sigma)^2 \\
&= \text{const.} + (1-n)(\ln \sigma - \ln s) - (n/2)\exp[2(\ln s - \ln \sigma)]
\end{aligned}$$
$$(5.3\text{-}7)$$

Any change in the data translates this log-likelihood function along the $\ln \sigma$ axis a distance exactly equal to the change in $\ln s$. Thus, $\ln \ell(\ln \sigma|s)$ is data-translated in the coordinate $\phi(\sigma) = \ln \sigma$, as illustrated by the curves in Figure 5.3. Consequently, the noninformative prior for this problem is

$$p(\ln \sigma) = \text{const.} \qquad (5.3\text{-}8)$$

Equating the probability elements $p(\sigma)|d\sigma|$ and $p(\ln \sigma)|d\ln \sigma|$, one then obtains

$$p(\sigma) = p(\ln \sigma)\,|d\ln \sigma/d\sigma| \propto (\text{const.})\sigma^{-1} \qquad (5.3\text{-}9)$$

as the noninformative prior for $\sigma$. The resulting posterior density function,

$$p(\sigma|s) \propto \sigma^{-n-1} \exp[-(n/2)(s/\sigma)^2] \qquad (5.3\text{-}10)$$

is used in Chapter 6.

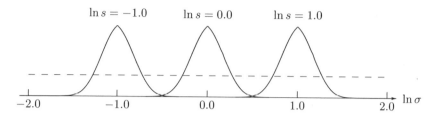

**Figure 5.3:** Noninformative prior (dashed line) for the normal $\ln \sigma$, and likelihood curves $\ell(\ln \sigma | s)$ after 10 observations.

The priors in Eqs. (5.3-2) and (5.3-8) would be improper (not normalizable to unit total probability) if applied over unbounded ranges. In practice this does not happen, since the parameters are restricted by (i) the range of significant likelihood, (ii) the physical or mathematical requirements of the model, and (iii) the number range of the computer used. Expressions like Eq. (5.3-2) and (5.3-8) should be regarded as "locally uniform" priors, and made proper by limiting the ranges of their arguments.

## 5.4 JEFFREYS PRIOR FOR ONE-PARAMETER MODELS

Jeffreys (1961) gave a formal method for constructing noninformative priors from the expectation of the log-likelihood function

$$L(\boldsymbol{\theta}|\boldsymbol{y}) \equiv \ln \ell(\boldsymbol{\theta}|\boldsymbol{y}) \equiv \ln p(\boldsymbol{y}|\boldsymbol{\theta}) \tag{5.4-1}$$

when $\boldsymbol{\theta}$ has a continuous range of values. His method is presented here for models with a single parameter $\theta$; the development for a multiparameter vector $\boldsymbol{\theta}$ will be given in Section 5.5.

Let $\theta_M$ denote the mode (initially unknown) of the log-likelihood function $L(\theta|\boldsymbol{y})$ for the single parameter $\theta$. A second-order Taylor expansion of $L$ around $\theta_M$, for a given model and pattern of experiments, gives

$$L(\theta|\boldsymbol{y}) \approx \ln p(\boldsymbol{y}|\theta_M) + \left[\frac{\partial^2 \ln p(\boldsymbol{y}|\theta)}{\partial \theta^2}\bigg|_{\theta_M}\right] \frac{(\theta - \theta_M)^2}{2!} \tag{5.4-2}$$

after use of the vanishing of $\partial L/\partial \theta$ at $\theta_M$. The expectation of this expansion over the predicted distribution of $\boldsymbol{y}$ is

$$\begin{aligned} \mathrm{E}[L(\theta|\theta_M)] &\approx \mathrm{E}[\ln p(\boldsymbol{y}|\theta_M)] + \mathrm{E}\left[\frac{\partial^2 \ln p(\boldsymbol{y}|\theta)}{\partial \theta^2}\bigg|_{\theta_M}\right] \frac{(\theta - \theta_M)^2}{2!} \\ &\approx K - \mathcal{I}_n(\theta_M)(\theta - \theta_M)^2/2! \end{aligned} \tag{5.4-3}$$

Here $K$ is a constant for a given model and pattern of $n$ experiments, and $\mathcal{I}_n(\theta_M)$ is the value at $\theta_M$ of the function

$$\mathcal{I}_n(\theta) \equiv \mathrm{E}\left[-\frac{\partial^2 \ln p(\boldsymbol{y}|\theta)}{\partial \theta^2}\right] \qquad (5.4\text{-}4)$$

that Fisher (1935) introduced as an information measure for experimental designs. The initial goal, as in the previous section, is to find a function $\phi(\theta)$ whose noninformative prior density $p(\phi)$ is constant. Jeffreys did this by use of the transformations

$$\mathcal{I}_n(\phi) = \mathcal{I}_n(\theta)\left|\frac{d\theta}{d\phi}\right|^2 \qquad (5.4\text{-}5)$$

$$p(\phi) = p(\theta)\left|\frac{d\theta}{d\phi}\right| \qquad (5.4\text{-}6)$$

which hold at the mode $\theta_M$ as long as $d\theta/d\phi$ is finite and nonzero.

Setting $\mathcal{I}_n(\phi) = 1$ in Eq. (5.4-5) leads to the differential equation

$$\frac{d\phi}{d\theta} = \pm\sqrt{\mathcal{I}_n(\theta)} \qquad (5.4\text{-}7)$$

and yields the following version of Eq. (5.4-3):

$$\mathrm{E}[L(\phi|\phi_M)] = K - [1](\phi - \phi_M)^2/2! \qquad (5.4\text{-}8)$$

To obtain a noninformative prior, we require that the interval $|\phi - \phi_M| < C$ have the same expected posterior probability content, wherever its center $\phi_M$ may turn out to be when data are analyzed. In other words, the integral $\int p(\phi)\exp[K - (\phi - \phi_M)^2/2]\,d\phi$ from $\phi_M - C$ to $\phi_M + C$ should be independent of $\phi_M$, so that $p(\phi)$ should be a constant. Equation (5.4-6) then gives

$$p(\theta) \propto \sqrt{\mathcal{I}_n(\theta)} \qquad (5.4\text{-}9)$$

which is Jeffreys prior for one-parameter models.

### Example 5.6. Jeffreys Prior for the Normal Mean

Taking logarithms in Eq. (5.1-12) gives

$$L(\theta|\sigma, \boldsymbol{y}) = \text{const.} - \frac{1}{2\sigma^2}\sum_{u=1}^{n}(y_u - \theta)^2 \qquad (5.4\text{-}10)$$

Hence,

$$\frac{\partial^2 L(\theta|\sigma, \boldsymbol{y})}{\partial \theta^2} = -\frac{n}{\sigma^2} = \text{const.} \qquad (5.4\text{-}11)$$

and

$$\mathcal{I}_n(\theta) = \text{const.} \quad \text{for all } \theta \qquad (5.4\text{-}12)$$

Since $\mathcal{I}_n(\theta)$ is independent of $\theta$ here, the Jeffreys prior $p(\theta)$ is uniform. This result agrees with Eq. (5.3-2).

**Example 5.7. Jeffreys Prior for the Normal Standard Deviation**

Taking logarithms in Eq. (5.3-4) gives

$$L(\sigma|s) = \text{const.} - n \ln \sigma - \frac{ns^2}{2\sigma^2} \qquad (5.4\text{-}13)$$

Hence,

$$\frac{\partial L}{\partial \sigma} = -\frac{n}{\sigma} + \frac{ns^2}{\sigma^3}$$

$$\frac{\partial^2 L}{\partial \sigma^2} = \frac{n}{\sigma^2} - \frac{3ns^2}{\sigma^4}.$$

Insertion of $\sigma^2$ as the expectation of $s^2$ in the latter equation gives the information measure

$$\mathcal{I}_n(\sigma) = -\frac{n}{\sigma^2} + \frac{3n}{\sigma^2} = \frac{2n}{\sigma^2} \qquad (5.4\text{-}14)$$

Application of Eq. (5.4-9) then gives the noninformative prior

$$p(\sigma) \propto \sqrt{2n/\sigma^2} \propto \sigma^{-1} \qquad (5.4\text{-}15)$$

which agrees with Eq. (5.3-7).

## 5.5 JEFFREYS PRIOR FOR MULTIPARAMETER MODELS

Two classes of parameters are needed in models of observations: *location parameters* $\boldsymbol{\theta}_l$ to describe expected response values and *scale parameters* $\boldsymbol{\theta}_s$ to describe distributions of errors. Jeffreys treated $\boldsymbol{\theta}_l$ and $\boldsymbol{\theta}_s$ separately in deriving his noninformative prior; this was reasonable since the two types of parameters are unrelated *a priori*. Our development here will parallel that given by Box and Tiao (1973, 1992), which provides a fuller discussion. The key result of this section is Eq. (5.5-8).

When data are obtained, we assume that the log-likelihood function will have a local maximum (mode) at some point $\boldsymbol{\theta} = \boldsymbol{\theta}_M$. A second-order Taylor expansion of $L(\boldsymbol{\theta})$ around that point will give

$$L(\boldsymbol{\theta}|\boldsymbol{y}) \approx \ln p(\boldsymbol{y}|\boldsymbol{\theta}_M) + \frac{1}{2!}(\boldsymbol{\theta} - \boldsymbol{\theta}_M)^T \left\{ \frac{\partial^2 \ln p(\boldsymbol{y}|\boldsymbol{\theta})}{\partial \theta_i \partial \theta_j} \bigg|_{\boldsymbol{\theta}_M} \right\} (\boldsymbol{\theta} - \boldsymbol{\theta}_M) \qquad (5.5\text{-}1)$$

The expectation of this expansion over the distribution of $\boldsymbol{y}$ predicted by the model is

$$\mathrm{E}[L(\boldsymbol{\theta}|\boldsymbol{\theta}_M)] \approx \mathrm{E}[\ln p(\boldsymbol{y}|\boldsymbol{\theta}_M)] + \frac{1}{2!}(\boldsymbol{\theta} - \boldsymbol{\theta}_M)^T \left\{ \mathrm{E}\left[ \frac{\partial^2 \ln p(\boldsymbol{y}|\boldsymbol{\theta}_M)}{\partial \theta_i \partial \theta_j} \bigg|_{\boldsymbol{\theta}_M} \right] \right\} (\boldsymbol{\theta} - \boldsymbol{\theta}_M)$$

$$\approx K - \frac{1}{2!}(\boldsymbol{\theta} - \boldsymbol{\theta}_M)^T \mathcal{I}_n(\boldsymbol{\theta}_M)(\boldsymbol{\theta} - \boldsymbol{\theta}_M). \qquad (5.5\text{-}2)$$

Here $\boldsymbol{\mathcal{I}}_n(\boldsymbol{\theta}_M)$ is the value at $\boldsymbol{\theta}_M$ of Fisher's (1935) information matrix

$$\boldsymbol{\mathcal{I}}_n(\boldsymbol{\theta}) \equiv \left\{ -\text{E}\left[ \frac{\partial^2 \ln p(\boldsymbol{y}|\boldsymbol{\theta})}{\partial \theta_i \partial \theta_j} \right] \right\} \tag{5.5-3}$$

for the given model and pattern of $n$ experiments. If a maximum-likelihood point $\boldsymbol{\theta}_M$ is found, then the quadratic term in Eq. (5.5-2) for the estimable parameters will be strictly negative, so that the matrix $\boldsymbol{\mathcal{I}}_n(\boldsymbol{\theta}_M)$ for those parameters will be positive definite. Then the constant-expected-likelihood contours given by Eq. (5.5-2) will be ellipsoidal.

Jeffreys (1961) investigated noninformative priors for $\boldsymbol{\theta}$ by use of the relations

$$\boldsymbol{\mathcal{I}}_n(\boldsymbol{\phi}) = \left[ \frac{\partial \boldsymbol{\theta}}{\partial \boldsymbol{\phi}} \right] \boldsymbol{\mathcal{I}}_n(\boldsymbol{\theta}) \left[ \frac{\partial \boldsymbol{\theta}}{\partial \boldsymbol{\phi}} \right]^T \tag{5.5-4}$$

$$p(\boldsymbol{\phi}) = p(\boldsymbol{\theta}) \left| \frac{\partial \boldsymbol{\theta}}{\partial \boldsymbol{\phi}} \right| \tag{5.5-5}$$

which hold at $\boldsymbol{\theta}_M$ as long as $\partial\boldsymbol{\theta}/\partial\boldsymbol{\phi}$ is neither zero nor infinite there.

The matrix $\boldsymbol{\mathcal{I}}_n(\boldsymbol{\phi})$ then inherits positive definiteness from $\boldsymbol{\mathcal{I}}_n(\boldsymbol{\theta})$, and the $\boldsymbol{\phi}$-analog of Eq. (5.5-2) gives the following log-likelihood contours,

$$(\boldsymbol{\phi} - \boldsymbol{\phi}_M)^T \boldsymbol{\mathcal{I}}_n(\boldsymbol{\phi}_M)(\boldsymbol{\phi} - \boldsymbol{\phi}_M) = C \geq 0 \tag{5.5-6}$$

with included volumes proportional to $\sqrt{|C\boldsymbol{\mathcal{I}}_n^{-1}(\boldsymbol{\phi}_M)|}$ in $\boldsymbol{\phi}$-space.

Of special interest are those mappings $\boldsymbol{\phi}(\boldsymbol{\theta})$ for which the determinant $|\boldsymbol{\mathcal{I}}_n(\boldsymbol{\phi})|$ is independent of $\boldsymbol{\phi}$, so that the volume enclosed by the contour of Eq. (5.5-6) for any positive $C$ is independent of $\boldsymbol{\phi}_M$. A uniform $p(\boldsymbol{\phi})$ then provides a noninformative prior, because the expected posterior probability content within any $C$-contour is then independent of $\boldsymbol{\phi}_M$. For constant $|\boldsymbol{\mathcal{I}}_n(\boldsymbol{\phi})|$ one also finds

$$|\boldsymbol{\mathcal{I}}_n(\boldsymbol{\theta})| \left| \frac{\partial \boldsymbol{\theta}}{\partial \boldsymbol{\phi}} \right|^2 = \text{const.} \tag{5.5-7}$$

by taking the determinant of each member in Eq. (5.5-4). Insertion of this result, and a uniform prior density $p(\boldsymbol{\phi})$, into Eq. (5.5-5) gives finally

$$p(\boldsymbol{\theta}) \propto p(\boldsymbol{\phi}) \left| \frac{\partial \boldsymbol{\phi}}{\partial \boldsymbol{\theta}} \right|$$

$$\propto \sqrt{|\boldsymbol{\mathcal{I}}_n(\boldsymbol{\theta})|} \tag{5.5-8}$$

as Jeffreys multiparameter noninformative prior.

**Example 5.8. Jeffreys Prior for the Normal Mean and $\sigma$**

The log-likelihood function in this case is obtained from Eq. (5.3-4):

$$L(\mu, \sigma | \boldsymbol{y}) = \text{const.} - n \ln \sigma - \frac{1}{2\sigma^2} \sum_{u=1}^{n} (y_u - \mu)^2 \qquad (5.5\text{-}9)$$

Here $\mu$ is $\theta_l$ and $\sigma$ is $\theta_s$. The Jeffreys priors for these parameters are available from Examples 5.6 and 5.7, with their arguments properly restricted as required by the prior independence of $\theta_l$ and $\theta_s$. Inserting those results into Eq. (5.5-9), we get the joint prior

$$p(\mu, \sigma) \propto \sqrt{(\text{const.})/\sigma^2} \propto \sigma^{-1} \qquad (5.5\text{-}10)$$

The posterior density expression is then

$$p(\mu, \sigma | \boldsymbol{y}) \propto \sigma^{-n-1} \exp\left[ -\frac{1}{2\sigma^2} \sum_{u=1}^{n} (y_u - \mu)^2 \right] \qquad (5.5\text{-}11)$$

as recommended by Jeffreys (1961) and by Box and Tiao (1973, 1992).

---

The following comments are offered on the use of Eq. (5.5-8):

(1) Any model linear in its location parameters $\boldsymbol{\theta}_l$ has a uniform Jeffreys prior $p(\boldsymbol{\theta}_l)$ over the permitted range of $\boldsymbol{\theta}_l$. This condition occurred in Examples 5.6 and 5.8, where the model $y_u = \mu + \varepsilon_u$ with location parameter $\mu$ was used. The Jeffreys prior $p(\boldsymbol{\theta}_l)$ is likewise uniform for any model nonlinear in $\boldsymbol{\theta}_l$, over the useful range of its linearized Taylor expansion that we provide in Chapters 6 and 7.

(2) Any unknown scale parameters should be treated by formula (5.5-8), with subdivision into independent groups when appropriate, as illustrated for the elements of $\boldsymbol{\Sigma}$ in Table 7.2. The priors thus obtained for normal variances $\sigma^2$ and covariance matrices $\boldsymbol{\Sigma}$ are consistent with the estimates and predictors found by Geisser and Cornfield (1963), Aitchison and Dunsmore (1975), and Murray (1977) by independent methods.

Comment (1) is consistent with normal practice in parameter estimation. Use of the detailed $p(\boldsymbol{\theta}_l)$ in nonlinear parameter estimation would take much extra effort, since fourth derivatives of the expectation model would then appear. Comment (2) is implemented in Chapters 6 and 7 and in the package GREGPLUS.

## 5.6 SUMMARY

This chapter has introduced Bayes' theorem and applied it to some simple estimation problems. Prior distributions based on much information and on little information have been described, and the resulting posterior distributions have been shown.

The following chapters and the package GREGPLUS apply these principles to practical models and various data structures. Least squares, multiresponse estimation, model discrimination, and process function estimation are presented there as special forms of Bayesian estimation.

## PROBLEMS

### 5.A Probability That a Die Is Loaded

Four dice of similar appearance are provided for a game of chance. Three of the dice are fair, and one is loaded; their score distributions are as follows:

| Score, $Z_i$ | 1 | 2 | 3 | 4 | 5 | 6 |
|---|---|---|---|---|---|---|
| $p(Z_i)$, fair dice | 1/6 | 1/6 | 1/6 | 1/6 | 1/6 | 1/6 |
| $p(Z_i)$, loaded die | 1/12 | 1/12 | 1/12 | 1/12 | 1/12 | 7/12 |

A die is chosen at random from the four and is tossed eight times, giving the following scores:
$$y_1 \ldots y_8 = \quad 3 \quad 6 \quad 6 \quad 5 \quad 6 \quad 1 \quad 6 \quad 4$$

Let $\theta = 0$ denote choosing of a fair die, and let $\theta = 1$ denote choosing of the loaded die. The following calculations are desired:
(a) The prior probabilities $p(\theta = 0)$ and $p(\theta = 1)$.
(b) The likelihoods $\ell(\theta = 0 | y_u)$ and $\ell(\theta = 1 | y_u)$ for each of the eight events.
(c) The posterior probabilities $p(\theta = 0 | y_1, \ldots, y_n)$ and $p(\theta = 1 | y_1, \ldots, y_n)$ for $n = 1, \ldots, 8$, normalized to unit total probability for each $n$.

### 5.B Standardized Likelihood of Normal Mean

Verify the development of Eq. (5.2-2) from Eq. (5.1-15). Proceed as follows:
(a) Integrate Eq. (5.1-15) over the range $(-\infty, +\infty)$ of $\theta$ with the aid of the formula

$$\int_{-\infty}^{\infty} \exp(-\alpha^2 x^2) dx = (1/\alpha)\sqrt{\pi} \tag{5.B-1}$$

(b) Use the result of (a) as a divisor to standardize Eq. (5.1-15).

### 5.C Normal Posterior Density Functions

(a) Combine Eqs. (5.2-1) and (5.2-2) according to Bayes' theorem to obtain Eq. (5.2-3).
(b) Verify Eq. (5.2-4) by finding the mode of Eq. (5.2-3).
(c) Verify Eq. (5.2-5) by equating the coefficient of $\theta^2$ in Eq. (5.2-3) to that of a normal density function with variance $\overline{\sigma}_i^2$.

## 5.D Parameter Estimates and Probabilities

(a) Use Eqs. (5.2-4) and (5.2-5) to compute the mode $\widehat{\theta}_B$ and the standard deviation $\overline{\sigma}_B$ of Investigator B's posterior distribution of $\theta$ after the first and the hundredth observation in Example 5.3.

(b) Compute and interpret the cumulative prior probability

$$\Pr_B(\theta > 850) \equiv \int_{850}^{\infty} p_B(\theta)\, d\theta \qquad (5.D-1)$$

by use of a table of the (cumulative) normal distribution or of the related function $\mathrm{erf}(x)$.

(c) Compute the probabilities $\Pr_B(\theta > 850|\sigma, \boldsymbol{y})$ at $n = 1$ and $n = 100$ by expressing $p_B(\theta|\sigma, \boldsymbol{y})$ as a normal density function (see Eqs. (5.2-4,5)) and using a table of the (cumulative) normal distribution.

## 5.E Computation of Likelihood and Posterior Density

In Example 5.3, suppose that the first two observations are $y_1 = 870$ and $y_2 = 850$.

(a) Compute the mode $\widehat{\theta}_i$ and variance $\overline{\sigma}^2$ of each investigator's posterior distribution $p_i(\theta|\sigma, y_1, y_2)$.

(b) Compute and plot the likelihood $\ell(\theta|y_1, y_2)$ and the posterior density $p_A(\theta|\sigma, y_1, y_2)$.

## REFERENCES and FURTHER READING

Aitchison, J., and I. R. Dunsmore, *Statistical Prediction Analysis*, Cambridge University Press, Cambridge (1975).

Bayes, T. R., An essay towards solving a problem in the doctrine of chances, *Phil. Trans. Roy. Soc. London*, **53**, 370–418 (1763). Reprinted in *Biometrika*, **45**, 293–315 (1958).

Berger, J. O., *Statistical Decision Theory and Bayesian Analysis*, 2nd edition, Springer, New York (1988).

Berry, D. A., and B. W. Lindgren, *Statistics – Theory and Methods*, Brooks/Cole, Pacific Grove, CA (1990).

Box, G. E. P., Sampling and Bayes' inference in scientific modelling and robustness (with Discussion), *J. Roy. Statist. Soc. A*, **143**, 383–430 (1980).

Box, G. E. P., and G. C. Tiao, *Bayesian Inference in Statistical Analysis*, Addison-Wesley, Reading, MA (1973). Reprinted by Wiley, New York (1992).

Fisher, R. A., On the mathematical foundations of theoretical statistics, *Phil. Trans. Roy. Soc. London A*, **222**, 309–368 (1922).

Fisher, R. A., *Design of Experiments*, Oliver and Boyd, London (1935).

Gauss, C. F., *Theoria motus corporum coelestium in sectionibus conicis solem ambientium*, Perthas et Besser, Hamburg (1809); *Werke*, **7**, 240–254. Translated as *Theory of Motion of the Heavenly Bodies Moving about the Sun in Conic Sections* by C. H. Davis. Little, and Brown, Boston (1857); Dover, New York (1963).

Geisser, S., and J. Cornfield, Posterior distributions for multivariate Normal parameters, *J. Roy. Statist. Soc. B*, **25**, 368–376 (1963).

Jeffreys, H., *Theory of Probability*, 3rd edition, Clarendon Press, Oxford (1961).

Laplace, P. S., Supplément au Mémoire Sur les approximations des fonctions qui sont fonctions de trés-grands nombres, *Mémoires de l'Académie des sciences des Paris*, (1810) 565.

Lee, P. M., *Bayesian Statistics: An Introduction*, Oxford University Press, New York (1989).

Meeden, G., Betting against a Bayesian bookie, *J. Am. Statist. Assoc.*, **76**, 202–204 (1981).

Murray, G. D., A note on the estimation of probability density functions, *Biometrika*, **64**, 150–152 (1977).

Stigler, S. M., Thomas Bayes's Bayesian Inference, *J. Roy. Statist. Soc. A*, **145**, 250–258 (1982).

Stigler, S. M., *History of Statistics*, Harvard University Press, Cambridge, MA (1986).

Zellner, A., *An Introduction to Bayesian Inference in Econometrics*, Wiley, New York (1971).

# Chapter 6

# Process Modeling

# with Single-Response Data

The development of a process model typically goes through several stages, outlined in Figure 4.1 and elaborated in this chapter and the next. Typical stages include model formulation; collection of data from new experiments or existing sources; parameter estimation; testing and discrimination of various postulated models; and extensions of the database with sequentially designed experiments. Computational aids are presented in this book for several of these tasks, but prudent investigators will also use diagnostic plots, physical and chemical clues, and experience as aids in checking the data and constructing good candidate models.

The statistical investigation of a model begins with the estimation of its parameters from observations. Chapters 4 and 5 give some background for this step. For single-response observations with independent normal error distributions and given relative precisions, Bayes' theorem leads to the famous method of least squares. Multiresponse observations need more detailed treatment, to be discussed in Chapter 7.

Gauss (1799) gave the first published least-squares solution, but without describing his method, which he identified later as least squares (Plackett 1972, pp. 240 and 246). Legendre (1805) presented least squares as a curve-fitting method, and Gauss (1809) derived the method from elegant probabilistic arguments. Later, setting probabilities aside, Gauss (1823, 1828) showed that linear least squares gives the parameter estimates with least variance among all linear combinations of the observations. Controversy ensued from Gauss' book of 1809, in which he referred to least squares as *principium nostrum* ("our principle") and said he had used it since 1795; Legendre refused to share credit for the discovery. Fascinating accounts of this dispute and the history of least squares are given by Plackett (1972), Stigler (1981, 1986), and G. W. Stewart (1995).

This chapter uses Gauss' 1809 treatment of nonlinear least squares (submitted in 1806, but delayed by the publisher's demand that it be translated into Latin). Gauss weighted the observations according to their precision, as we do in Sections 6.1 and 6.2. He provided *normal equations* for parameter estimation, as we do in Section 6.3, with iteration for models nonlinear in the parameters. He gave efficient algorithms for the parameter estimates and their variances; we follow his lead in Sections 6.4 and 6.6. Developments since his time have included goodness-of-fit testing (Section 6.5), interval estimation for parameters and functions (Section 6.6), model discrimination (Section 6.6), and design of the next experiment for optimal estimation and/or model discrimination (Section 6.7). These developments are outlined in this chapter and implemented conveniently in the software package GREGPLUS; see Appendix C.

## 6.1 THE OBJECTIVE FUNCTION $S(\boldsymbol{\theta})$

An objective function $S(\boldsymbol{\theta})$ is presented here for use in Bayesian estimation of the parameter vector $\boldsymbol{\theta}$ in a mathematical model

$$
\begin{aligned}
y_u &= f_u(\boldsymbol{\xi}_u, \boldsymbol{\theta}) + \varepsilon_u \\
&= f_u(\boldsymbol{\theta}) + \varepsilon_u \qquad u = 1, \dots, n
\end{aligned}
\tag{6.1-1}
$$

of single-response events $u = 1, \dots, n$. Such data give one response value $y_u$ per event at settings $\boldsymbol{\xi}_u$ of the independent variables. A value $y_u$ may be directly observed or may be a chosen function of observations from that event; in either case, corrections for any known causes of systematic error should be included in $y_u$. Each value $y_u$ is then modeled as an expectation function $f_u(\boldsymbol{\xi}_u, \boldsymbol{\theta})$ plus an independent random error $\varepsilon_u$ with probability density

$$
p(\varepsilon_u | \sigma_u) = \frac{1}{\sqrt{2\pi}\sigma_u} \exp\left[-\frac{\varepsilon_u^2}{2\sigma_u^2}\right] \qquad u = 1, \dots, n
\tag{6.1-2}
$$

based on the central limit theorem of Chapter 4 and on the postulate that the proposed model form $f(\boldsymbol{\xi}_u, \boldsymbol{\theta})$ is true. The predictive probability density[1] for the $u$th event is then

$$
p(y_u | \boldsymbol{\theta}, \sigma_u) = \frac{1}{\sqrt{2\pi}\sigma_u} \exp\left\{-\frac{[y_u - f_u(\boldsymbol{\theta})]^2}{2\sigma_u^2}\right\} \qquad u = 1, \dots, n
\tag{6.1-3}
$$

and the predictive probability density for the full data vector $\boldsymbol{y}$ is the product of these independent functions:

$$
p(\boldsymbol{y} | \boldsymbol{\theta}, \sigma_1, \dots, \sigma_n) = \left[\prod_{u=1}^{n} \frac{1}{\sqrt{2\pi}\sigma_u}\right] \exp\left\{-\sum_{u=1}^{n} \frac{[y_u - f_u(\boldsymbol{\theta})]^2}{2\sigma_u^2}\right\}
\tag{6.1-4}
$$

---

[1] This density times $d\varepsilon$ is the probability that a prospective observation $y_u$ at $\boldsymbol{\xi}_u$ will fall between $f(\boldsymbol{\xi}_u, \boldsymbol{\theta}) - d\varepsilon/2)$ and $f(\boldsymbol{\xi}_u, \boldsymbol{\theta}) + d\varepsilon/2$ if the model form is true.

Relative weights are assigned to the values $y_u$ in the forms

$$w_u = \sigma^2/\sigma_u^2 \qquad u = 1, \ldots, n \tag{6.1-5}$$

where $\sigma^2$ is the variance of observations of unit weight and $\sigma_u^2$ is the variance assigned to observations at $\xi_u$. Unit weights ($w_u = 1$) are appropriate for data of uniform precision; other weightings are discussed in Section 6.2. Equation (6.1-4) then takes the form

$$p(\boldsymbol{y}|\boldsymbol{\theta}, \sigma, \boldsymbol{w}) = \left[ \prod_{u=1}^{n} \frac{\sqrt{w_u}}{\sqrt{2\pi}\sigma} \right] \exp \left[ -\frac{S(\boldsymbol{\theta})}{2\sigma^2} \right] \tag{6.1-6}$$

with the weighted sum-of-squares function

$$\begin{aligned} S(\boldsymbol{\theta}) &= \sum_{u=1}^{n} \left( \sqrt{w_u} \, [y_u - f_u(\boldsymbol{\theta})] \right)^2 \\ &= \sum_{u=1}^{n} [Y_u - F_u(\boldsymbol{\theta})]^2 \\ &= \sum_{u=1}^{n} E_u(\boldsymbol{\theta})^2 \end{aligned} \tag{6.1-7}$$

Here the notations

$$Y_u = \sqrt{w_u}\, y_u; \qquad F_u(\boldsymbol{\theta}) = \sqrt{w_u}\, f_u(\boldsymbol{\theta}); \qquad E_u(\boldsymbol{\theta}) = Y_u - F_u(\boldsymbol{\theta}) \tag{6.1-8}$$

have reduced $S(\boldsymbol{\theta})$ to a simple sum of squares of the weighted errors $E_u(\boldsymbol{\theta})$. Correspondingly, Eq, (6.1-3) yields the following predictive probability density for a weighted observation $Y_u$,

$$p(Y_u|\boldsymbol{\theta}, \sigma) = \frac{p(y_u|\boldsymbol{\theta}, \sigma)}{\sqrt{w_u}} = (\sqrt{2\pi}\sigma)^{-1} \exp \left[ -\frac{E_u^2}{2\sigma^2} \right] \qquad u = 1, \ldots, n \tag{6.1-9}$$

and for a vector $Y_1, \ldots, Y_n$ of weighted independent observations, it gives

$$p(\boldsymbol{Y}|\boldsymbol{\theta}, \sigma) = (\sqrt{2\pi}\sigma)^{-n} \exp \left[ -\frac{S(\boldsymbol{\theta})}{2\sigma^2} \right] \tag{6.1-10}$$

When $\boldsymbol{Y}$ is given instead of $\boldsymbol{\theta}$ and $\sigma$, we call this function the *likelihood*,

$$\ell(\boldsymbol{\theta}, \sigma|\boldsymbol{Y}) = (\sqrt{2\pi}\sigma)^{-n} \exp \left[ -\frac{S(\boldsymbol{\theta})}{2\sigma^2} \right] \tag{6.1-11}$$

in the manner of Eq. (5.1-6).

To apply Bayes' theorem, we need a prior probability density for the unknowns, $\boldsymbol{\theta}$ and $\sigma$. Treating $\boldsymbol{\theta}$ and $\sigma$ as independent *a priori* and $p(\boldsymbol{\theta})$ as uniform over the permitted range of $\boldsymbol{\theta}$, we obtain the joint prior density

$$p(\boldsymbol{\theta}, \sigma) \propto \begin{cases} \sigma^{-1} & \text{for permitted values of } \boldsymbol{\theta} \\ 0 & \text{otherwise} \end{cases} \qquad (6.1\text{-}12)$$

consistent with the Jeffreys prior of Eq. (5.5-12). Multiplication of the likelihood $\ell(\boldsymbol{\theta}, \sigma | \boldsymbol{Y})$ by this prior density, in accordance with Bayes' theorem, then gives the posterior density

$$p(\boldsymbol{\theta}, \sigma | \boldsymbol{Y}) \propto \sigma^{-(n+1)} \exp\left[ -\frac{S(\boldsymbol{\theta})}{2\sigma^2} \right] \qquad (6.1\text{-}13)$$

over the permitted range of $\boldsymbol{\theta}$. This probability density takes its largest value at the least sum of squares $S(\boldsymbol{\theta})$ in the permitted range of $\boldsymbol{\theta}$. Of course, one should examine not only the least-squares $\boldsymbol{\theta}$ value (the so-called *point estimate* $\widehat{\boldsymbol{\theta}}$), but also its uncertainty as indicated by measures such as a 95% posterior probability interval for each estimated parameter $\theta_i$. The measures provided by GREGPLUS are described in Section 6.6, along with corresponding results for auxiliary functions $\phi_i(\boldsymbol{\theta})$ that the user may define.

An alternate argument for minimizing $S(\boldsymbol{\theta})$ is to maximize the function $\ell(\boldsymbol{\theta}, \sigma | \boldsymbol{Y})$ given in Eq. (6.1-10). This *maximum likelihood* approach, advocated by Fisher (1925), gives the same point estimate $\widehat{\boldsymbol{\theta}}$ as does the posterior density function in Eq. (6.1-13). The posterior density function is essential, however, for calculating posterior probabilities for regions of $\boldsymbol{\theta}$ and for rival models, as we do in later sections of this chapter.

The permitted region of $\boldsymbol{\theta}$ can take various forms. Our package GREGPLUS uses a rectangular region

$$\text{BNDLW}(i) \leq \theta_i \leq \text{BNDUP}(i) \qquad i = 1, \ldots, \text{NPAR} \qquad (6.1\text{-}14)$$

for a model containing NPAR parameters.

Many methods are available for least-squares calculations. Models linear in $\boldsymbol{\theta}$ allow direct solutions; other models need iteration. The choice of iteration method depends on one's goal. For a mere curve-fit of the data, a direct search procedure such as that of Powell (1965) or of Nelder and Mead (1965) may suffice. But to determine the most important parameters and their most probable values, a method based on derivatives of $S(\boldsymbol{\theta})$ is essential and is followed here.

## 6.2 WEIGHTING AND OBSERVATION FORMS

Uniform weighting (known as *simple least squares*) is appropriate when the expected variances of the observations are equal and is commonly used when these values are unknown. GREGPLUS uses this weighting when called with JWT $= 0$; the values $\sqrt{w_u} = 1$ are then provided automatically.

Observations done with uniform *relative* precision are fairly common. For example, reported catalytic rate data often include adjustments of reaction temperature, catalyst activity, and other variables to standard values. These combined adjustments amount to a correction factor $C_u$ for the catalyst mass in each event $u = 1, \ldots, n$. Then the total rate of production of any product P in a reactor experiment can be expressed as

$$\mathrm{R}_{\mathrm{P}u} \equiv \left( \frac{F \Delta x}{CW} \right)_u = f(\boldsymbol{\xi}_u, \boldsymbol{\theta}) + \varepsilon_u \quad u = 1, \ldots, n \qquad (6.2\text{-}1)$$

Here $F$ is the total molar feed rate to the reactor, $\Delta x$ is the production of the product P in moles per mole of feed, and $W$ is the catalyst mass.

The function $y_u = \ln \mathrm{R}_{\mathrm{P}u}$ has the following variance according to Eq. (4.D-4), with the errors in $\ln F_u$, $\ln \Delta x_u$, $\ln C_u$, and $\ln W_u$ regarded as random independent variables:

$$\mathrm{Var}(\ln \mathrm{R}_{\mathrm{P}u}) = \mathrm{Var}(\ln F_u) + \mathrm{Var}(\ln \Delta x_u) + \mathrm{Var}(\ln C_u) + \mathrm{Var}(\ln W_u) \quad (6.2\text{-}2)$$

Therefore, the weight $w_u$ for each observation $y_u = \ln(\mathrm{R})_{\mathrm{P}u}$ satisfies

$$w_u = \frac{\sigma^2}{\mathrm{Var}(\ln F_u) + \mathrm{Var}(\ln \Delta x_u) + \mathrm{Var}(\ln C_u) + \mathrm{Var}(\ln W_u)} \qquad (6.2\text{-}3)$$

according to Eq. (6.1-5). Simple weights $w_u = 1$ for all $u$ are thus appropriate for observations modeled by Eq. (6.1-1) if the terms in Eq. (6.2-3) are independent of $u$. A somewhat more detailed weighting is used in Example 6.2.

The remaining sections of this chapter are implemented in the package GREGPLUS, described and demonstrated in Appendix C.

## 6.3 PARAMETRIC SENSITIVITIES; NORMAL EQUATIONS

Gauss (1809) used a Newton-like iteration scheme to minimize $S(\boldsymbol{\theta})$ for nonlinear models; a single iteration suffices for linear models. He approximated the departures of the data from a model as linear expansions $\widetilde{E}_u(\boldsymbol{\theta})$ around the starting point $\boldsymbol{\theta}^k$ of the current iteration:

$$E_u(\boldsymbol{\theta}) \approx \widetilde{E}_u(\boldsymbol{\theta}) \equiv Y_u - F_u(\boldsymbol{\theta}^k) - \sum_{r=1}^{p} \left[ \frac{\partial F_u}{\partial \theta_r} \bigg|_{\boldsymbol{\theta}^k} \right] (\theta_r - \theta_r^k)$$

$$\equiv E_u(\boldsymbol{\theta}^k) - \sum_{r=1}^{p} X_{ur}(\theta_r - \theta_r^k) \qquad (u = 1, \ldots, n) \qquad (6.3\text{-}1)$$

Here $X_{ur}$ denotes $\partial F_u / \partial \theta_r$ evaluated at $\boldsymbol{\theta}^k$. This set of equations can be expressed concisely as

$$\boldsymbol{E}(\boldsymbol{\theta}) \approx \widetilde{\boldsymbol{E}}(\boldsymbol{\theta}) \equiv \boldsymbol{E}^k - \boldsymbol{X}(\boldsymbol{\theta} - \boldsymbol{\theta}^k) \qquad (6.3\text{-}2)$$

in which $\boldsymbol{E}(\boldsymbol{\theta})$ is the column vector of functions $E_u(\boldsymbol{\theta})$, $\widetilde{\boldsymbol{E}}(\boldsymbol{\theta})$ is the local linear approximation to $\boldsymbol{E}(\boldsymbol{\theta})$, and $\boldsymbol{X}$ is the local matrix of *parametric sensitivities* of the model. The resulting approximation for the error sum of squares is

$$
\begin{aligned}
\widetilde{S}(\boldsymbol{\theta}) &= \widetilde{\boldsymbol{E}}^T(\boldsymbol{\theta})\widetilde{\boldsymbol{E}}(\boldsymbol{\theta}) \\
&= [\boldsymbol{E}^k - \boldsymbol{X}(\boldsymbol{\theta} - \boldsymbol{\theta}^k)]^T[\boldsymbol{E}^k - \boldsymbol{X}(\boldsymbol{\theta} - \boldsymbol{\theta}^k)] \\
&= \boldsymbol{E}^{kT}\boldsymbol{E}^k - 2(\boldsymbol{\theta} - \boldsymbol{\theta}^k)^T\boldsymbol{X}^T\boldsymbol{E}^k + (\boldsymbol{\theta} - \boldsymbol{\theta}^k)^T\boldsymbol{X}^T\boldsymbol{X}(\boldsymbol{\theta} - \boldsymbol{\theta}^k)
\end{aligned} \tag{6.3-3}
$$

and its gradient $\partial\widetilde{S}/\partial\boldsymbol{\theta}^T$ is

$$
\nabla\widetilde{S}(\boldsymbol{\theta}) = \boldsymbol{0} - 2\boldsymbol{X}^T\boldsymbol{E}^k + 2\boldsymbol{X}^T\boldsymbol{X}(\boldsymbol{\theta} - \boldsymbol{\theta}^k) \tag{6.3-4}
$$

Setting $\nabla\widetilde{S} = \boldsymbol{0}$ gives the equation system

$$
\boldsymbol{X}^T\boldsymbol{X}(\boldsymbol{\theta}^* - \boldsymbol{\theta}^k) = \boldsymbol{X}^T\boldsymbol{E}^k \tag{6.3-5}
$$

for the locus of level points $\boldsymbol{\theta}^*$ of the current function $\widetilde{S}(\boldsymbol{\theta})$. If the matrix $\boldsymbol{X}^T\boldsymbol{X}$ is nonsingular, then it is also positive definite (see Problem 6.A) and the locus is the global minimum of $\widetilde{S}(\boldsymbol{\theta})$; this outcome is favored by use of well-designed experiments and by absence of unnecessary parameters. A method will be given in Section 6.4 to find a local minimum $\widetilde{S}$ within the useful range of Eq. (6.3-2).

The system (6.3-5) can also be written as

$$
\boldsymbol{X}^T[\boldsymbol{E}^k - \boldsymbol{X}(\boldsymbol{\theta}^* - \boldsymbol{\theta}^k)] = \boldsymbol{0} \tag{6.3-6}
$$

Thus, the correction vector $[\boldsymbol{E}^k - \boldsymbol{X}(\boldsymbol{\theta}^* - \boldsymbol{\theta}^k)]$ is orthogonal (normal) to each column vector of $\boldsymbol{X}$. The rows of Eq. (6.3-5) are accordingly known as the *normal equations* of the given problem.

If the model functions $F_u(\boldsymbol{\theta})$ all are linear in $\boldsymbol{\theta}$, then the sensitivity matrix $\boldsymbol{X}$ is constant and a single application of Eq. (6.3-5) will give the least-squares solution. In practice, $\boldsymbol{X}$ is often far from constant, making iteration necessary as described in the following section.

One can extend Eq. (6.3-1) to second order to get a fuller quadratic function $\widetilde{S}(\boldsymbol{\theta})$. This gives the *full Newton* equations of the problem, whereas Eqs. (6.3-3)–(6.3-5) are called the *Gauss-Newton* equations. Either form can be expressed as a symmetric matrix expansion

$$
\widetilde{S}(\boldsymbol{\theta}) = S(\boldsymbol{\theta}^k) + \left[ (\boldsymbol{\theta} - \boldsymbol{\theta}^k)^T \vdots -1 \right]
\begin{bmatrix}
\boldsymbol{A}_{\theta\theta} & \vdots & \boldsymbol{A}_{\theta L} \\
\cdots\cdots\cdots\cdots \\
\boldsymbol{A}_{\theta L}^T & \vdots & 0
\end{bmatrix}
\begin{bmatrix}
\boldsymbol{\theta} - \boldsymbol{\theta}^k \\
\cdots\cdots \\
-1
\end{bmatrix} \tag{6.3-7}
$$

with submatrices defined as follows

$$A_{\theta L} = -\frac{1}{2}\frac{\partial S}{\partial \boldsymbol{\theta}}\bigg|_{\boldsymbol{\theta}^k} = \boldsymbol{X}^T \boldsymbol{E}^k \quad \text{(Always)} \tag{6.3-8}$$

$$A_{\theta\theta} = \frac{1}{2}\frac{\partial^2 \widetilde{S}}{\partial\boldsymbol{\theta}\partial\boldsymbol{\theta}^T} = \begin{cases} \boldsymbol{X}^T\boldsymbol{X} & \text{(Gauss-Newton)} \\ \boldsymbol{X}^T\boldsymbol{X} - \left\{\dfrac{1}{2}\displaystyle\sum_{u=1}^{n} E_u \dfrac{\partial^2 F_u}{\partial\theta_r\partial\theta_s}\bigg|_{\boldsymbol{\theta}^k}\right\} & \text{(full Newton)} \end{cases} \tag{6.3-9}$$

Here the sum within the curly brackets denotes a square matrix with the indicated elements. The normal equations[1] in either form give

$$A_{\theta\theta}(\boldsymbol{\theta} - \boldsymbol{\theta}^k) = A_{\theta L} \tag{6.3-10}$$

for the locus of level points of the quadratic function $\widetilde{S}$. Maxima and saddle points of $\widetilde{S}(\boldsymbol{\theta})$ are possible for the full Newton equations, because positive definiteness of $A_{\theta\theta}$ is no longer guaranteed when the curly-bracketed term of Eq. (6.3-9) is included; however, GREGPLUS avoids such points by the solution method described below. Dennis, Gay, and Welsch (1981) give an adaptive algorithm that approximates the full Newton matrix $A_{\theta\theta}$ by use of results from successive iterations.

The derivatives $F_{ur}$ are called the *first-order parametric sensitivities* of the model. Their direct computation via Newton's method is implemented in Subroutines DDAPLUS (Appendix B) and PDAPLUS. Finite-difference approximations are also provided as options in GREGPLUS to produce the matrix $A$ in either the Gauss-Newton or the full Newton form; these approximations are treated in Problems 6.B and 6.C.

The useful range of Eq. (6.3-7) can often be enhanced by writing the model in a different form. For example, the equation

$$\ln k_j = \ln A_j - E_j/RT \tag{6.3-11}$$

yields better-conditioned normal equations when rewritten in the form

$$\ln k_j = \ln k_{jB} + \frac{E_j}{R}\left[\frac{1}{T_B} - \frac{1}{T}\right] \tag{6.3-12}$$

used by Arrhenius (1889) and recommended by Box (1960). This parameterization is demonstrated in Example C.3, along with the robust form provided by Ratkowsky (1985) for heterogeneous reaction rate expressions.

---

[1] Since $A_{\theta\theta}$ is real and symmetric, it belongs to the class of *normal* matrices [Stewart (1973), p. 288]. Thus, Eqs. (6.3-10) are properly called *normal equations*, whichever form of Eq. (6.3-9) is used.

## 6.4 CONSTRAINED MINIMIZATION OF $S(\boldsymbol{\theta})$

Models linear in $\boldsymbol{\theta}$, with unconstrained parameters, can be fitted directly by solving Eq. (6.3-5). Efficient algorithms and software for such problems are available [Lawson and Hanson (1974, 1995); Dongarra, Bunch, Moler, and Stewart (1979); Anderson et al., (1992)], and will not be elaborated here. We will focus on nonlinear models with bounded parameters, which are common in chemical kinetics and chemical reaction engineering.

Models nonlinear in $\boldsymbol{\theta}$ need careful treatment. Direct iteration with Eq. (6.3-10) often fails because of the limited range of the expansions $\widetilde{E}_u(\boldsymbol{\theta})$. Gauss-Newton iteration schemes with steps adjusted by line search work well [Booth, Box, Muller and Peterson (1958); Hartley (1961, 1964); Box and Kanemasu (1972, 1984); Bard (1974); Bock (1981)] when $\boldsymbol{A}_{\theta\theta}$ is well-conditioned and $\boldsymbol{\theta}$ unrestricted, but give difficulty otherwise.

Levenberg (1944) and Marquardt (1963) augmented $\boldsymbol{X}^T\boldsymbol{X}$ in Eq. (6.3-5) with a positive diagonal matrix $\lambda\boldsymbol{d}$, thus obtaining a nonsingular coefficient matrix for each iteration while shortening each correction step $\Delta\boldsymbol{\theta}$. This approach has the difficulty that the true rank of the problem is concealed until near the end, when $\lambda$ is reduced toward zero in an effort to recover Eq. (6.3-5). If $\boldsymbol{X}^T\boldsymbol{X}$ then turns out to be singular, the original problem is indeterminate (having too many parameters or inappropriate data) and the method gives no guidance for dealing with this.

Difficulties of this sort can be avoided by minimizing $\widetilde{S}(\boldsymbol{\theta})$ of Eq. (6.3-7) over a trust region of $\boldsymbol{\theta}$ in each iteration. A line search can be used to adjust the correction vector $\Delta\boldsymbol{\theta}^k$ and the trust-region dimensions for the next iteration. Such a method is outlined below and used in GREGPLUS.

### 6.4.1 The Quadratic Programming Algorithm GRQP

A rectangular *trust region* of the form

$$|\theta_i - \text{PARB}(i)| \leq \begin{cases} |\text{CHMAX}(i)| & \text{if } \text{CHMAX}(i) \geq 0.0; \\ |\text{CHMAX}(i) * \text{PARB}(i)| & \text{if } \text{CHMAX}(i) < 0.0 \\ & \text{and } \text{PARB}(i) \neq 0.0; \\ \sqrt{0.1 S(\boldsymbol{\theta}^k)/A_{ii,\text{ref}}} & \text{if } \text{CHMAX}(i) < 0.0 \\ & \text{and } \text{PARB}(i) = 0.0 \\ & \text{and } A_{ii,\text{ref}} \neq 0.0; \\ 0.0 & \text{otherwise} \end{cases}$$

$$i = 1, \ldots, \text{NPAR} \qquad (6.4\text{-}1)$$

is used in Subroutine GRQP as the working range of the quadratic expansion $\widetilde{S}(\boldsymbol{\theta})$ for the current iteration. Here PARB(i) and $A_{ii,\text{ref}}$ are the starting values of PAR(i) and $A_{ii}$ for the iteration. The values CHMAX(i) are specified initially by the user and are adjustable by GREGPLUS at the end of each iteration.

Subroutine GRQP minimizes the quadratic function $\widetilde{S}(\boldsymbol{\theta})$ of Eq. (6.3-7) over the *feasible region*

$$C_{i-} \leq \theta_i \leq C_{i+}, \qquad i = 1, \ldots, \text{NPAR} \qquad (6.4\text{-}2)$$

of parameter values that satisfy both (6.1-14) and (6.4-1). By working in this region, GRQP gives stable solutions even when the system (6.3-10) is ill-conditioned or singular.

Three classes of parameters are considered in GRQP:

1.  The basis set $\boldsymbol{\theta}_e$ of *estimated parameters*, obtained by solving the corresponding rows of Eq. (6.3-10). Nonzero entries $i$ in the vector LBAS identify these parameters, whose current values are stored in the vector PAR at the end of each transformation.
2.  A set $\boldsymbol{\theta}_c$ of actively *constrained parameters*, which currently lie at a limit of Eq. (6.4-2). Nonzero entries $i$ or $-i$ in the vector LBOX identify any such parameters and the active limit (upper or lower) for each.
3.  The set $\boldsymbol{\theta}_f$ of *free parameters* (all others). These parameters are candidates for transfer into $\boldsymbol{\theta}_e$ or $\boldsymbol{\theta}_c$ in the next transformation.

Each step in the minimization begins with a search for a descent variable $\theta_i$ among the candidate set composed of $\boldsymbol{\theta}_f$ plus any members of $\boldsymbol{\theta}_c$ that lie at an uphill limit. The candidate $\theta_i$ with the largest test divisor $D_i \equiv |A_{ii}|/A_{ii,\text{ref}}$ is chosen, provided[1] that $D_i$ exceeds the threshold tolerance ADTOL and that the test values $D_j = 1/(A^{jj} A_{jj,\text{ref}})$ for the resulting inverse matrix all exceed ADTOL. When such a candidate is found, one of the following moves is made:

(1)  If possible, bring $\theta_i$ into the basis set $\boldsymbol{\theta}_e$ as the pivotal variable for row $i$ and column $i$, by applying the modified Gauss-Jordan method described at the end of Section E.3 to the full $\boldsymbol{A}$-matrix, including the last row and column. The requirements for such a move are: LBAS($i$) must be zero; $A_{ii}$ must be positive (to ensure a descent of $\widetilde{S}$); $D_i$ and all resulting $D_j$ must exceed ADTOL; and every parameter must remain within the permitted region of Eq. (6.4-2).
(2)  If a step of Type 1 would cross a limit of the permitted range of $\theta_i$, move the solution to that limit, accordingly updating the last row and column of the $\boldsymbol{A}$-matrix. Then $\theta_i$ becomes a member of the constrained set $\boldsymbol{\theta}_c$. No changes occur in the other $\boldsymbol{A}$-matrix elements.
(3)  If basis parameter $\theta_j$ would be the first to reach a limit of its permitted range in a step of Type 1, shorten the step to take $\theta_j$

---

[1] The criterion $D_i > \text{ADTOL}$ corresponds to ones given by Efroymson (1960) and by Stewart (1987). The criteria corresponding to $D_j > \text{ADTOL}$ were given by Frane (1977).

exactly to that limit. Then remove $\theta_j$ from the basis set $\boldsymbol{\theta}_e$ by doing a Gauss-Jordan step with the inverse matrix element $A^{jj}$ as pivot, and record $\theta_j$ in LBOX as a member of the constrained set $\boldsymbol{\theta}_c$.

(4)  When each remaining free parameter $\theta_i$ meets the requirements for a step of Type 1, except that $D_i \leq$ ADTOL, GRQP moves those parameters individually to least-squares values, starting with the one that reduces $\widetilde{S}$ most. These moves alter only the last row and column of $\boldsymbol{A}$.

This algorithm continues until no more such moves are available or until $2p$ steps are done, whichever happens first.

The transformed $\boldsymbol{A}$-matrix quadratic expansion of $\widetilde{S}$ in terms of the set $\boldsymbol{\theta}_e$ of current basis parameters. The submatrix $\boldsymbol{A}^{ee}$ is the inverse of the current basis matrix. The basis determinant $|\boldsymbol{A}_{ee}|$ is the product of the pivots $A_{ii}$ that were used in steps of Type 1, divided by the pivots $A^{ii}$ that were used in steps of Type 3. The use of strictly positive pivots for the Gauss-Jordan transformations ensures a positive definite submatrix $\boldsymbol{A}^{ee}$, hence a positive determinant $|\boldsymbol{A}_{ee}|$ throughout the minimization.

Choosing the basis candidates in order of their current test divisors $D_i = A_{ii}/A_{ii,\mathrm{ref}}$ selects those parameters for which the normal equations are best conditioned. This order of selection may sometimes exclude a parameter that is important on other grounds. Such a parameter might be made estimable by providing additional experiments (preferably selected by the determinant criterion in Section 6.6), or by using an ADTOL value smaller than the excluded test divisor $D_i$, or by rewriting the model in a better form.

### 6.4.2 The Line Search Algorithm GRS1

GRS1 has two main tasks in each iteration $k$: (1) to determine if the estimation is completed, and (2) to prepare for another iteration when needed.

GRS1 begins with a calculation of $S$ at the point $\boldsymbol{\theta}^{(k+1)} = \boldsymbol{\theta}^k + \Delta\boldsymbol{\theta}^k$ computed by GRQP. If LINMOD was set to 1 in the current call of GREG-PLUS, indicating the model to be linear with unbounded parameters, $S$ and $\boldsymbol{\theta}$ are accepted as final upon completion of this calculation.

If LINMOD is zero, the model is treated as nonlinear, and a test is made for convergence of the parameter estimation. Convergence is normally declared, and final statistics of Section 6.6 are reported, if no free parameter is at a CHMAX limit and if $|\Delta\theta_i^k|$ for each basis parameter is less than one-tenth of the 95% posterior probability half-interval calculated from Eq. (6.6-10). If the number of basis parameters equals the number of events, leaving no degrees of freedom for the estimation, this interval criterion is replaced by $|\Delta\theta_i^k| \leq$ RPTOL $* |\theta_i^k|$.

If convergence is not declared, the correction $\Delta\boldsymbol{\theta}$ is halved as many times as needed (up to a maximum of 10) to find a point with $S(\boldsymbol{\theta}) < S(\boldsymbol{\theta}^k)$.

This process is called a *weak line search*.

Sometimes a test value of $S$ cannot be calculated, because some values in the model call go out of range or fail to converge. Subroutine GRS1 keeps the computation going by halving $\Delta\boldsymbol{\theta}$ and trying again whenever return occurs from the user's subroutine MODEL with a nonzero value of the integer IERR. To use this feature, the user should include code in Subroutine MODEL to return the value IERR=1 whenever failure is imminent in the current call.

Before starting another iteration, any trust-region values CHMAX($i$) that were active in the quadratic programming solution may be enlarged according to the line search results. Then control is returned to subroutine GR1 for a calculation of matrix $\boldsymbol{A}$ at the latest line search point.

Any parameter $\theta_i$ that lies at a bound of Eq. (6.1-13) at the end of the minimization has its constrained least-squares value there. Such a parameter is invariant to small changes in the data, since its value is determined by the bound.

Any parameter $\theta_i$ that is free at the end of the minimization is declared indeterminate, as its test divisor $D_i$ is not above the rounding threshold value ADTOL described in Section 6.4.1.

### 6.4.3 Final Expansions Around $\widehat{\boldsymbol{\theta}}$

Several expansions of $\widetilde{S}$ around the least-squares solution are summarized here for later use. The expansion

$$
\begin{aligned}
\widetilde{S}(\boldsymbol{\theta}) &= \widehat{S} + (\boldsymbol{\theta}_e - \widehat{\boldsymbol{\theta}}_e)^T \left[ \widehat{\boldsymbol{A}}_{ee} \right] (\boldsymbol{\theta}_e - \widehat{\boldsymbol{\theta}}_e) \\
&= \widehat{S} + (\boldsymbol{\theta}_e - \widehat{\boldsymbol{\theta}}_e)^T \left[ \widehat{\boldsymbol{A}}^{ee} \right]^{-1} (\boldsymbol{\theta}_e - \widehat{\boldsymbol{\theta}}_e)
\end{aligned}
\tag{6.4-3}
$$

involves the full vector $\boldsymbol{\theta}_e$ of estimated parameters, with the vectors $\boldsymbol{\theta}_c$ and $\boldsymbol{\theta}_f$ fixed at their exit values. The first version here comes from Eq. (6.3-7) applied at $\boldsymbol{\theta}^k = \widehat{\boldsymbol{\theta}}$; the second uses the inverse matrix $\widehat{\boldsymbol{A}}^{ee}$ computed in the final iteration.

Shorter expansions of $\widetilde{S}$ are obtainable by use of the final inverse matrix $\widehat{\boldsymbol{A}}^{ee}$. Taking an $n_a$-parameter test vector $\boldsymbol{\theta}_a$ from $\boldsymbol{\theta}_e$, and taking the other basis parameters at least-squares values conditional on $\boldsymbol{\theta}_a$, one obtains the expansion

$$
\widetilde{S}(\boldsymbol{\theta}_a) = \widehat{S} + (\boldsymbol{\theta}_a - \widehat{\boldsymbol{\theta}}_a)^T [\widehat{\boldsymbol{A}}^{aa}]^{-1} (\boldsymbol{\theta}_a - \widehat{\boldsymbol{\theta}}_a)
\tag{6.4-4}
$$

based on the corresponding subset of elements of the inverse matrix $\widehat{\boldsymbol{A}}^{ee}$. This expansion reduces to

$$
\widetilde{S}(\theta_i) = \widehat{S} + (\theta_i - \widehat{\theta}_i)^2 / \widehat{A}^{ii}
\tag{6.4-5}
$$

for a single test parameter and to Eq. (6.4-3) when $n_a = p$.

## 6.5 TESTING THE RESIDUALS

Once $S(\boldsymbol{\theta})$ has been minimized for a given model and data collection, the assumptions of the calculation should be tested. In this step, one examines the *weighted residuals*

$$\widehat{E}_u \equiv Y_u - \widehat{Y}_u = Y_u - F_u(\widehat{\boldsymbol{\theta}}) \tag{6.5-1}$$

of the least-squares solution, looking for errors in the model or in the data. This process of criticism is based on common sense and on sampling theory, as discussed by Box (1980) and by Draper and Smith (1981).

One looks first for unlikely values or trends of the residuals by plotting or tabulating them against time, expected values, or experimental variables. Such plots or tabulations may reveal errors in the data, the expectation model, or the weightings used. The largest weighted residuals are most useful in identifying errors. Histograms and normal probability plots of the residuals [Daniel (1976); Box, Hunter, and Hunter (1978); Draper and Smith (1981)] give additional perspectives.

Systematic trends of the residuals may suggest alterations of the model or the weighting. Systematic differences of parameters fitted to different parts of the data can give similar information, as Hunter and Mezaki (1964) showed in their study of model building for methane oxidation.

When possible, one should also test the residual sum of squares against its predictive probability distribution. If the weighted residuals are distributed in the manner assumed in Section 6.1, then the standardized sum of squares $\chi^2 \equiv \widehat{S}/\sigma^2$ has the predictive probability density

$$p(\chi^2|\nu) = \frac{|\chi|^{\nu-2}\exp(-\chi^2/2)}{2^{\nu/2}\Gamma(\nu/2)} \tag{6.5-2}$$

with $\nu = n - p$ degrees of freedom. The corresponding probability of obtaining, in random sampling of observations, a ratio $\widehat{S}/\sigma^2$ greater than the realized value $\chi^2$ is

$$\alpha(\chi^2,\nu) = \int_{\chi^2}^{\infty} \frac{Z^{(\nu-2)/2}\exp(-Z/2)}{2^{\nu/2}\Gamma(\nu/2)}dZ \tag{6.5-3}$$

and is reported by GREGPLUS whenever it is called with $\sigma$ known. The fitted model should be rejected if the calculated probability $\alpha(\chi^2,\nu)$ is unacceptably small — say, less than 0.01; otherwise, it may be provisionally accepted. The function $1-\alpha(\chi^2,\nu) = P(\chi^2,\nu)$ is called the $\chi^2$ distribution.

A more powerful criterion of goodness of fit is the $F$-test, pioneered by Fisher (1925), of the variance ratio

$$F = \frac{S_L/\nu_L}{S_E/\nu_E} = \frac{s_L^2}{s_E^2} \tag{6.5-4}$$

Here $S_E$ is an experimental error sum of squares with $\nu_E$ degrees of freedom, normally determined from $\nu_E$ independent differences of replicate observations or residuals. $S_L$ is a lack-of-fit sum of squares, given by

$$S_L = \widehat{S} - S_E \qquad (6.5\text{-}5)$$

if the replicate data are included in $\widehat{S}$, and having

$$\nu_L = n - p - \nu_E \qquad (6.5\text{-}6)$$

lack-of-fit degrees of freedom.

To see if the model fits the data acceptably, one tests the hypothesis that the residuals in $S_L$ and $S_E$ are samples from normal distributions with equal variances and expected value zero. On this hypothesis, the predictive probability of obtaining a sample variance ratio $f$ larger than the realized value $F$ is

$$\Pr\left[f > F\right] = \alpha(F, \nu_L, \nu_E) \qquad (6.5\text{-}7)$$

The function on the right comes from the work of Sir Ronald Fisher (1925). It is given by

$$\alpha(F, \nu_1, \nu_2) = \frac{1}{\mathrm{B}\left(\nu_1/2, \nu_2/2\right)} \int_F^\infty \left(\frac{\nu_1}{\nu_2}\right)^{\nu_1/2} \mathrm{f}^{(\nu_1-2)/2}\left(1 + \frac{\nu_1}{\nu_2}\mathrm{f}\right)^{-(\nu_1+\nu_2)/2} d\mathrm{f}$$

$$(6.5\text{-}8)$$

and its predictive probability distribution goes from 1 to 0 as $F$ goes from 0 to $\infty$. The hypothesis is rejected if the probability $\alpha$ is judged to be unacceptably small — say, less than 0.01.

GREGPLUS calculates $\alpha(F, \nu_1, \nu_2)$ for this test whenever replicates are identified in the input; see Examples 6.1, 6.2, and Appendix C. The user can then decide whether the reported probability $\alpha$ is large enough for acceptance of the model.

The literature on residual analysis is extensive. Good treatments are given by Daniel (1976), Box, Hunter, and Hunter (1978), Draper and Smith (1981), Guttman, Wilks, and Hunter (1982), and Chambers et al. (1983). Guttman, Wilks, and Hunter include lucid developments of the predictive densities of $\chi^2$ and $F$.

## 6.6 INFERENCES FROM THE POSTERIOR DENSITY

The posterior density function, Eq. (6.1-13), expresses all the information available about the postulated model from the data $Y$ and the prior density function, Eq. (6.1-12). In this section we present some of the major results obtainable from this function.

Let $\widehat{\boldsymbol{\theta}}_e$ denote the value of $\boldsymbol{\theta}_e$ on completion of the minimization of $S(\boldsymbol{\theta})$. This solution corresponds to a mode (local maximum) of the posterior density $p(\boldsymbol{\theta}_e | Y)$. There will be just one mode if the model is linear in the parameters, but models nonlinear in $\boldsymbol{\theta}$ may have additional modes and one

should watch for these. Useful tests for this purpose include plots of the
$S$-contours (very effective for simple models) and least-squares calculations
with various starting points and/or different bounds on the parameters.
Some kinds of multiplicity can be removed by reparameterization, as in the
study by Stewart and Mastenbrook (1983b) on pulmonary modeling with
sums of Gaussian functions, or by including appropriate physical bounds
on the parameters.

Once the principal mode is found, the posterior probability content of
the adjoining region of $\boldsymbol{\theta}$ is computable (in principle) by numerical integration. This approach, however, is seldom feasible for multiparameter models.
The needed integrals are expressible more concisely via the quadratic expansions $\widetilde{S}$ given in Eqs. (6.4-3) to (6.4-5), which are exact for models linear
in $\boldsymbol{\theta}$.

### 6.6.1 Inferences for the Parameters

Use of Eq. (6.4-3) to approximate $S(\boldsymbol{\theta})$ in Eq. (6.1-13) gives

$$p(\boldsymbol{\theta}_e, \sigma | \boldsymbol{Y}) \propto \sigma^{-(n+1)} \exp\left\{ -\frac{1}{2\sigma^2} \left[ \widehat{S} + (\boldsymbol{\theta}_e - \widehat{\boldsymbol{\theta}}_e)^T \widehat{\boldsymbol{A}}_{ee} (\boldsymbol{\theta}_e - \widehat{\boldsymbol{\theta}}_e) \right] \right\} \quad (6.6\text{-}1)$$

for the joint posterior density of the estimable quantities, $\boldsymbol{\theta}_e$ and $\sigma$. If $\sigma$
were known, this would be a normal (Gaussian) density function for $\boldsymbol{\theta}_e$;
however, in practice, $\sigma$ is generally unknown.

It would be possible to estimate $\boldsymbol{\theta}$ and $\sigma$ jointly via Eq. (6.6-1); however, this is not the usual procedure. The standard procedure is to integrate
Eq. (6.1-13) over the permitted range of $\sigma$, thus obtaining the *marginal* posterior density function

$$\begin{aligned}
p(\boldsymbol{\theta}_e | \boldsymbol{Y}) &= \int_0^\infty p(\boldsymbol{\theta}_e, \sigma | \boldsymbol{Y}) \, d\sigma \\
&\propto \int_0^\infty \sigma^{-(n+1)} \exp\left[ -\frac{\widetilde{S}(\boldsymbol{\theta})}{2\sigma^2} \right] d\sigma
\end{aligned} \quad (6.6\text{-}2)$$

and removing the "nuisance parameter" $\sigma$ from the problem. W. S. Gossett,
writing under the pseudonym "Student" (1908), pioneered this approach.

Introduction of a new variable, $\phi = 1/(2\sigma^2)$, into the latter integral
gives

$$\begin{aligned}
p(\boldsymbol{\theta}_e | \boldsymbol{Y}) &\propto -\int_\infty^0 \phi^{(n-2)/2} \exp\left[ -\phi \widetilde{S}(\boldsymbol{\theta}) \right] d\phi \\
&\propto \Gamma(n/2) \left[ \widetilde{S}(\boldsymbol{\theta}) \right]^{-n/2} \\
&\propto \left[ 1 + \frac{\widetilde{S}(\boldsymbol{\theta}) - \widehat{S}}{\widehat{S}} \right]^{-n/2}
\end{aligned} \quad (6.6\text{-}3)$$

Insertion of Eq. (6.4-3) gives

$$p(\boldsymbol{\theta}_e|\boldsymbol{Y}) \propto \left[1 + \frac{(\boldsymbol{\theta}_e - \widehat{\boldsymbol{\theta}}_e)^T \hat{A}_{ee}(\boldsymbol{\theta}_e - \widehat{\boldsymbol{\theta}}_e)}{\widehat{S}}\right]^{-n/2} \tag{6.6-4}$$

as the *marginal* posterior density for $\boldsymbol{\theta}_e$. This result, exact for models linear in $\boldsymbol{\theta}$, represents the constant-density contours as ellipsoids.

The marginal posterior density for a $n_a$-parameter subset $\boldsymbol{\theta}_a$ of $\boldsymbol{\theta}_e$ is obtained by integrating Eq. (6.6-4) over any other estimated parameters, as described in Problem 6.E. When the ranges of the parameters are unrestricted, one obtains

$$p(\boldsymbol{\theta}_a|\boldsymbol{Y}) \propto \left[1 + \frac{(\boldsymbol{\theta}_a - \widehat{\boldsymbol{\theta}}_a)^T \left[\hat{A}^{aa}\right]^{-1}(\boldsymbol{\theta}_a - \widehat{\boldsymbol{\theta}}_a)}{\widehat{S}}\right]^{-(n-p+n_a)/2} \tag{6.6-5}$$

which reduces for $n_a = 1$ to the marginal posterior density expression

$$p(\theta_i|\boldsymbol{Y}) \propto \left[1 + \frac{(\theta_i - \widehat{\theta}_i)^2}{\widehat{S}\hat{A}^{ii}}\right]^{-(n-p+1)/2} \tag{6.6-6}$$

for each estimated parameter. Several consequences of these results are summarized here.

1. Each of these density functions has its maximum at the least-squares values of its included parameters. The resulting distributions of parameters are known as *t-distributions*. Such distributions were first investigated by Gossett (1908) for single-response problems of quality control at the Guinness brewery in Dublin. The covariance matrix of the estimated parameter vector $\boldsymbol{\theta}_e$ in Eq. (6.6-4) is

$$\text{cov}\,\boldsymbol{\theta}_e = \hat{A}^{ee}\widehat{S}/(n-p) \tag{6.6-7}$$

and the variance of the individual parameter estimate $\theta_{ei}$ is

$$\text{Var}\,\theta_{ei} = \sigma^2_{\theta_{ei}} = \hat{A}^{ii}\widehat{S}/(n-p) \tag{6.6-8}$$

The normalized covariances $\hat{A}^{ij}/(\hat{A}^{ii}\hat{A}^{jj})^{1/2}$ are reported by GREGPLUS for all members of the estimated parameter set $\boldsymbol{\theta}_e$. Several examples of these are given in Appendix C.

2. The posterior probability that $\theta_i$ exceeds a given value $\theta_{i,\text{ref}}$ is

$$\alpha(\nu, t_i) = \frac{1}{B(1/2, \nu/2)\sqrt{\nu}} \int_{t_i}^{\infty} \left[1 + \frac{t_1^2}{\nu}\right]^{-(\nu+1)/2} dt_1 \tag{6.6-9}$$

evaluated with $\nu = n - p$ and $t_i = (\theta_{i,\text{ref}} - \widehat{\theta}_i)/\sigma_{\theta_i}$. The function $1 - \alpha(\nu, t)$ is the famous $t$-distribution.

3. The highest-posterior-density (HPD) interval with probability content $(1 - \alpha)$ for any estimated parameter $\theta_i$ is

$$-t(\alpha/2, \nu) \leq \frac{(\theta_i - \widehat{\theta}_i)}{\sigma_{\theta_i}} \leq t(\alpha/2, \nu) \qquad (6.6\text{-}10)$$

with $\nu = (n - p)$. A popular choice is $\alpha = 0.05$, which gives a 95% probability HPD interval. GREGPLUS reports this interval for each parameter in the set $\boldsymbol{\theta}_e$.

4. The HPD region with probability content $(1-\alpha)$ for a parameter subset $\boldsymbol{\theta}_a$ of $\boldsymbol{\theta}_e$ is given implicitly by

$$\frac{[\widetilde{S}(\boldsymbol{\theta}_a) - S(\widehat{\boldsymbol{\theta}}_a)]/n_a}{\widehat{S}/(n - p)} \leq F(n_a, n - p, \alpha) \qquad (6.6\text{-}11)$$

The boundary shape is exact, since the posterior density of Eq. (6.1-13) is uniform there, but the probability content $(1 - \alpha)$ is exact only if the expectation model is linear in $\boldsymbol{\theta}$. This region can be further approximated as

$$\frac{(\boldsymbol{\theta}_a - \widehat{\boldsymbol{\theta}}_a)^T [\widehat{\boldsymbol{A}}^{aa}]^{-1}(\boldsymbol{\theta}_a - \widehat{\boldsymbol{\theta}}_a)/n_a}{\widehat{S}/(n - p)} \leq F(n_a, n - p, \alpha) \qquad (6.6\text{-}11a)$$

by use of Eq. (6.4-4). The resulting boundary is ellipsoidal when $n_a \geq 2$ and encloses an $n_a$-dimensional volume proportional to $|\widehat{\boldsymbol{A}}^{aa}|^{1/2}$. The volume feature is important for experimental design, as will be seen in Section 6.6. This region reduces to the line interval of Eq. (6.6-10) when $n_\alpha = 1$.

The last two equations involve ratios of mean squares: a numerator mean square based on $n_a$ degrees of freedom and a denominator mean square based on $(n - p)$ degrees of freedom. The probability $\alpha$ satisfies Eq. (6.5-8), with $\nu_1 = n_\alpha$ and $\nu_2 = n - p$; thus, the same integral encountered there in a sampling problem arises here in Bayesian estimation of an HPD region.

### 6.6.2 Inferences for Predicted Functions

Frequently one wants to analyze the posterior density of some result calculated from a fitted model. Such information is obtainable by a transformation of the posterior density, as outlined below and illustrated in Example C.2 of Appendix C. This analysis parallels that of Clarke (1987).

Consider a vector $\boldsymbol{\phi}(\boldsymbol{\theta}_e) = \phi_1(\boldsymbol{\theta}_e), \ldots, \phi_p(\boldsymbol{\theta}_e)$ of auxiliary functions of the estimated parameter vector $\boldsymbol{\theta}_e$. Suppose that these functions are differentiable and linearly independent, so that the matrix $\boldsymbol{J} = \partial\boldsymbol{\phi}/\partial\boldsymbol{\theta}_e^T$ is nonsingular over the considered region of $\boldsymbol{\theta}_e$. Then the vector $\boldsymbol{\phi}(\boldsymbol{\theta}_e)$ has the posterior probability density

$$
\begin{aligned}
p(\boldsymbol{\phi}|\boldsymbol{Y}) &= \left| \frac{\partial\boldsymbol{\theta}_e}{\partial\boldsymbol{\phi}^T} \right| p(\boldsymbol{\theta}_e|\boldsymbol{Y}) \\
&= |\boldsymbol{J}|^{-1} p(\boldsymbol{\theta}_e|\boldsymbol{Y})
\end{aligned}
\tag{6.6-12}
$$

First-order Taylor expansions of the auxiliary functions around $\widehat{\boldsymbol{\theta}}_e$ and $\hat{\boldsymbol{\phi}}$ give

$$
(\boldsymbol{\theta}_e - \widehat{\boldsymbol{\theta}}_e) = \hat{\boldsymbol{J}}^{-1}(\boldsymbol{\phi} - \hat{\boldsymbol{\phi}})
\tag{6.6-13}
$$

so that Eq. (6.4-3) can be rewritten as

$$
\begin{aligned}
\widetilde{S}(\boldsymbol{\phi}) &= \widehat{S} + (\boldsymbol{\phi} - \hat{\boldsymbol{\phi}})^T \hat{\boldsymbol{J}}^{-T} \left[\hat{\boldsymbol{A}}_{ee}\right] \hat{\boldsymbol{J}}^{-1}(\boldsymbol{\phi} - \hat{\boldsymbol{\phi}}) \\
&= \widehat{S} + (\boldsymbol{\phi} - \hat{\boldsymbol{\phi}})^T \left[\hat{\boldsymbol{\mathcal{A}}}_{ee}\right](\boldsymbol{\phi} - \hat{\boldsymbol{\phi}}) \\
&= \widehat{S} + (\boldsymbol{\phi} - \hat{\boldsymbol{\phi}})^T \left[\hat{\boldsymbol{\mathcal{A}}}^{ee}\right]^{-1}(\boldsymbol{\phi} - \hat{\boldsymbol{\phi}})
\end{aligned}
\tag{6.6-14}
$$

Here $\hat{\boldsymbol{J}}^{-T}$ denotes the transpose of $\hat{\boldsymbol{J}}^{-1}$, and

$$
\hat{\boldsymbol{\mathcal{A}}}_{ee} = \hat{\boldsymbol{J}}^{-T}\hat{\boldsymbol{A}}_{ee}\hat{\boldsymbol{J}}^{-1}; \quad \text{thus,} \quad \hat{\boldsymbol{\mathcal{A}}}^{ee} = \hat{\boldsymbol{J}}\hat{\boldsymbol{A}}^{ee}\hat{\boldsymbol{J}}^T
\tag{6.6-15,16}
$$

by the product inversion rule $(\boldsymbol{ABC})^{-1} = \boldsymbol{C}^{-1}\boldsymbol{B}^{-1}\boldsymbol{A}^{-1}$ of Problem A.E in Appendix A.

The parametric inferences in Section 6.6.1 can now be taken over directly. One simply replaces $\hat{\boldsymbol{A}}$ by $\hat{\boldsymbol{\mathcal{A}}}$ and elements of $\boldsymbol{\theta}_e$ by corresponding elements of $\boldsymbol{\phi}$. Thus, Eqs. (6.6-7) and (6.6-8) give the covariance matrix

$$
\text{Cov } \boldsymbol{\phi} = \hat{\boldsymbol{\mathcal{A}}}^{-1}\widehat{S}/(n-p)
\tag{6.6-17}
$$

for the estimated function vector $\boldsymbol{\phi}$ and the variances

$$
\text{Var } \phi_i = \sigma_{\phi_i}^2 = \hat{\boldsymbol{\mathcal{A}}}^{ii}\widehat{S}/(n-p)
\tag{6.6-18}
$$

for the estimated auxiliary functions $\phi_i$.

A vector $\boldsymbol{\phi}_a$ of $n_a \le p$ functions can also be treated, noting that Eq. (6.6-16) needs only the corresponding rows of $\boldsymbol{J}$. Let

$$
\hat{\boldsymbol{J}}_a \equiv \left[\partial\boldsymbol{\phi}_a/\partial\boldsymbol{\theta}_e^T\right]
\tag{6.6-19}
$$

denote the resulting $(n_a \times p)$ transformation matrix. Then the matrix $\hat{\boldsymbol{\mathcal{A}}}^{aa}$ can be calculated as

$$
\hat{\boldsymbol{\mathcal{A}}}^{aa} = \hat{\boldsymbol{J}}_a\hat{\boldsymbol{A}}^{ee}\hat{\boldsymbol{J}}_a^T
\tag{6.6-20}
$$

from the inverse matrix $\hat{\boldsymbol{A}}^{ee}$ described in Section 6.4.1. The matrix $\hat{\boldsymbol{A}}^{aa}$ will be nonsingular when computed by GREGPLUS, which selects a set of linearly independent rows of $\boldsymbol{J}$. In particular, one obtains

$$\text{Var } \phi_i = [\partial\phi_i/\partial\boldsymbol{\theta}_e^T]\hat{\boldsymbol{A}}^{ee}[\partial\phi_i/\partial\boldsymbol{\theta}_e]\widehat{S}/(n-p) \qquad (6.6\text{-}21)$$

for the variance of the auxiliary function reported by GREGPLUS as PHI(i).

### 6.6.3 Discrimination of Rival Models by Posterior Probability

Candidate expectation models $M_1, \ldots, M_J$ are to be tested on a common data vector $\boldsymbol{Y}$ of weighted observations in $n$ independent events and ranked in order of their posterior probabilities on this data set. Prior probabilities $p(M_1), \ldots, p(M_J)$, commonly equal, are assigned to the models without examining the data. These probabilities need not add up to 1, since only their ratios affect the ranking of the models.

*Case I. Variance known*

For model $M_j$, let $p(\boldsymbol{Y}|M_j, \sigma)$ denote the predictive probability density of prospective data $\boldsymbol{Y}$ predicted in the manner of Eq. (6.1-10), with $\sigma$ known. Bayes' theorem then gives the *posterior* probability of model $M_j$ conditional on *given* $\boldsymbol{Y}$ and $\sigma$ as

$$p(M_j|\boldsymbol{Y}, \sigma) = p(M_j)p(\boldsymbol{Y}|M_j, \sigma)/C \qquad (6.6\text{-}22)$$

in which C is a normalization constant, equal for all the models.

Introducing a vector $\boldsymbol{\theta}_{je}$ of estimable parameters for model $M_j$, and integrating Eq. (6.6-22) over the permitted range of $\boldsymbol{\theta}_{je}$, gives the posterior probability of model $Mj$ in the form

$$p(M_j|\boldsymbol{Y}, \sigma) = p(M_j) \int p(\boldsymbol{Y}|\boldsymbol{\theta}_{je}, M_j, \sigma)p(\boldsymbol{\theta}_{je}|M_J)d\boldsymbol{\theta}_{je}/C \qquad (6.6\text{-}23)$$

This relation can be expressed more naturally in the form

$$p(M_j|\boldsymbol{Y}, \sigma) = p(M_j) \int \ell(\boldsymbol{\theta}_{je}|\boldsymbol{Y}, M_j, \sigma)p(\boldsymbol{\theta}_{je}|M_j)d\boldsymbol{\theta}_{je}/C \qquad (6.6\text{-}24)$$

by rewriting $p(\boldsymbol{Y}|\boldsymbol{\theta}_{je}, M_j, \sigma)$ as a likelihood function for $\boldsymbol{\theta}_{je}$ conditional on the given $\boldsymbol{Y}$, $M_j$, and $\sigma$.

We treat the prior density $p(\boldsymbol{\theta}_{je}|M_j)$ as uniform over the region of appreciable likelihood. This approximation, consistent with Eq. (6.1-13), reduces Eq. (6.6-24) to

$$p(M_j|\boldsymbol{Y}, \sigma) = p(M_j)p(\boldsymbol{\theta}_{je}|M_j) \int \ell(\boldsymbol{\theta}_{je}|\boldsymbol{Y}, \sigma)d\boldsymbol{\theta}_{je}/C \qquad (6.6\text{-}25)$$

without restricting the model dependence of the priors.

The posterior probability $p(M_j|\boldsymbol{Y},\sigma)$, being based on data, has a sampling distribution over conceptual replications of the experimental design. We require this distribution to have expectation $p(M_j)$ whenever model $M_j$ is true. With this sensible requirement, Eq. (6.6-25) yields

$$p(M_j|\boldsymbol{Y},\sigma) = p(M_j)\frac{\int \ell(\boldsymbol{\theta}_{je}|\boldsymbol{Y},\sigma)d\boldsymbol{\theta}_{je}}{\mathrm{E}\left[\int \ell(\boldsymbol{\theta}_{je}|\boldsymbol{Y},\sigma)d\boldsymbol{\theta}_{je}\right]} \qquad (6.6\text{-}26)$$

for each candidate model $M_j$.

Evaluation of the likelihood via Eq. (6.1-11) gives

$$p(M_j|\boldsymbol{Y},\sigma) = \frac{\int \exp\{-[\widehat{S}_j + (\boldsymbol{\theta}_{je} - \widehat{\boldsymbol{\theta}}_{je})^T \hat{\boldsymbol{A}}_{ee}(\boldsymbol{\theta}_{je} - \widehat{\boldsymbol{\theta}}_{je})]/2\sigma^2\}d\boldsymbol{\theta}_{je}}{\mathrm{E}\left[\int \exp\{-[\widehat{S}_j + (\boldsymbol{\theta}_{je} - \widehat{\boldsymbol{\theta}}_{je})^T \hat{\boldsymbol{A}}_{ee}(\boldsymbol{\theta}_{je} - \widehat{\boldsymbol{\theta}}_{je})]/2\sigma^2\}d\boldsymbol{\theta}_{je}\right]}$$
$$(6.6\text{-}27)$$

in which the integrations cover the permitted region of $\boldsymbol{\theta}_{je}$. This reduces to

$$p(M_j|\boldsymbol{Y},\sigma) = \frac{\exp[-\widehat{S}_j/2\sigma^2]}{\mathrm{E}\left[\exp(-\widehat{S}_j/2\sigma^2)\right]} \qquad (6.6\text{-}28)$$

because the integrals of the functions of $\boldsymbol{\theta}_{je}$ factor out. Since $\widehat{S}_j/\sigma^2$ has the predictive density in Eq. (6.5-2) whenever the weighted errors are distributed as in Eq. (6.1-9), the expectation in Eq. (6.6-28) is

$$\mathrm{E}\left[\exp(-\widehat{S}_j/2\sigma^2)\right] = \int_0^\infty \exp(-\chi^2/2)p(\chi^2|\nu_j)d\chi^2$$
$$= \int_0^\infty \exp(-\chi^2)\frac{(\chi^2)^{(\nu_j-2)/2}}{2^{\nu_j/2}\Gamma(\nu_j/2)}d\chi^2$$
$$= 2^{-\nu_j/2} \qquad \text{with } \nu_j = n - p_j \qquad (6.6\text{-}29)$$

Insertion of this result in Eq. (6.6-28) gives

$$p(M_j|\boldsymbol{Y},\sigma) = p(M_j)2^{\nu_j/2}\exp(-\widehat{S}_j/2\sigma^2)$$
$$\propto p(M_j)2^{-p_j/2}\exp(-\widehat{S}_j/2\sigma^2) \qquad j = 1,\dots J \qquad (6.6\text{-}30)$$

for the posterior probability of candidate model $j$.

The model with the largest posterior probability is normally preferred, but before acceptance it should be tested for goodness of fit, as in Section 6.5. GREGPLUS automatically performs this test and summarizes the results for all the candidate models if the variance $\sigma^2$ has been specified.

*Case II. Variance unknown but estimable from the data*

Suppose that $\sigma^2$ is unknown but that there is genuine replication of some observations or residuals. Let these replications provide a variance estimate $s^2 = S_E/\nu_E$ having $\nu_E$ degrees of freedom. Subtraction of the constant $S_E$ from $\widehat{S}_j$ in each of Eqs. (6.6-30) gives the proportional expressions

$$p(M_j|\boldsymbol{Y},\sigma) \propto p(M_j)2^{-p_j/2}\exp(-S_{Lj}/2\sigma^2) \qquad j = 1,\ldots J \qquad (6.6\text{-}31)$$

in which $S_{Lj} \equiv \widehat{S}_j - S_E$ is the "lack-of-fit" sum of squares for model $M_j$; see Section 6.5. The "nuisance parameter" $\sigma$ can be removed by integrating it out, in the manner of Eq. (6.6-2), to obtain a marginal posterior density

$$p(M_j|\boldsymbol{Y}) = \int_0^\infty p(M_j|\boldsymbol{Y},\sigma)p(\sigma|S_E,\nu_E)d\sigma \qquad (6.6\text{-}32)$$

From Box and Tiao (1973, 1992, p. 100) we obtain the conditional density function

$$p(\sigma|S_E,\nu_E) = \frac{2}{2^{\nu_E/2}\Gamma(\nu_E/2)} \frac{S_E^{\nu_E/2}}{\sigma^{\nu_E+1}} \exp(-S_E/2\sigma^2) \qquad (6.6\text{-}33)$$

Insertion of this function into Eq. (6.6-32) then gives the relative posterior probabilities of the candidate models:

$$p(M_j|\boldsymbol{Y}) \propto p(M_j)2^{-p_j/2} \int_0^\infty \frac{1}{\sigma^{\nu_E+1}} \exp[-(S_E + S_{Lj})/2\sigma^2]d\sigma$$

$$\propto p(M_j)2^{-p_j/2}\widehat{S}_j^{-\nu_E/2} \qquad j = 1,\ldots J \qquad (6.6\text{-}34)$$

For large $\nu_E$ this result corresponds to Eq. (6.6-30), but with the sample estimate $s^2 = S_E/\nu_E$ replacing $\sigma^2$. The strength of the discrimination increases exponentially with the number of replicate degrees of freedom, $\nu_E$.

The factor $2^{-p_j/2}$ in Eqs. (6.6-30) and (6.6-34) is an *a priori* penalty for the number of parameters estimated in candidate model $j$. It offsets the closer fit expected when parameters are added to an already adequate model. Thus, the expectation of the posterior probability is not altered by adding parameters to an adequate model, whether the variance $\sigma^2$ is known or is estimated from the data. Omission of this penalty would improperly favor overparameterization.

Equations (6.6-30) and (6.6-34) were presented by Box and Henson (1969, 1970) and extended by Stewart, Henson, and Box (1996), with several applications. The development given here is more direct, and includes models with parameters restricted by simple bounds, as in Eq. (6.1-13).

Before accepting a model, one should test its goodness of fit, as outlined in Section 6.5 and implemented in GREGPLUS. Examples of model discrimination and criticism are given in Appendix C.

## 6.7 SEQUENTIAL PLANNING OF EXPERIMENTS

Test patterns for models linear in the parameters can be planned in advance and are highly developed for polynomial response-surface models. Experimental designs of this type are well treated, for instance, in Box and Draper (1987) and in Box, Hunter and Hunter (1978, 2004).

Models nonlinear in $\boldsymbol{\theta}$ can be studied economically in a sequential manner once an exploratory test pattern has been run. Computer simulations based on the current information can then be used to select the next test condition from a set of candidate conditions. The resulting economy of experimental effort can be substantial, as noted by Rippin (1988).

There is a risk in starting model-based planning too soon. These strategies choose tests in regions appropriate for the known models, and so may exclude experiments that could lead to better models. To minimize this risk, one could use a regular grid of candidate conditions without repeating any of them. Hybrid strategies, in which the set of candidate experiments is modified as the work proceeds, are also useful.

The selection strategy depends on the goals of the investigator. Normally, one or more of the following goals is chosen:

1. Optimal estimation of parameters.
2. Optimal estimation of functions of the parameters, such as operating states or expected economic return.
3. Optimal discrimination among models.
4. Building a better model.

Recommended strategies for these goals are described below.

### 6.7.1 Planning for Parameter Estimation

The current parameter estimates are obtained from the posterior density function, which we approximated in Section 6.6 by expansion via the matrix $\hat{\boldsymbol{A}}_{ee}$ or its inverse. Expected values of these matrices after additional observations are computable from simulations of those augmented data sets. The preferred candidate for the next event can then be selected according to the desired criterion of estimation.

*Determinant* criteria aim to minimize the volume of an HPD region in parameter space. For the full estimable parameter set $\boldsymbol{\theta}_e$, one selects to maximize the determinant $|\hat{\boldsymbol{A}}_{ee}|$, thus minimizing the volume of the joint HPD region according to Eq. (6.6-4). For optimal estimation of a parameter subset $\boldsymbol{\theta}_a$, one selects to maximize the determinant $|\hat{\boldsymbol{A}}_{aa}|$, thus minimizing the volume of the joint HPD region for the subset $\boldsymbol{\theta}_a$ according to Eq. (6.6-11a).

The full determinant strategy was introduced by Box and Lucas (1959) and applied sequentially by Box and Hunter (1965); the subset strategy was introduced by Michael Box (1971). Experimental designs of these types are called *D-optimal* [St. John and Draper (1975)]; the literature on them is extensive.

Two *shape criteria* for the HPD region have been proposed for planning experiments for parameter estimation. Hosten (1974) advocated selection to maximize the smallest eigenvalue of $A$, thus giving the HPD region a rounder shape. Pritchard and Bacon (1975) selected the next experiment to reduce the interdependence of the parameter estimates. Reparameterization can give such results more easily, as shown by Agarwal and Brisk (1985) and by Ratkowsky (1985).

A *trace criterion* for precise estimation was introduced by Pinto, Lobão, and Monteiro (1990). Here one selects the next experiment to minimize $\sum_i A^{ii}$, the trace of the inverse matrix $A^{ee}$. This quantity is the sum of the eigenvalues of $A^{ee}$ and is proportional to the sum of the variances of the parameter estimates. Pinto, Lobão, and Monteiro (1991) showed that selection to minimize a weighted trace, $\sum_i \hat{\theta}_i^2 A^{ii}$, makes the *relative* variances of the parameters more nearly equal. (A simple implementation of the latter criterion is to reparameterize the model with the logarithms of the original parameters; then minimization of $\mathrm{tr}A^{ee}$ accomplishes the desired result). Finally, the trace criterion is readily applied to any subset $\theta_a$ of the parameters by including just the corresponding diagonal elements of $A^{ee}$ in the trace calculation.

The determinant criterion and the trace criterion are well regarded, and both are provided as options in GREGPLUS. Example C.1 applies the full determinant criterion to a three-parameter model.

## 6.7.2 Planning for Auxiliary Function Estimation

Suppose that the goal of an investigation is to predict the yield of a process at $p$ or fewer chosen conditions, $p$ being the number of parameters to be estimated. This task amounts to estimating a vector $\phi_a(\theta)$ of functions of the parameters. The optimal strategies for this task parallel those just given for parameter estimation, the only difference being that one uses matrix $\hat{\mathcal{A}}^{aa}$ of Eq. (6.6-20) instead of $\hat{A}^{aa}$. Example C.2 illustrates the use of this criterion to select the candidate next event that gives the smallest inference-region volume for a pair of auxiliary functions.

Different choices of conditions can occur in function-based planning than in parameter-based planning when a criterion of order $n_a$ less than $p$ is used. However, the criteria $|\hat{\mathcal{A}}_{ee}|$ (for function estimation) and $|\hat{A}_{ee}|$ (for parameter estimation) choose identically when $n_a = p$, since $|\hat{\mathcal{A}}|$ and $|\hat{A}|$ are then proportional at each value of $\hat{\theta}$. The proportionality relation, $|\hat{\mathcal{A}}_{ee}| = |\hat{A}_{ee}|/|\hat{J}|^2$, comes from the determinants of the two members of Eq. (6.6-15).

## 6.7.3 Planning for Model Discrimination

Early strategies for model discrimination called for experiments at locations where the predictions of the leading candidate models differed most. Such a criterion was given by Hunter and Reiner (1965) for discrimination between two models, and has been generalized to pairwise discrimination among multiple models [Atkinson and Fedorov (1975); Dumez and Froment (1976); Buzzi-Ferraris and co-workers (1983, 1990)]. Such criteria have the defect that the best models are identified last; thus, much of the effort is spent on eliminating inferior models.

Box and Hill (1967) gave a more natural criterion: that the next event be designed for maximum expected information gain (entropy decrease). They used the entropy function

$$S = - \sum_{j=1}^{|\text{nmod}|} \Pi_{jn} \ln \Pi_{jn} \tag{6.7-1}$$

given by Shannon (1948), shown here for discrimination among models $1, \ldots, |\text{nmod}|$. Here $\Pi_{jn}$ is the posterior probability share for Model $j$ after $n$ events; thus, the quality of every candidate model is included in the entropy function. $|\text{nmod}|$ is the number of candidate models. In GREGPLUS, the sign of nmod indicates how the events will be selected: for discrimination if nmod is negative or for combined discrimination and estimation if nmod is positive.

Box and Hill (1967) derived the following expression for the expectation $R$ of information gain (entropy decrease) obtainable by adding an observation value $y_n$ to the database:

$$R = \sum_{j=1}^{n} \Pi_{jn-1} \int p_j \ln \frac{p_j}{\sum_{k=1}^{|\text{nmod}|} \Pi_{kn-1} p_k(y_n)} \, dy_n \tag{6.7-2}$$

Here the indices $i$ and $m$ used by Box and Hill are replaced by $j$ and $|\text{nmod}|$, to save them for other use in Chapter 7. In their design criterion, Box and Hill actually used an upper bound $D$ on the entropy decrease, which is more easily computed than the expectation $R$ itself.

The observation $y_n$ was regarded as normally distributed, with a known variance $\sigma^2$ and expectation $y_n^{(j)} = E_j(y_n)$ under Model $j$, giving a probability density

$$p(y_n) = p(y_n|\eta, \sigma) = \frac{1}{\sqrt{2\pi}\sigma} \exp\left\{ -\frac{1}{\sigma^2}(y_n - y_n^{(j)})^2 \right\} \tag{6.7-3}$$

for the prospective observation $y_n$. Then a local linearization of Model $j$ with respect to its parameters gave the predictive probability density of $y_n$ under that model,

$$p_j(y_n|\sigma) = \frac{1}{\sqrt{2\pi(\sigma^2 + \sigma_j^2)}} \exp\left\{ -\frac{\{y_n - y_n^{(j)}\}^2}{2(\sigma^2 + \sigma_j^2)} \right\} \tag{6.7-4}$$

as shown by Box and Hill (1967).

Reilly (1970) gave an improved criterion that the next event be designed to maximize the *expectation*, $R$, of information gain rather than the upper bound $D$. His expression for $R$ with $\sigma$ known is included in GREGPLUS and extended to unknown $\sigma$ by including a posterior probability density $p(\sigma|s, \nu_e)$ based on a variance estimate $s^2$ with $\nu_e$ error degrees of freedom. The extended $R$ function thus obtained is an expectation over the posterior distributions of $y_n$ and $\sigma$,

$$R = \sum_{j=1}^{|\text{nmod}|} \Pi_{jn-1} \int_{y_n} \int_{\sigma} \Pi_j(y_n|\sigma) \ln \frac{p_j(y_n|\sigma)}{\sum_{k=1}^{|\text{nmod}|} \Pi_{kn-1}p_k(y_n|\sigma)} \, dy_n p(\sigma|s, \nu_e) \, d\sigma$$

(6.7-5)

and is computed efficiently in GREGPLUS by use of adaptive quadrature subroutines from Press et al. (1992).

### 6.7.4 Combined Discrimination and Estimation

The sequential procedures discussed above are appropriate when the investigator wants *either* to estimate parameters *or* to discriminate among rival models. However, often an investigator wants to do both. A design procedure for this purpose should emphasize discrimination until one model is gaining favor and should emphasize parameter estimation after that point. The criterion should guard against premature choice of a model, with consequent waste of effort on parameter estimation for an inferior model.

Hill, Hunter, and Wichern (1968) suggested a weighted criterion

$$C = w_1 D + w_2 E \tag{6.7-6}$$

for selection of the next event; here $D$ and $E$ are measures of discrimination and estimation. GREGPLUS defines $D$ as the greatest of the $R$ values for the candidates and defines $E$ as

$$E = \sum_{j=1}^{|\text{nmod}|} \Pi_{jn} \begin{cases} \det A & \text{if JNEXT.gt.0} \\ \text{tr} A & \text{if JNEXT.lt.0} \end{cases} \tag{6.7-7}$$

(here JNEXT is the signed number of candidate conditions for the next event), while the weights are defined as

$$w_1 = |\text{nmod}|(1 - \Pi_{jn})/(|\text{nmod}| - 1)]^{\lambda} \tag{6.7-8}$$

$$w_2 = 1 - w_1 \tag{6.7-9}$$

Here we have modified the formulas of Hill, Hunter, and Wichern (1968) by using Eq. (6.7-2) for $R$ in place of the upper bound derived by Box and Hill (1967). We have correspondingly modified the discrimination measure

$D$ in the weighted criterion $C$ and have provided a choice between two estimation measures $E$, distinguished by the sign of JNEXT, as shown in Eq. (6.7-7).

The exponent $\lambda$ is set at 2 by GREGPLUS to emphasize discrimination initially as Hill, Hunter and Wichern did.

Sequential strategies for discrimination can be risky if applied before a thorough enumeration of plausible model forms. Important regions of experimentation may then be missed, and the kinds of models needed to describe those regions may go undiscovered. Factorial experimentation at the outset is desirable to allow inductive building of good model forms.

### 6.7.5 Planning for Model Building

Model building should normally begin with a factorial experimental pattern to give impartial coverage of the accessible region. Such designs are described in Box, Hunter and Hunter (1978) and in Box and Draper (1987). When several candidate models have been constructed, one can begin sequential discrimination and estimation via Eq. (6.7-4) to find a preferred model and estimate its parameters. To broaden the coverage of the accessible region, one can do the sequential selections from unexplored points of a uniform grid, as advocated by Rippin (1988).

## 6.8 EXAMPLES

Two examples are provided here to illustrate nonlinear parameter estimation, model discrimination, and analysis of variance.

### Example 6.1. Discrimination Between Series and Parallel Models

Suppose that some data are available for the concentration [B] of species B as a function of time in a constant-volume, isothermal batch reactor containing chemical species A, B, and C. Sixteen simulated replicate pairs of observations similar to those considered by Box and Coutie (1956) are given in Table 6.1 for the initial condition $[A]_0 = 1$, $[B]_0 = [C]_0 = 0$.

Two rival models are postulated:

Model 1: Consecutive first-order reactions $A \xrightarrow{k_1} B \xrightarrow{k_2} C$ give

$$[B] = \frac{k_1}{k_2 - k_1}\{\exp(-k_1 t) - \exp(-k_2 t)\}$$

Model 2: Parallel first-order reactions $A \underset{k_2'}{\overset{k_1'}{\rightleftharpoons}} B$ and $A \xrightarrow{k_3'} C$ give

$$[B] = \frac{k_1'}{\lambda_2 - \lambda_3}\{\exp(-\lambda_3 t) - \exp(-\lambda_2 t)\}$$

$$\lambda_2 = (p+q)/2 \qquad p = k_1' + k_2' + k_3'$$
$$\lambda_3 = (p-q)/2 \qquad q = (p^2 - 4k_2' k_3')^{1/2}$$

| t, min. | [B] | t, min. | [B] |
|---|---|---|---|
| 10 | .192, .140 | 160 | .407, .464 |
| 20 | .144, .240 | 180 | .439, .380 |
| 30 | .211, .161 | 200 | .387, .393 |
| 40 | .423, .308 | 220 | .362, .324 |
| 60 | .406, .486 | 240 | .269, .293 |
| 80 | .421, .405 | 260 | .240, .424 |
| 100 | .457, .519 | 280 | .269, .213 |
| 120 | .505, .537 | 300 | .297, .303 |
| 140 | .558, .581 | 320 | .271, .223 |

**Table 6.1:** Simulated Data for Example 6.1.

Fitting these models to the data by nonlinear least squares yields the results in Table 6.2. Model 1 fits the data better than Model 2 and has a posterior probability share of 0.987. The posterior probability shares $\Pi(M_j|\boldsymbol{Y})$ are calculated here from Eq. (6.6-33) with $p(M_1) = p(M_2)$ and normalization to a sum of 1. Analyses of variance for both models are given in Table 6.3; there is a significant lack of fit for Model 2. These results are as expected, since the data were constructed from Model 1 with simulated random errors. All these calculations were done with GREGPLUS.

### Example 6.2. Discrimination of Models for Codimer Hydrogenation

Tschernitz et al. (1946) investigated the kinetics of hydrogenation of mixed isooctenes (codimer) over a supported nickel catalyst. Their article gives 40 unreplicated observations (reproduced in Table 6.5) of the hydrogenation rate. The independent variables investigated were the catalyst temperature $T$ and the partial pressures $p_H$, $p_U$, and $p_S$ of hydrogen, unsaturates, and saturated hydrocarbons. Eighteen rival models (shown in Table 6.4) of the reaction mechanism were formulated and fitted by least squares to the experimental data. These models and data are reanalyzed here.

For this new analysis, the observations were expressed as values of the response function $y \equiv \ln \mathcal{R}$, where $\mathcal{R}$ is the hydrogenation rate in mols per hour per unit mass of catalyst, adjusted to a standard level of catalyst activity. Variances were assigned to these adjusted observations according to the formula

$$\sigma_y^2 = (0.00001/\Delta n_{\mathrm{RI}})^2 + (0.1)^2 \qquad (6.8\text{-}1)$$

based on assigned standard deviations of 0.00001 for the observations of refractive index difference $\Delta n_{\mathrm{RI}}$ (used in calculating the reactant conversions) and 0.1 for the catalyst activity corrections. The activity variance, $(0.1)^2$, proved to be dominant in this formula; thus, the weights $w_u = 1/(\sigma_y^2)_u$ were nearly equal for the 40 events.

The 18 models investigated by Tschernitz et al. are expressed in Table 6.4 as expectation models for $\ln \mathcal{R}$. Each coefficient $k_i(T)$ or $K_i(T)$

| Model $j$ | $\hat{k}_{j1}$ | $\hat{k}_{j2}$ | $\hat{k}_{j3}$ | $\widehat{S}_j$ | $\Pi(M_j|Y)$ |
|---|---|---|---|---|---|
| 1 | 0.0121 ±0.0015 | 0.00644 ±0.00053 | | 0.1142 | 0.987 |
| 2 | 0.0158 ±0.0034 | 0.0078 ±0.0050 | 0.0078 ±0.0050 | 0.1748 | 0.013 |

**Table 6.2:** Estimates for Example 6.1.

ANOVA for Model 1:

| Source | Sum of Squares | Deg. of Freedom | Mean Square |
|---|---|---|---|
| Residuals | .114243 | 34 | |
| Lack of Fit | .070335 | 16 | $s_1^2 = .004396$ |
| Pure Error | .043908 | 18 | $s^2 = .002439$ |

$F_1 = s_1^2/s^2 = 1.80$
$Q(1.80|16,18) = 0.115$

ANOVA for Model 2:

| Source | Sum of Squares | Deg. of Freedom | Mean Square |
|---|---|---|---|
| Residuals | .174806 | 33 | |
| Lack of Fit | .130898 | 15 | $s_2^2 = .008727$ |
| Pure Error | .043908 | 18 | $s^2 = .002439$ |

$F_2 = s_2^2/s^2 = 3.58$
$Q(3.58|15,18) = 0.006$

**Table 6.3:** Analyses of Variance for Example 6.1.

denotes an Arrhenius function centered at the mean $1/T$ value of $1/538.9$ $K^{-1}$. Functions of the form $\exp[\theta_{iA} + \theta_{iB}(1/538.9 - 1/T)]$ were tried initially so that every model could be fairly tested, with nonnegative $k_i(T)$ and $K_i(T)$. This form often needed huge magnitudes of $\theta_{iA}$ and/or $\theta_{iB}$ to describe the unimportance of one or more individual rate or equilibrium coefficients; therefore, the form $\theta_{iA}\exp[\theta_{iB}(1/538.9 - 1/T)]$ with nonnegative rate coefficients $\theta_{iA}$ was preferred.

Each model $f_j$ was fitted to the data by weighted least squares, using our program GREGPLUS and the nonnegativity constraints $\theta_{iA} \geq 0$. The data for all temperatures were analyzed simultaneously under these constraints, as advocated by Blakemore and Hoerl (1963) and emphasized by Pritchard and Bacon (1975). The resulting nonlinear parameter estimation calculations were readily handled by GREGPLUS.

To estimate the experimental variance, a 30-parameter polynomial in the four independent variables was constructed and fitted to the data in the same manner. Reduced versions of this polynomial were then tested until a minimum value of the residual mean square, $\widehat{S}/(n - p_j)$, was found with 23 terms, giving the estimate $\nu_E = 40 - 23 = 17$ for the "pure error" degrees of freedom. The resulting residual sum of squares, $\widehat{S} = 60.8714$, correspondingly approximates the pure error sum of squares $S_E$.

Our tests of the candidate models are summarized in Table 6.5, with the posterior probability shares $\Pi(M_j|\mathbf{Y})$ calculated from Eq. (6.6-33) and the prior probabilities $p(M_j) = 1$. These probabilities indicate a strong preference for Model h, with Models d and g next best and with negligible probabilities for the other models. Model h also gives the best result on the goodness-of-fit test, with a sampling probability of 0.095 which considerably exceeds the values for the other candidates. Thus, Model h gives the most acceptable fit yet obtained for these data as well as the highest posterior probability of being true.

Our choice of model differs from that of Tschernitz et al. (1946), who preferred Model d over Model h on the basis of a better fit. The difference lies in the weightings used. Tschernitz et al. transformed each model to get a linear least-squares problem (a necessity for their desk calculations) but inappropriately used weights of 1 for the transformed observations and response functions. For comparison, we refitted the data with the same linearized models, but with weights $w_u$ derived for each model and each event according to the variance expression in Eq. (6.8-1) for $\ln \mathcal{R}$. The residual sums of squares thus found were comparable to those in Table 6.5, confirming the superiority of Model h among those tested.

## 6.9 SUMMARY

Data plots are very useful at the start of a modeling study. They are helpful in assessing the quality of the data and can be useful in explorations of potential model forms.

Graphical methods suffice for approximate parameter estimation when the models considered are simple and the parameters are few. Least squares gives closer estimates, especially for multiparameter models.

Modern minimization algorithms can handle highly nonlinear models directly. Nevertheless, it helps to express a model in a form linear in $\boldsymbol{\theta}$ when this can be done, with proper weightings for the transformed functions and observations. Tschernitz et al. (1946) used linearization to get direct solutions in their comprehensive study of surface reaction rate expressions; that approach can work well when applied with appropriate weighting and a constrained estimation algorithm to avoid unacceptable parameter values. Later workers showed that reparameterizations of models to achieve approximate linearity in $\boldsymbol{\theta}$ led to better interval estimates as well as easier minimization [Ratkowsky (1985); Bates and Watts (1988); Espie and Machietto (1988)]. Such transformations are, unfortunately, less effective for treating data obtained at several temperatures because of the nonlinear influence of the energy parameters on rate and equilibrium coefficients.

It is helpful to choose a parameterization that eliminates one or more off-diagonal elements of the matrix $\boldsymbol{X}^T \boldsymbol{X}$ in Eq. (6.3-9). For example, the

$$f_a(\boldsymbol{\xi}, \boldsymbol{\theta}_a) = \ln\left[k_a(T)p_H / (1 + K_U(T)p_U + K_S(T)p_S))\right]$$

$$f_b(\boldsymbol{\xi}, \boldsymbol{\theta}_b) = \ln\left[k_b(T)p_U / (1 + K_H(T)p_H + K_S p_S)\right]$$

$$f_c(\boldsymbol{\xi}, \boldsymbol{\theta}_c) = \ln\left[k_c(T)p_H p_U / (1 + K_H(T)p_H + K_U(T)p_U)\right]$$

$$f_d(\boldsymbol{\xi}, \boldsymbol{\theta}_d) = \ln\left[k_d(T)p_H p_U / (1 + K_H(T)p_H + K_U(T)p_U + K_S(T)p_S)^2\right]$$

$$f_e(\boldsymbol{\xi}, \boldsymbol{\theta}_e) = \ln\left[k_e(T)p_H / (1 + K_U(T)p_U + K_S(T)p_S)^2\right]$$

$$f_f(\boldsymbol{\xi}, \boldsymbol{\theta}_f) = \ln\left[k_f(T)p_U / (1 + \sqrt{K_H(T)p_H} + K_S(T)p_S)\right]$$

$$f_g(\boldsymbol{\xi}, \boldsymbol{\theta}_g) = \ln\left[k_g(T)p_H p_U / (1 + \sqrt{K_H(T)p_H} + K_U(T)p_U)\right]$$

$$f_h(\boldsymbol{\xi}, \boldsymbol{\theta}_h) = \ln\left[k_h(T)p_H p_U / (1 + \sqrt{K_H(T)p_H} + K_U(T)p_U + K_S(T)p_S)^3\right]$$

$$f_i(\boldsymbol{\xi}, \boldsymbol{\theta}_i) = \ln\left[k_i(T)p_H / (1 + K_S(T)p_S)\right]$$

$$f_j(\boldsymbol{\xi}, \boldsymbol{\theta}_j) = \ln\left[k_j(T)p_H p_U / 1 + K_H(T)p_H)\right]$$

$$f_k(\boldsymbol{\xi}, \boldsymbol{\theta}_k) = \ln\left[k_k(T)p_H p_U / (1 + K_H(T)p_H + K_S(T)p_S)\right]$$

$$f_l(\boldsymbol{\xi}, \boldsymbol{\theta}_l) = \ln\left[k_l(T)p_H / (1 + K_S(T)p_S)^2\right]$$

$$f_m(\boldsymbol{\xi}, \boldsymbol{\theta}_m) = \ln\left[k_m(T)p_H p_U / (1 + \sqrt{K_H(T)p_H})\right]$$

$$f_n(\boldsymbol{\xi}, \boldsymbol{\theta}_n) = \ln\left[k_n(T)p_H p_U / (1 + \sqrt{K_H(T)p_H} + K_S(T)p_S)^2\right]$$

$$f_o(\boldsymbol{\xi}, \boldsymbol{\theta}_o) = \ln\left[k_o(T)p_H p_U / (1 + K_U(T)p_U + K_S(T)p_S)\right]$$

$$f_p(\boldsymbol{\xi}, \boldsymbol{\theta}_p) = \ln\left[k_p(T)p_H p_U / (1 + K_U(T)p_U)\right]$$

$$f_q(\boldsymbol{\xi}, \boldsymbol{\theta}_q) = \ln\left[k_q(T)p_U / (1 + K_S(T)p_S)\right]$$

$$f_r(\boldsymbol{\xi}, \boldsymbol{\theta}_r) = \ln\left[k_r(T)p_H p_U\right]$$

**Table 6.4:** Expectation Models for Response $y = \ln\mathcal{R}$ in Example 6.2 [Adapted from Tschernitz et al. (1946)].

linear model

$$f_u = \beta_1 + x_u\beta_2 \qquad u = 1, \ldots, n \qquad (6.9\text{-}1)$$

with parameters $\beta_1$ and $\beta_2$ is better written as

$$f_u = \theta_1 + (x_u - \overline{x})\theta_2 \qquad (6.9\text{-}2)$$

unless the mean value $\overline{x}$ is small relative to the limits of $x_u$. Equation (6.9-2) has the advantage that $\theta_1$ represents the mean of the data, whereas $\beta_1$ may be a considerably extrapolated value. Furthermore, Eq. (6.9-2) makes the Gauss-Newton matrix $\boldsymbol{X}^T\boldsymbol{X}$ for $\theta_1$ and $\theta_2$ diagonal, so that the posterior distributions of $\theta_1$ and $\theta_2$ are uncorrelated according to Eq. (6.6-7). This principle is commonly utilized in linear regression programs by working with independent variables $(x_{ui} - \overline{x}_i)$ measured from the weighted mean

| Model $j$ | Residual Sum of Squares | No. of Parameters Estimated | Posterior Probability Share $\Pi(M_j\|\boldsymbol{Y})$ | Test Ratio $F_j$ | Degrees of Freedom Lack of Fit | Degrees of Freedom Exptl. Error | Sampling Probability of Larger $F_j$ |
|---|---|---|---|---|---|---|---|
| a | 970.2 | 4 | 0.000 | 13.4 | 19 | 17 | 9.2D-7 |
| b | 2175.6 | 3 | 0.000 | 29.5 | 20 | 17 | 1.7D-9 |
| c | 279.4 | 6 | 0.015 | 3.6 | 17 | 17 | 5.9D-3 |
| d | 192.2 | 8 | 0.177 | 2.4 | 15 | 17 | 4.0D-2 |
| e | 1013.8 | 4 | 0.000 | 14.0 | 19 | 17 | 6.5D-7 |
| f | 2175.6 | 3 | 0.000 | 29.5 | 20 | 17 | 1.7D-9 |
| g | 210.2 | 6 | 0.165 | 2.5 | 17 | 17 | 3.6D-2 |
| h | 165.2 | 8 | **0.641** | 1.9 | 15 | 17 | **9.5D-2** |
| i | 970.2 | 4 | 0.000 | 13.4 | 19 | 17 | 9.2D-7 |
| j | 844.7 | 4 | 0.000 | 11.5 | 19 | 17 | 2.8D-6 |
| k | 826.2 | 6 | 0.000 | 12.6 | 17 | 17 | 1.8D-6 |
| l | 1013.8 | 4 | 0.000 | 14.0 | 19 | 17 | 6.5D-7 |
| m | 824.8 | 4 | 0.000 | 11.2 | 19 | 17 | 3.4D-6 |
| n | 788.8 | 6 | 0.000 | 12.0 | 17 | 17 | 2.6D-6 |
| o | 420.1 | 6 | 0.000 | 5.9 | 17 | 17 | 3.3D-4 |
| p | 496.4 | 2 | 0.000 | 6.2 | 21 | 17 | 2.8D-4 |
| q | 2139.2 | 3 | 0.000 | 29.0 | 20 | 17 | 2.0D-9 |
| r | 925.4 | 2 | 0.000 | 11.5 | 21 | 17 | 2.4D-6 |

**Table 6.5:** Testing of Kinetic Models for Data of Tschernitz et al. (1946)

of the data. The same principle led Box (1960) and Krug, Hunter, and Grieger (1976) to recommend Eq. (3.4-7) as an improved parameterization for the Arrhenius equation.

Interval estimates for nonlinear models are usually approximate, since exact calculations are very difficult for more than a few parameters. But, as our colleague George Box once said, "One needn't be excessively precise about uncertainty." In this connection, Donaldson and Schnabel (1987) found the Gauss-Newton normal equations to be more reliable than the full Newton equations for computations of confidence regions and intervals.

The error model in Eq. (6.1-2) is regarded as unbiased; that is, the expected value of $\varepsilon_u$ is assumed to be zero. For consistency with this model, all known systematic errors should be corrected out of the data. This is often done before submitting the data, though with proper care the adjustments can be done by including the appropriate constraints (material balances, etc.) in the computational model. In the latter case, the parameters are calculated with the optimally adjusted values of $\boldsymbol{\xi}_u$ and $y_u$ for each experiment, thus rectifying errors in the independent variables. This idea originated with Gauss [see also Deming (1943)], and is known in the recent literature as the *error-in-variables* method [Anderson, Abrams, and Grens

(1978), Kim, Edgar, and Bell (1991)]. A simple way of doing the latter type of calculation is to treat some observed values as parameters; such a calculation is described by Young and Stewart (1992).

More detailed error models are available in which occasional large errors are given a separate distribution with small probability and large variance. Robust estimation procedures based on such models are available [Fariss and Law (1979); Box (1980)] and provide an objective means of dealing with unusually large residuals (known as *outliers*). Though this method of identification of questionable data is helpful, one should not reject any observation without further evidence. Outliers may give valuable clues for improving the model, as in the discoveries of Neptune and Pluto through anomalies in the orbits calculated for other planets.

The weighting of the observations is important, and can be done best when replicate data are available to estimate the variance of $y$ at several test conditions. Otherwise, the weighting may need to be based on prior experience, or on a theoretical analysis of the precisions of the observations, as in Example 6.2. If transformed observations are used as $y_u$, the corresponding transformed variances $\sigma_u^2$ must be used. Choosing $w_u = 1$ without regard for the form of $y_u$ is a common error that continues to occur in the literature.

An unnecessary difficulty in weighting occurs if a model is fitted first in its isothermal form and the resulting parameters are then fitted as functions of temperature, e.g., $k_j(T)$ or $\ln k_j(T)$. If this method is used, the second stage should be followed by a simultaneous fit of the original data for all temperatures. The simultaneous calculation will fit the original data better, as Blakemore and Hoerl (1963) showed for Model d of Table 6.4.

Least squares has played a prominent role in the chemical engineering literature, especially since the advent of automatic computation. Some further references to this literature and to least-squares algorithms are included at the end of this chapter. Multiresponse data require a more detailed error model and will be treated in Chapter 7.

## 6.10 NOTATION

| | |
|---|---|
| $A$ | matrix in Eq. (6.3-7) for current iteration |
| $\hat{A}$ | matrix $A$ at $\hat{\theta}$ |
| $\hat{A}^{ee}$ | inverse of submatrix $\hat{A}_{ee}$ for estimated parameters |
| $\hat{\mathcal{A}}^{ee}$ | inverse of submatrix $\hat{\mathcal{A}}_{ee}$ for estimated functions |
| $\hat{A}^{aa}$ | submatrix of $\hat{A}^{ee}$ for parameter subset $\theta_a$ |
| $A_{ij}, A^{ij}$ | elements of $A$ and $A^{ee}$, respectively |
| $E_u(\theta)$ | weighted error variable, Eq. (6.1-8) |
| $\widehat{E}_u$ | $= E_u(\hat{\theta})$, weighted residual, Eq. (6.5-1) |
| $f(\xi_u, \theta)$ | $= f_u(\theta)$, expectation model for $y_u$ |
| $F_u(\theta)$ | $\sqrt{w_u} f_u(\theta)$, expectation model for $Y_u$ |
| $F$ | variance ratio in Eq. (6.5-4) |

$\widehat{\boldsymbol{J}}_a$            transformation matrix in Eq. (6.6-19)

$n$            number of observations $y_u$ or $Y_u$

$n_a$            number of parameters in subset $\boldsymbol{\theta}_a$ of $\boldsymbol{\theta}_e$

$p$            number of parameters estimated from Eq. (6.3-10)

$p(\boldsymbol{\theta}, \sigma)$            prior density for $\boldsymbol{\theta}$ and $\sigma$

$p(\boldsymbol{\theta}, \sigma | \boldsymbol{Y})$            posterior density for $\boldsymbol{\theta}$ and $\sigma$

$p(\boldsymbol{\theta}_a | \boldsymbol{Y})$            marginal posterior density for subset $\boldsymbol{\theta}_a$ of $\boldsymbol{\theta}_e$

$p(\boldsymbol{\theta}_e | \boldsymbol{y})$            marginal posterior density for estimated parameters

$p(M_j)$            prior probability for Model $j$

$p(M_j | \boldsymbol{Y}, \sigma)$            posterior probability for Model $j$ with $\sigma$ known, Eq. (6.6-30)

$p(M_j | \boldsymbol{Y})$            posterior probability for Model $j$ with $\sigma$ estimated from $\boldsymbol{Y}$, Eq. (6.6-34)

$S(\boldsymbol{\theta})$            sum of squares of weighted errors, Eqs. (6.1-7)

$\widetilde{S}(\boldsymbol{\theta})$            local quadratic approximation to $S(\boldsymbol{\theta})$, Eq. (6.3-7) and (6.4-3)

$\Delta \widetilde{S}^k$            increment of $\widetilde{S}$ computed by GRQP in iteration $k$

$S_E$            pure error sum of squares, in Eq. (6.5-5)

$S_L$            lack-of-fit sum of squares, Eq. (6.5-5)

$s^2$            sample estimate of variance $\sigma^2$

$u$            event index in Eq. (6.1-1)

$w_u$            weight of observation $y_u$, Eq. (6.1-5)

$\boldsymbol{X}$            matrix of parametric sensitivities $\partial F_u / \partial \theta_r$

$\boldsymbol{\xi}_u$            vector of experimental settings for event $u$

$\boldsymbol{y}$            vector of observations $y_u$

$\boldsymbol{Y}$            vector of weighted observations $\sqrt{w_u} y_u$

$\widehat{Y}_u$            $= F_u(\widehat{\boldsymbol{\theta}})$, modal estimate of $Y_u$

*Greek letters*

$\alpha$            cumulative sampling probability (Eqs. (6.5-7)) or posterior probability (Eqs. (6.6-9) to (6.6-12))

$\varepsilon_u$            error $y_u - f_u(\boldsymbol{\theta})$, a random variable until $y_u$ is known

$\boldsymbol{\theta}$            parameter vector in Eq. (6.1-1)

$\boldsymbol{\theta}_c$            vector of parameters actively constrained by Eq. (6.4-2)

$\boldsymbol{\theta}_e$            vector of parameters estimated from Eq. (6.3-10)

$\boldsymbol{\theta}_f$            vector of free (undetermined) parameters

$\widehat{\boldsymbol{\theta}}$            modal (least-squares) estimate of $\boldsymbol{\theta}$

$\theta_r$            element $r$ of $\boldsymbol{\theta}$

$\Delta \boldsymbol{\theta}^k$            parameter step computed by GRQP in iteration $k$

$\nu_1, \nu_2$            degrees of freedom $\nu_L$ and $\nu_E$ in Eq. (6.5-8)

$\sigma^2$            variance of observations of unit weight

$\sigma_u^2$            variance assigned to observation $y_u$; see Eq. (6.1-5)

$\sigma_{\theta_r}^2$            variance of marginal posterior distribution of $\theta_r$

$\phi$            vector of auxiliary functions

*Subscripts*

$a$            for a subset of estimated parameters or functions

| $c$ | for actively constrained parameters |
|---|---|
| $e$ | for estimated parameters |
| $f$ | for free parameters |
| $L$ | last row or column of matrix $\boldsymbol{A}$ |
| $u$ | for event $u$ |
| $\theta$ | derivatives with respect to $\boldsymbol{\theta}$ |

*Superscripts*

| $aa, ee$ | inverse submatrix |
|---|---|
| $-1$ | inverse matrix |
| $T$ | transpose of vector or matrix |
| $\hat{}$ | modal (least-squares) value |

*Functions and operations*

| $\equiv$ | preceding symbol is defined by the following expression |
|---|---|
| $\lvert \boldsymbol{A} \rvert$ | determinant of $\boldsymbol{A}$ |
| $\propto$ | proportionality sign |
| $\prod$ | product |
| $\sum$ | summation |

## PROBLEMS

### 6.A Properties of the Gauss-Newton Normal Equations

Here the properties asserted in Section 6.3 for the Gauss-Newton normal equations are to be verified.

(a) A real, $k \times k$ symmetric matrix $\boldsymbol{B}$ is *positive semidefinite* if $\boldsymbol{z}^T \boldsymbol{B} \boldsymbol{z} \geq 0$ for all real $k$-vectors $\boldsymbol{z}$. Show that $\boldsymbol{X}^T \boldsymbol{X}$ is positive semidefinite for any real $n \times k$ matrix $\boldsymbol{X}$ by expressing $\boldsymbol{z}^T \boldsymbol{X}^T \boldsymbol{X} \boldsymbol{z}$ as a sum of squares.

(b) A real, $k \times k$ symmetric matrix $\boldsymbol{B}$ is *positive definite* if $\boldsymbol{z}^T \boldsymbol{B} \boldsymbol{z} > 0$ for all real nonzero $k$-vectors $\boldsymbol{z}$. Extend the result of (a) to show that $\boldsymbol{X}^T \boldsymbol{X}$ is positive definite if and only if the columns of $\boldsymbol{X}$ are linearly independent.

(c) Use the results of (a) and (b) to show that $\widetilde{S}(\boldsymbol{\theta})$ of Eq. (6.3-3) takes its minimum value on the solution locus of Eq. (6.3-5). Under what conditions does this proof also hold for the true sum-of-squares function, $S(\boldsymbol{\theta})$?

### 6.B Optimal Stepsize for Finite-Difference Sensitivities

In the forward-difference approximation

$$\left. \frac{\partial F_u}{\partial \theta_r} \right|_{\boldsymbol{\theta}^k} \approx \frac{F(\boldsymbol{\theta}^k + h_r \boldsymbol{I}_r) - F(\boldsymbol{\theta}^k)}{(\theta_r^k + h_r) - \theta^k} \tag{6.B-1}$$

for the first-order parametric sensitivities $X_{ur}$, it is convenient to use the same set of steplengths $h_r$ for $u = 1, \ldots, n$. Here $\boldsymbol{I}_r$ is column $r$ of the unit matrix of order $p$. As shown by Bard (1974), approximately optimal steplengths for the next iteration can be computed by use of the current expansion of the objective function. One writes

$$\widetilde{S}(\boldsymbol{\theta} + \boldsymbol{I}_r h_r) = S(\boldsymbol{\theta}^k) + \left[\frac{\partial S}{\partial \theta_r}\Big|_{\boldsymbol{\theta}^k}\right] h_r + \frac{1}{2!}\left[\frac{\partial^2 S}{\partial \theta_r^2}\Big|_{\boldsymbol{\theta}^k}\right] h_r^2$$

$$= S(\boldsymbol{\theta}^k) - 2A_{rL}h_r + A_{rr}h_r^2 \qquad (6.\text{B-}2)$$

and chooses $h_r$ by minimizing the sum of the truncation error $\epsilon_T$ and rounding error $\epsilon_R$ in the difference quotient

$$\frac{S(\boldsymbol{\theta}^k + \boldsymbol{I}_r H_r) - S(\boldsymbol{\theta}^k)}{H_r} \approx \frac{\partial S}{\partial \theta_r}\Big|_{\boldsymbol{\theta}^k} \qquad (6.\text{B-}3)$$

Here $H_r$ is the machine representation of the denominator in Eq. (6.B-1), and is maintained nonzero by proper choice of $h_r$.

(a) The truncation error $\epsilon_T$ is estimated as the contribution of the second-derivative term in Eq. (6.B-2) to the difference quotient in Eq. (6.B-3). Show that this gives

$$\epsilon_T \approx A_{rr}H_r \qquad (6.\text{B-}4)$$

(b) With $H_r$ defined as above, the rounding error in the difference quotient is dominated by the subtraction in the numerator of Eq. (6.B-3). GREGPLUS uses the estimate $\epsilon_M S(\boldsymbol{\theta}^k)$ for the rounding error of each numerator term, thereby obtaining the root-mean-square estimate

$$\epsilon_R \approx \sqrt{2}\epsilon_M |S(\boldsymbol{\theta}^k)/H_r| \qquad (6.\text{B-}5)$$

for the rounding error of the quotient in Eq. (6.B-3).

(c) Show that the minimum of $(\epsilon_T + \epsilon_R)$ with respect to $H_r$ occurs when $\epsilon_T = \epsilon_R$ if $H_r$ is approximated as a continuous variable. From this result, obtain the step size expression

$$H_{r,\text{opt}} \approx \sqrt{\sqrt{2}\epsilon_M |S(\boldsymbol{\theta}^k)|/|A_{rr}|} \qquad \text{for } |A_{rr}| > 0 \qquad (6.\text{B-}6)$$

This result is used by GREGPLUS for those parameters that are marked with nonzero DEL(r) values to request divided-difference computations of parametric sensitivities. The same expression is used in the multiresponse levels of GREGPLUS, where negative values of $S(\boldsymbol{\theta})$ can occur.

## 6.C Full Newton Version of Normal Equations

The submatrices $\boldsymbol{A}_{\theta\theta}$ and $\boldsymbol{A}_{\theta L}$ in Eq. (6.3-7) can be computed in full Newton form by numerical differencing of $S(\boldsymbol{\theta})$. For these calculations, $S$ is represented as a quadratic expansion

$$\widetilde{S} = S(\boldsymbol{\theta}^k) - 2\sum_{r=1}^{\text{NPAR}} A_{rL}(\theta_r - \theta_r^k) + \sum_{r=1}^{\text{NPAR}}\sum_{s=r}^{p} A_{rs}(\theta_r - \theta_r^k)(\theta_s - \theta_s^k) \quad (6.\text{C-}1)$$

(a) Give expressions for the matrix elements $A_{rr}$ and $A_{rL}$ in terms of partial derivatives of $S$.

(b) Give an expression for the minimum number of evaluations of $S$ required to compute all the coefficients in Eq. (6.C-1) for a $p$-parameter model. For this calculation, assume that each partial derivative is approximated at $\boldsymbol{\theta}^k$ by a corresponding finite-difference quotient.

(c) Test your result from (b) against the number of function calls reported by GREGPLUS when solving the first example of Appendix C, with IDIF reset to 2. Note that each line search point takes one function call, and that second-order differencing is not used until the minimum $S$ is nearly reached.

## 6.D Decomposition of Benzazide

Newman, Lee and Garett (1947) reported the following data on isothermal decomposition of benzazide in dioxane solution at 75°C:

| $t$ | 0.20 | 0.57 | 0.92 | 1.22 | 1.55 | 1.90 |
|---|---|---|---|---|---|---|
| $V$ | 12.62 | 30.72 | 44.59 | 52.82 | 60.66 | 68.20 |

| $t$ | 2.25 | 2.63 | 3.05 | 3.60 | 4.77 | 5.85 | $\infty$ |
|---|---|---|---|---|---|---|---|
| $V$ | 73.86 | 78.59 | 82.02 | 86.29 | 91.30 | 93.59 | 95.20 |

Here $t$ is the time of reaction in hours and $V$ is the volume of $N_2$ evolved, expressed as ml at 0°C and 1 atm.

(a) Estimate the parameters $\theta_1 = V_\infty$ and $\theta_2 = k$ in the model

$$V = V_\infty[1 - \exp(-kt)] \qquad (6.\text{D-1})$$

using GREGPLUS at LEVEL $= 10$, with $\boldsymbol{\theta}^0 = [100 \quad 1.0]^T$ as the starting guess and with the default settings from GRPREP. Assume the $t$ values to be error-free and assume equal variances for all the observations of $V$. Include test statements in your subroutine MODEL to detect unusable argument values and return control to the calling program in such cases with IERR $= 1$. Give 95% HPD interval estimates of the form $a \pm b$ for $\theta_1$ and $\theta_2$.

(b) Do corresponding calculations with analytic derivatives from MODEL and compare the results.

(c) Repeat (b) with other starting guesses $\boldsymbol{\theta}^0$ to test the convergence and the possibility of other modal solutions.

(d) Plot the residuals of your best solution versus $V$ and comment on any interesting features.

## 6.E Posterior Densities for Parameter Subsets

We wish to derive from Eq. (6.6-1) the marginal posterior density

$$p(\boldsymbol{\theta}_a|\boldsymbol{Y}) \propto \int_\sigma \int_{\boldsymbol{\theta}_b} p(\boldsymbol{\theta}_e, \sigma|\boldsymbol{Y})\, d\boldsymbol{\theta}_b d\sigma \qquad (6.\text{E-1})$$

for a subset $\boldsymbol{\theta}_a$ of the parameter set $\boldsymbol{\theta}_e$. Here $\boldsymbol{\theta}_b$ consists of the $n_b = n_e - n_a$ estimated parameters not counted in $\boldsymbol{\theta}_a$, and the integrals span the full ranges of $\sigma$ and $\boldsymbol{\theta}_b$.

(a) Consider the expansion

$$
\tilde{S}(\boldsymbol{\theta}) = \hat{S} + \begin{pmatrix} \boldsymbol{x}_a \\ \vdots \\ \boldsymbol{x}_b \end{pmatrix}^T \begin{bmatrix} \boldsymbol{A}_{aa} & \vdots & \boldsymbol{A}_{ab} \\ \cdots\cdots\cdots\cdots \\ \boldsymbol{A}_{ba} & \vdots & \boldsymbol{A}_{bb} \end{bmatrix} \begin{pmatrix} \boldsymbol{x}_a \\ \vdots \\ \boldsymbol{x}_b \end{pmatrix}
\tag{6.E-2}
$$

obtained by partitioning Eq. (6.4-3), with $\boldsymbol{x}_a = (\boldsymbol{\theta}_a - \hat{\boldsymbol{\theta}}_a)$ and $\boldsymbol{x}_b = (\boldsymbol{\theta}_b - \hat{\boldsymbol{\theta}}_b)$. Show that the following expansion is equivalent

$$
\begin{aligned}
\tilde{S}(\boldsymbol{\theta}) = \hat{S} &+ \boldsymbol{x}_a^T (\boldsymbol{A}_{aa} - \boldsymbol{\alpha}^T \boldsymbol{A}_{bb} \boldsymbol{\alpha}) \boldsymbol{x}_a \\
&+ (\boldsymbol{x}_a^T \boldsymbol{\alpha}^T + \boldsymbol{x}_b^T) \boldsymbol{A}_{bb} (\boldsymbol{\alpha} \boldsymbol{x}_a + \boldsymbol{x}_b)
\end{aligned}
\tag{6.E-3}
$$

and determine the required expression for the $n_b \times n_a$ matrix $\boldsymbol{\alpha}$.

(b) Using the results of part (a) and Example A.1, rewrite Eq. (6.E-3) as

$$
\tilde{S}(\boldsymbol{\theta}) = \hat{S} + \boldsymbol{x}_a^T (\boldsymbol{A}^{aa})^{-1} \boldsymbol{x}_a + \boldsymbol{z}_b^T \boldsymbol{A}_{bb} \boldsymbol{z}_b
\tag{6.E-4}
$$

in which $\boldsymbol{z}_b = \boldsymbol{\alpha} \boldsymbol{x}_a + \boldsymbol{x}_b$.

(c) Combine Eqs. (6.6-1), (6.E-1), and (6.E-4) to obtain

$$
p(\boldsymbol{\theta}_a | \boldsymbol{Y}) \propto \int_\sigma \sigma^{-n-1} \exp\left[ -\frac{\hat{S} + \boldsymbol{x}_a^T (\boldsymbol{A}^{aa})^{-1} \boldsymbol{x}_a}{2\sigma^2} \right] \int_{\boldsymbol{z}_b} \exp\left[ -\frac{\boldsymbol{z}_b^T \boldsymbol{A}_{bb} \boldsymbol{z}_b}{2\sigma^2} \right] d\boldsymbol{z}_b d\sigma
\tag{6.E-5}
$$

(d) Evaluate the integral over $\boldsymbol{z}_b$ by use of the normalization condition

$$
\frac{1}{(\sqrt{2\pi}\sigma)^{n_e - n_a}} |\boldsymbol{A}_{bb}|^{1/2} \int_{\boldsymbol{z}_b} \exp\left[ -\frac{\boldsymbol{z}_b^T \boldsymbol{A}_{bb} \boldsymbol{z}_b}{2\sigma^2} \right] d\boldsymbol{z}_b = 1
\tag{6.E-6}
$$

which adjusts the density function to unit total probability content. Thus, reduce Eq. (6.E-5) to

$$
p(\boldsymbol{\theta}_a | \boldsymbol{Y}) \propto \int_0^\infty \sigma^{(n_e - n_a - n - 1)} \exp\left[ -\frac{\hat{S} + \boldsymbol{x}_a^T (\boldsymbol{A}^{aa})^{-1} \boldsymbol{x}_a}{2\sigma^2} \right] d\sigma
\tag{6.E-7}
$$

(e) Treat Eq. (6.E-7) by the method in Eq. (6.6-3) to obtain Eq. (6.6-5).

## 6.F Calibration of a Thermistor

Meyer and Roth (1972) give the following data on a thermistor's resistance, R, as a function of temperature $t$:

| $R_u$, $\Omega$ | 34780 | 28610 | 23650 | 19630 | 16370 | 13720 | 11540 | 9744 |
|---|---|---|---|---|---|---|---|---|
| $t_u$, C | 50 | 55 | 60 | 65 | 70 | 75 | 80 | 85 |

| $R_u$, $\Omega$ | 8261 | 7030 | 6005 | 5147 | 4427 | 3820 | 3307 | 2872 |
|---|---|---|---|---|---|---|---|---|
| $t_u$, C | 90 | 95 | 100 | 105 | 110 | 115 | 120 | 125 |

The following expectation model is proposed to represent $\ln R$

$$f = \theta_1 + [\theta_2/(T + \theta_3)] \qquad \text{(6.F-1)}$$

as a function of the absolute temperature T in degrees K.

(a) Estimate the parameters with GREGPLUS at Level = 10, starting from $\theta^0 = (0.0, 1.0D4, 0.0)^T$ and CHMAX(i) = −0.1. Use uniform weighting of the observations expressed as $y_u = \ln R_u$. Report the normalized covariance matrix for the parameter estimates, the number of iterations required, and the 95% HPD interval for each parameter.

(b) Plot the residuals versus $t$. Comment on any interesting features, including the suitability of the weighting.

## 6.G Reparameterization of Thermistor Calibration

Consider the following reparameterized version of Eq. (6.F-1):

$$f = \theta_1 + \theta_2 \left[ \frac{1}{T + \theta_3} - \frac{1}{400 + \theta_3} \right] \qquad \text{(6.G-1)}$$

(a) Do calculations corresponding to those of Problem 6.F, using this model in place of Eq. (6.F-1). A good starting guess for the new $\theta_1$ can be found from the data at $T \approx 400$ K.

(b) Compare the covariance matrix, the test divisors $D_i$, and the rate of progress toward the minimum with the corresponding results of Problem 6.F. Discuss.

## 6.H Nuclear Spin Relaxation in Block Copolymers

Okamoto, Cooper, and Root (1992) derived the following expectation model of coupled spin relaxation of $^1$H and $^{13}$C nuclei for their nuclear magnetic resonance experiments with block copolymers:

$$\frac{I_z}{I_{eq}} = 1 - \exp\left[-\frac{\tau + t}{T_{1H}}\right] \qquad \text{(6.H-1)}$$

$$\frac{S_z}{S_{eq}} = 1 - \exp\left[-\frac{t}{T_{1C}}\right] \qquad \text{(6.H-2)}$$

$$+ 2.0\varepsilon \frac{T_{1H}}{T_{1H} - T_{1C}} \exp\left[-\frac{\tau}{T_{1H}}\right] \left\{ \exp\left[-\frac{t}{T_{1C}}\right] - \exp\left[-\frac{t}{T_{1H}}\right] \right\}$$

Here $I_z$ and $S_z$ are the axial magnetizations of $^1$H and $^{13}$C nuclei, respectively. The theory requires $0 \le \varepsilon \le 1$. The parameters to be estimated are $\theta_1 = S_{eq}$, $\theta_2 = T_{1C}$, $\theta_3 = T_{1H}$, and $\theta_4 = \varepsilon$. The following observations of $S_z$ as a function of $t$ and $\tau$ were obtained for a particular sample:

| | $\tau = 0$ | 0.3 | 0.5 | 1.0 | 3.0 | 5.0 |
|---|---|---|---|---|---|---|
| $t = 0.3$ | $S_{z,obs} = 6.8$ | 5 | 4.4 | 3.5 | 3.4 | 3.4 |
| 0.5 | 7.1 | 5.9 | 5.3 | 4.7 | 4.25 | 4.25 |
| 1.0 | 6.3 | 5.8 | 5.4 | 5.3 | 4.9 | 4.9 |
| 2.0 | 5.05 | 5.05 | 5.05 | 5.05 | 5.05 | 5.05 |

(a) Analyze the data at $\tau = 5.0$ according to the large-$\tau$ asymptotic form of Eq. (6.H-2) to get approximate values of $S_{eq}$ and $T_{1C}$.

(b) Analyze the data at $t = 0.3$ according to Eq. (6.H-2) and the results of part (a) to get approximate values of $T_{1H}$ and $\varepsilon$.

(c) Estimate the four parameters from all 24 observations using GREG-PLUS with LEVEL $= 10$ and weights $w_u = 1$, starting from the initial parameter values obtained in (a) and (b). Report the 95% HPD interval and probability $\Pr(\theta_i > 0|Y)$ for each parameter and the standard deviation of the weighted observations.

(d) Plot the residuals versus $t$ at each reported value of $\tau$. Comment on any interesting features, including the suitability of the weighting.

## 6.I Pulse Chromatography

The expectation model

$$f = \theta_1 \frac{1}{\sqrt{2\pi H u v t}} \exp\left[-\frac{(t - L/(uv))^2}{2Ht/(uv)}\right] + \theta_4 + \theta_5[t - L/(uv)] \qquad (6.\text{I-1})$$

is to be tested against a packed-tube chromatography experiment of Atha-lye, Gibbs, and Lightfoot (1992). Here $f$ is the expected solute analyzer reading at the column exit, $\theta_1$ is a corresponding measure of the solute pulse mass per unit fluid cross-section, $H$ is the height per theoretical plate, $u$ is the solute partition coefficient, $v$ is the interstitial fluid velocity (superficial velocity divided by bed porosity), and $L$ is the column length. The last two terms of the model function give a linear representation of the analyzer background reading. Some of the analyzer data $y(t)$ are given here for an experiment with a pulse of bovine hemoglobin in a packed tube of 2.5 cm ID and 11.5 cm length, with bed porosity $\epsilon_b = 0.36$ and a flow rate of 0.01 ml/s:

| $t$, s | 2971 | 3041 | 3111 | 3181 | 3251 | 3321 | 3391 | 3461 |
|---|---|---|---|---|---|---|---|---|
| $y$ | .0025 | .0021 | .0028 | .0042 | .0071 | .0103 | .0162 | .0300 |

| $t$, s | 3531 | 3601 | 3671 | 3741 | 3811 | 3881 | 3951 | 4021 |
|---|---|---|---|---|---|---|---|---|
| $y$ | .0509 | .0861 | .1240 | .1604 | .1769 | .1706 | .1458 | .1116 |

| $t$, s | 4091 | 4161 | 4231 | 4301 | 4371 | 4441 | 4511 | 4581 |
|---|---|---|---|---|---|---|---|---|
| $y$ | .0755 | .0466 | .0292 | .0164 | .0092 | .0047 | .0015 | .0013 |

From these data, the parameters $\theta_1$, $\theta_2 = u$, $\theta_3 = H$, $\theta_4$ and $\theta_5$ of Eq. (6.I-1) are to be estimated. Proceed as follows:

(a) The exponential function in Eq. (6.I-1) has its mode at the time $t_{max} = L/(uv)$. Apply this relation to the data to obtain a starting value of $\theta_2$.

(b) The time interval between inflection points of the function $f(t)$ corresponds roughly to $2\sqrt{Ht_{max}^2/L}$, estimated by approximating the exponential term as a normal error density with a variance calculated at $t = t_{max}$. Apply this relation and the result of (a) to obtain a starting value of $\theta_3$.

(*c*) Find a starting value of $\theta_1$ from the result of (*b*) and the data at the largest $y$ value.
(*d*) Estimate the 95% HPD intervals for the five parameters, using GREG-PLUS with LEVEL = 10 and starting with $\theta_4$ and $\theta_5$ equal to zero.
(*e*) Plot the residuals versus $t$. Comment on any interesting features, including the suitability of the weighting.

**6.J Rosenbrock's Function**
The quartic function

$$S(\boldsymbol{\theta}) = 100(\theta_2 - \theta_1^2)^2 + (1 - \theta_1)^2 \qquad (6.\text{J-}1)$$

given by Rosenbrock (1960) is used here as a test of Subroutine GREG-PLUS. The minimization of $S(\boldsymbol{\theta})$ is equivalent to a least-squares calculation with "data" $y_1 = y_2 = 0$, weights $w_1 = w_2 = 1$, and expectation functions

$$f_1 = \theta_2 - \theta_1^2; \qquad f_2 = 1 - \theta_1 \qquad (6.\text{J-}2)$$

(*a*) Minimize $S(\boldsymbol{\theta})$ with GREGPLUS, using LEVEL = 10 and $\boldsymbol{\theta}^{0T} = (-1.2, 1.0)$. Use LISTS = 3 to obtain the line search points.
(*b*) Graph the points $(\theta_1, \theta_2)$ visited by GREGPLUS. On the same diagram, show the point of least $S$ at each $\theta_1$.

**REFERENCES and FURTHER READING**

Agarwal, A. K., and M. L. Brisk, Sequential experimental design for precise parameter estimation. 1. Use of reparametrization. 2. Design criteria, *Ind. Eng. Chem Process Des. Dev.*, **24**, 203–207, 207–210 (1985).
Anderson, D. F., Significance tests based on residuals, *Biometrika*, **58**, 139–148 (1971).
Anderson, E., Z. Bai, C. Bischof, J. Demmel, J. J. Dongarra, J. Du Croz, A. Greenbaum, S. Hammarling, A. McKenney, S. Ostrouchov, and D. Sorensen, *LAPACK Users' Guide*, SIAM, Philadelphia (1992).
Anderson, T. F., D. S. Abrams, and E. A. Grens II, Evaluation of parameters for nonlinear thermodynamic models, *AIChE J.*, **24**, 20–29 (1978).
Arrhenius, S., Ueber die Reaktionsgeschwindigkeit der Inversion von Rohrzucker durch Saeuren, *Z. Physik. Chem.*, **4**, 226–248 (1889).
Athalye, A. M., S. J. Gibbs, and E. N. Lightfoot, Predictability of chromatographic protein separations, *J. Chromatogr.*, **589**, 71–85 (1992).
Atkinson, A. C., and V. V. Fedorov, The design of experiments for discrimination between rival models, *Biometrika*, **62**, 57–70, (1975).
Bajramovic, R., and P. M. Reilly, An explanation for an apparent failure of the Box-Hill procedure for model discrimination, *Can. J. Chem. Eng.*, **55**, 82–86 (1971).
Bard, Y., *Nonlinear Parameter Estimation*, Academic Press, New York (1974).
Bard, Y., and L. Lapidus, Kinetics analysis by digital parameter estimation, *Catal. Rev.*, **2**, 67–112 (1968).

Bates, D. M., and D. G. Watts, *Nonlinear Regression Analysis*, Wiley, New York (1988).

Blakemore, J. W., and A. W. Hoerl, Fitting non-linear reaction rate equations to data, *Chem. Eng. Prog. Symp. Ser.*, **59**, 14–27 (1963).

Bock, H. G., Numerical treatment of inverse problems in chemical reaction kinetics, in *Modelling of Chemical Reaction Systems*, K. H. Ebert, P. Deuflhard, and W. Jäger, eds., Springer, New York (1981), 102–125.

Booth, G. W., G. E. P. Box, M. E. Muller, and T. I. Peterson, Forecasting by generalized regression methods, non-linear estimation, *IBM Share Program Package* **687**, International Business Machines Corp., New York (1958).

Box, G. E. P., Fitting empirical data, *Ann. N.Y. Acad. Sci.*, **86**, 792–816 (1960).

Box, G. E. P., Sampling and Bayes' inference in scientific modelling and robustness (with Discussion), *J. Roy. Statist. Soc. A*, **143**, 383–430 (1980).

Box, G. E. P., and G. A. Coutie, Application of digital computers in the exploration of functional relationships, *Proc. Institution Elec. Engineers*, **103**, Part B, supplement No. 1, 100–107 (1956).

Box, G. E. P., and N. R. Draper, *Empirical Model-Building and Response Surfaces*, Wiley, New York (1987).

Box, G. E. P., and T. L. Henson, Model Fitting and Discrimination, *Dept. of Statistics Tech. Rep.* **211**, Univ. of Wisconsin–Madison (1969).

Box, G. E. P., and T. L. Henson, Some aspects of mathematical modeling in chemical engineering, *Proc. Inaugural Conf. of the Scientific Computation Centre and the Institute of Statistical Studies and Research*, Cairo Univ. Press, Cairo (1970), 548.

Box, G. E. P., and W. J. Hill, Discrimination among mechanistic models, *Technometrics*, **9**, 57–71 (1967).

Box, G. E. P., and W. G. Hunter, Sequential design of experiments for nonlinear models, in *IBM Scientific Computing Symposium in Statistics*, October 21–23, 1963, 113–137 (1965).

Box, G. E. P., W. G. Hunter, and J. S. Hunter, *Statistics for Experimenters*, Wiley, New York (1978, 2004).

Box, G. E. P., and H. Kanemasu, Topics in model building. Part II. On linear least squares, University of Wisconsin, Statistics Department Technical Report No. 321, (1972).

Box, G. E. P., and H. Kanemasu, Constrained nonlinear least squares, in *Contributions to Experimental Design, Linear Models, and Genetic Statistics: Essays in Honor of Oscar Kempthorne*, K. Hinklemann, ed., Marcel Dekker, New York (1984), 297–318.

Box, G. E. P., and H. L. Lucas, Design of experiments in non-linear situations, *Biometrika*, **46**, 77–90 (1959).

Box, G. E. P., and G. C. Tiao, *Bayesian Inference in Statistical Analysis*, Addison-Wesley, Reading, MA (1973). Reprinted by Wiley, New York (1992).

Box, M. J., An experimental design criterion for precise estimation of a subset of the parameters in a nonlinear model, *Biometrika*, **58**, 149–153 (1971).

Bradshaw, R. W., and B. Davidson, A new approach to the analysis of heterogeneous reaction rate data, *Chem. Eng. Sci.*, **24**, 1519–1527 (1970).

Burke, A. L., T. A. Duever, and A. Penlidis, Model discrimination via designed experiments: Discriminating between the terminal and penultimate models on the basis of composition data, *Macromolecules*, **27**, 386–399 (1994).

Buzzi-Ferraris, G., and P. Forzatti, A new sequential experimental design procedure for discrimination among models, *Chem. Eng. Sci.*, **38**, 225–232 (1983).

Buzzi-Ferraris, G., P. Forzatti, and P. Canu, An improved version of a sequential design criterion of discriminating among rival response models, *Chem. Eng. Sci.*, **48**, 477–481 (1990).

Caracotsios, M., *Model Parametric Sensitivity Analysis and Nonlinear Parameter Estimation: Theory and Applications*, Ph. D. thesis, University of Wisconsin–Madison (1986).

Chambers, J. M., W. S. Cleveland, B. Kleiner, and P. A. Tukey, *Graphical Methods for Data Analysis*, Wadsworth, Belmont, CA (1983).

Clarke, G. P. Y., Approximate confidence limits for a parameter function in nonlinear regression, *J. Am. Statist. Assoc.*, **82**, 221-230 (1987).

Daniel, C., *Applications of Statistics to Industrial Experimentation*, Wiley, New York (1976).

Daniel, C., and F. S. Wood, *Fitting Equations to Data*, 2nd edition, Wiley, New York (1980).

Deming, W. E., *Statistical Adjustment of Data*, Wiley, New York (1943).

Dennis, J. E., Jr., D. M. Gay, and R. E. Welsch, An adaptive nonlinear least-squares algorithm, *ACM Trans. Math. Software*, **7**, 343–368 (1981).

Donaldson, J. R., and R. B. Schnabel, Computational experiences with confidence regions and confidence intervals for nonlinear least squares, *Technometrics*, **29**, 67–82 (1987).

Dongarra, J. J., J. R. Bunch, C. B. Moler, and G. W. Stewart, *LINPACK Users' Guide*, SIAM, Philadelphia (1979).

Draper, N. R., and H. Smith, *Applied Regression Analysis*, 2nd edition, Wiley, New York (1981).

Dumez, F. J., and G. F. Froment, Dehydrogenation of 1-butene into butadiene. Kinetics, catalyst coking, and reactor design, *Ind. Eng. Chem. Proc. Des. Def.*, **15**, 291–301 (1976).

Efroymson, M. A., Multiple regression analysis, in *Mathematical methods for digital computers*, Wiley, New York, (1960).

Espie, D. M., and S. Machietto, Nonlinear transformations for parameter estimation, *Ind. Eng. Chem. Res.*, **27**, 2175–2179 (1988).

Fariss, R. H., and V. J. Law, An efficient computational technique for generalized application of maximum likelihood to improve correlation of experimental data, *Comput. Chem. Eng.*, **3**, 95–104 (1979).

Atkinson, A. C., and V. V. Fedorov, The design of experiments for discrimination between rival models, *Biometrika*, **62**, 57–70, (1975).

Feng, C. F., V. V. Kostrov, and W. E. Stewart, Multicomponent diffusion of gases in porous solids. Models and experiments, *Ind. Eng. Chem. Fundam.*, **5**, 5–9 (1974).

Fisher, R. A., *Statistical Methods for Research Workers*, First edition, Oliver and Boyd, Edinburgh (1925).

Frane, J. W., A note on checking tolerance in matrix inversion and regression, *Technometrics*, **19**, 513–514 (1977).

Froment, G. F., and L. H. Hosten, Catalytic kinetics: modelling, in *Catalysis Science and Technology*, **2**, J. R. Anderson and M. Boudart, eds., Springer, New York, (1984), 97–170.

Froment, G. F., and R. Mezaki, Sequential discrimination and estimation procedures for rate modelling in heterogeneous catalysis, *Chem. Eng. Sci.*, **25**, 293–301 (1970).

Gauss, C. F., Letter to the Editor, Vermischte Nachrichten no. 3, *Allgemeine Geographische Ephemeridenz* **4**, 378 (1799).

Gauss, C. F., *Theoria motus corporum coelestium in sectionibus conicis solem ambientium*, Perthas et Besser, Hamburg (1809); *Werke*, **7**, 240–254. Translated by C. H. Davis as *Theory of Motion of the Heavenly Bodies Moving about the Sun in Conic Sections*. Little, Brown, Boston (1857); Dover, New York (1963).

Gauss, C. F., *Theoria combinationis observationum erroribus minimis obnoxiae: Pars prior; Pars posterior; Supplementum. Commentatines societatis regiae scientarium Gottingensis recentiores*, **5** (1823); **6** (1828). Translated by G. W. Stewart as *Theory of the Combination of Observations Least Subject to Errors, Part One, Part Two, and Supplement*, by the Society for Industrial and Applied Mathematics, Philadelphia (1995). A valuable Afterword by the translator is included.

Gossett, W. S. ("Student"), The probable error of a mean, *Biometrika*, **6**, 1-25 (1908).

Guttman, I., S. S. Wilkes, and J. S. Hunter, *Introductory Engineering Statistics*, 3rd edition, Wiley, New York (1982).

Hartley, H. O., The modified Gauss-Newton method for the fitting of non-linear regression functions by least squares, *Technometrics*, **3**, 269–280 (1961).

Hartley, H. O., Exact confidence regions for the parameters in non-linear regression laws, *Biometrika*, **51**, 347–353 (1964).

Hartley, H. O., and A. Booker, Non-linear least squares estimation, *Ann. Math. Statist.*, **36**, 638–650 (1965).

Hertzberg, T., and O. A. Asbjornsen, Parameter estimation in nonlinear differential equations, in *Computer Applications in the Analysis of Data and Plants*, Science Press, Princeton NJ (1977).

Hill, W. J., W. G. Hunter, and D. W. Wichern, A joint design criterion for the dual problem of model discrimination and parameter estimation, *Technometrics*, **10**, 145–160 (1968).

Hill, W. J., and W. G. Hunter, A note on designs for model discrimination: variance unknown case, *Technometrics*, **11**, 396–400 (1969).

Hill, W. J., and W. G. Hunter, Design of experiments for subsets of the parameters, *Technometrics*, **16**, 425–434 (1974).

Himmelblau, D. M., *Process Analysis by Statistical Methods*, Wiley, New York (1970).

Himmelblau, D. M., C. R. Jones, and K. B. Bischoff, Determination of rate constants for complex kinetics models, *Ind. Eng. Chem. Fundam.*, **6**, 539–543 (1967).

Hosten, L. H., A sequential experimental procedure for precise parameter estimation based on the shape of the joint confidence region, *Chem. Eng. Sci.*, **29**, 2247–2252 (1974).

Hsiang, T., and P. M. Reilly, A practical method for discriminating among mechanistic models, *Can. J. Chem. Eng.*, **48**, 865–871 (1971).

Huber, P. J., *Robust Statistics*, Wiley, New York (1981).

Hunter, W. G., W. J. Hill, and T. L. Henson, Design of experiments for precise estimation of all or some of the constants in a mechanistic model, *Can. J. Chem. Eng.*, **47**, 76–80 (1969).

Hunter, W. G., and R. Mezaki, A model-building technique for chemical engineering kinetics, *AIChE J.*, **10**, 315–322 (1964).

Hunter, W. G., and A. M. Reiner, Design for discriminating between two rival models, *Technometrics*, **7**, 307–323 (1965).

Kalogerakis, N., and R. Luus, Sequential experimental design of dynamic systems through the use of information index, *Can. J. Chem. Eng.*, **62**, 730–737 (1984).

Kim, I.-W., T. F. Edgar, and N. H. Bell, Parameter estimation for a laboratory water-gas-shift reactor using a nonlinear error-in-variables method, *Comput. Chem. Eng.*, **15**, 361–367 (1991).

Kittrell, J. R., Mathematical modelling of chemical reactions, *Adv. in Chem. Eng.*, **8**, 97–183, Academic Press, New York (1970).

Kittrell, J. R., and J. Erjavec, Response surface methods in heterogeneous kinetic modelling, *Ind. Eng. Chem.*, *Process Des. Devel.*, **7**, 321–327 (1968).

Kittrell, J. R., W. G. Hunter, and C. C. Watson, Obtaining precise parameter estimates for nonlinear catalytic rate models, *AIChE J.*, **12**, 5–10 (1966).

Krug, R. R., W. G. Hunter, and R. A. Grieger, Statistical interpretation of enthalpy-entropy compensation, *Nature*, **261**, 566–567 (1976).

Lawson, C. L., and R. J. Hanson, *Solving Least Squares Problems*, Prentice-Hall, Englewood Cliffs, NJ (1974); 2nd edition published by SIAM, Philadelphia (1995).

Legendre, A. M., *Nouvelles Méthodes pour la Determination des Orbites de Comètes*, Paris (1805).

Levenberg, K., A method for the solution of certain non-linear problems in least squares, *Quart. Appl. Math.*, **2**, 164–168 (1944).

Lumpkin, R. E., W. D. Smith, Jr., and J. M. Douglas, Importance of the structure of the kinetic model for catalytic reactions, *Ind. Eng. Chem. Fundam.*, **8**, 407–411 (1969).

Marquardt, D. W., An algorithm for least-squares estimation of nonlinear parameters, *J. Soc. Ind. Appl. Math.*, **11**, 431–441 (1963).

Meyer, R. R., and P. M. Roth, Modified damped least squares: an algorithm for non-linear estimation, *J. Inst. Maths. Applics.* **9**, 218–233 (1972).

Mezaki., R, N. R. Draper, and R. A. Johnson, On the violation of assumptions in nonlinear least squares by interchange of response and predictor variables, *Ind. Eng. Chem. Fundam.*, **12**, 251–253 (1973).

Nelder, J. A., and R. Mead, A simplex method for function minimization, *Computer J.*, **7**, 308–313 (1965).

Newman, M. S., S. H. Lee, and A. B. Garrett, *J. Am. Chem. Soc.*, **69**, 1, 113–116 (1947).

Nowak, U., and P. Deuflhard, Numerical identification of selected rate constants in large chemical reaction systems, *Appl. Numer. Anal.*, **1**, 59–75 (1985).

Okamoto, D. T., S. L. Cooper, and T. W. Root, Control of solid-state nuclear Overhauser enhancement in polyurethane block copolymers, *Macromolecules*, **25**, 3301 (1992).

Peterson, T. I., Kinetics and mechanism of naphthalene oxidation by nonlinear estimation, *Chem. Eng. Sci.*, **17**, 203–219 (1962).

Pinchbeck, P. H., The kinetic implications of an empirically fitted yield surface for the vapour-phase oxidation of naphthalene to phthalic anhydride, *Chem. Eng. Sci.*, **6**, 105–111 (1957).

Pinto, J. C., M. W. Lobão, and J. L. Monteiro, Sequential experimental design for parameter estimation: a different approach, *Chem. Eng. Sci.*, **45**, 883–892 (1990).

Pinto, J. C., M. W. Lobão, and J. L. Monteiro, Sequential experimental design for parameter estimation: analysis of relative deviations, *Chem. Eng. Sci.*, **46**, 3129–3138 (1991).

Plackett, R. L., Studies in the history of probability and statistics. XXIX. The discovery of least squares, *Biometrika*, **59**, 239–251 (1972).

Powell, M. J. D., A method for minimizing a sum of squares of non-linear functions without calculating derivatives, *Computer J.*, **7**, 303–307 (1965).

Press, W. H., S. A. Teukolsky, W. T. Vetterling, and B. P. Flannery, *Numerical Recipes in Fortran*, 2nd edition, Cambridge University Press, New York (1992).

Pritchard, D. J., and D. W. Bacon, Statistical assessment of chemical kinetic models, *Chem. Eng. Sci.*, **30**, 567–574 (1975).

Pritchard, D. J., and D. W. Bacon, Prospects for reducing correlations among parameter estimates in kinetic models, *Chem. Eng. Sci.*, **33**, 1539–1543 (1978).

Pritchard, D. J., J. Downie, and D. W. Bacon, Further consideration of heteroscedasticity in fitting kinetic models, *Technometrics*, **19**, 227–236 (1977).

Ralston, M. L., and R. Jennrich, DUD, a derivative-free algorithm for nonlinear least squares, *Technometrics*, **20**, 7–14 (1978).

Ratkowsky, D. A., *Nonlinear Regression Modelling: A Unified Practical Approach*, Marcel Dekker, New York (1983).

Ratkowsky, D. A., A statistically suitable general formulation for modelling catalytic chemical reactions, *Chem. Eng. Sci.*, **40**, 1623–1628 (1985).

Reilly, P. M., Statistical methods in model discrimination, *Can. J. Chem. Eng.*, **48**, 168–173 (1970).

Rippin, D. W. T., Statistical methods for experimental planning in chemical engineering, *Comput. Chem. Eng.*, **12**, 109–116 (1988).

Rosenbrock, H. H., An automatic method for finding the greatest or least value of a function, *Computer J.*, **3**, 175–184 (1960).

Schwedock, M. J., L. C. Windes, and W. H. Ray, Steady state and dynamic modelling of a packed bed reactor for the partial oxidation of methanol to formaldehyde–II. Experimental results compared with model predictions, *Chem. Eng. Commun.*, **78**, 45–71 (1989).

Seinfeld, J. H., Identification of parameters in partial differential equations, *Chem. Eng. Sci.*, **24**, 65–74 (1969).

Seinfeld, J. H., and L. Lapidus, *Mathematical Methods in Chemical Engineering. Volume III. Process Modelling, Estimation and Identification*, Prentice-Hall, Englewood Cliffs, NJ (1974).

Shannon, C. E., A mathematical theory of communication, *Bell System Tech. J.*, **27**, 373–423 and 623–656 (1948).

Sørensen, J. P., *Simulation, Regression and Control of Chemical Reactors by Collocation Techniques*, doctoral thesis, Danmarks tekniske Højskole, Lyngby (1982).

Steinberg, D. M., and W. G. Hunter, Experimental design: review and comment (with Discussion), *Technometrics*, **26**, 71–130 (1984).

Stewart, G. W., *Introduction to Matrix Computations*, Academic Press, New York (1973).

Stewart, G. W., Collinearity and least squares regression (with Discussion), *Stat. Sci.*, **2**, 68–100 (1987).

Stewart, G. W., Afterword, in C. F. Gauss, *Theory of the Combination of Observations Least Subject to Errors*, SIAM, Philadelphia (1995), 205–235.

Stewart, W. E., T. L. Henson, and G. E. P. Box, Model discrimination and criticism with single-response data, *AIChE J.*, **42**, 3055–3062 (1996).

Stewart, W. E., and S. M. Mastenbrook, Graphical analysis of multiple inert gas elimination data, *J. Appl. Physiol.: Respir., Environ. Exerc. Physiol.*, **55**, 32–36 (1983a); Errata, **56**, No. 6 (1984).

Stewart, W. E., and S. M. Mastenbrook, Parametric estimation of ventilation-perfusion ratio distributions, *J. Appl. Physiol.: Respir., Environ. Exerc. Physiol.*, **55**, 37–51 (1983b); Errata, **56**, No. 6 (1984).

Stewart, W. E., and J. P. Sørensen, Sensitivity and regression of multicomponent reactor models, *4th Int. Symp. Chem. React. Eng.*, DECHEMA, Frankfurt, **I**, 12–20 (1976).

Stewart, W. E., J. P. Sørensen, and B. C. Teeter, Pulse-response measurement of thermal properties of small catalyst pellets, *Ind. Eng. Chem. Fundam.*, **17**, 221–224 (1978); Errata, **18**, 438 (1979).

Stigler, S. M., Gauss and the invention of least squares, *Ann. Stat.*, **9**, 465–474 (1981).

Stigler, S. M., *History of Statistics*, Harvard University Press, Cambridge, MA (1986).

St. John, R. C., and N. R. Draper, D-optimality for regression designs: a review, *Technometrics*, **17**, 15–23 (1975).

Tan, H. S., D. W. Bacon, and J. Downie, Sequential statistical design strategy in a kinetic study of propylene oxidation, *Can. J. Chem. Eng.*, **67**, 397–404 (1989).

Tjoa, I.-B., and L. T. Biegler, Simultaneous solution and optimization strategies for parameter estimation of differential-algebraic equation systems, *Ind. Eng. Chem. Res.*, **30**, 376–385 (1991).

Tschernitz, J., S. Bornstein, R. B. Beckmann, and O. A. Hougen, Determination of the kinetics mechanism of a catalytic reaction, *Trans. Am. Inst. Chem. Engrs.*, **42**, 883–903 (1946).

Wang, B-C., and R. Luus, Increasing the size of region of convergence for parameter estimation through the use of shorter data-length, *Int. J. Control*, **31**, 947–972 (1980).

Young, T. C., and W. E. Stewart, Correlation of fractionation tray performance via a cross-flow boundary-layer model, *AIChE J.*, **38**, 592–602 (1992); Errata, **38**, 1302 (1992).

# Chapter 7
# Process Modeling
# with Multiresponse Data

Multiresponse experimentation is important in studies of complex systems and of systems observed by multiple methods. Chemical engineers and chemists use multiresponse experiments to study chemical reactions, mixtures, separation, and mixing processes; similar data structures occur widely in science and engineering. In this chapter we study methods for investigating process models with multiresponse data. Bayes' theorem now yields more general methods than those of Chapter 6, and Jeffreys rule, discussed in Chapter 5, takes increased importance.

The methods of Chapter 6 are not appropriate for multiresponse investigations unless the responses have known relative precisions and independent, unbiased normal distributions of error. These restrictions come from the error model in Eq. (6.1-2). Single-response models were treated under these assumptions by Gauss (1809, 1823) and less completely by Legendre (1805), co-discoverer of the method of least squares. Aitken (1935) generalized weighted least squares to multiple responses with a specified error covariance matrix; his method was extended to nonlinear parameter estimation by Bard and Lapidus (1968) and Bard (1974). However, least squares is not suitable for multiresponse problems unless information is given about the error covariance matrix; we may consider such applications at another time.

Bayes' theorem (Bayes 1763; Box and Tiao 1973, 1992) permits estimation of the error covariance matrix $\Sigma$ from a multiresponse data set, along with the parameter vector $\theta$ of a predictive model. It is also possible, under further assumptions, to shorten the calculations by estimating $\theta$ and $\Sigma$ separately, as we do in the computer package GREGPLUS provided in Athena. We can then analyze the goodness of fit, the precision of estimation of parameters and functions of them, the relative probabilities of alternative models, and the choice of additional experiments to improve a chosen information measure. This chapter summarizes these procedures and their implementation in GREGPLUS; details and examples are given in Appendix C.

Jeffreys (1961) advanced Bayesian theory by giving an unprejudiced prior density $p(\theta, \Sigma)$ for suitably differentiable models. His result, given in Chapter 5 and used below, is fundamental in Bayesian estimation.

Box and Draper (1965) took another major step by deriving a posterior density function $p(\boldsymbol{\theta}|\boldsymbol{Y})$, averaged over $\boldsymbol{\Sigma}$, for estimating a parameter vector $\boldsymbol{\theta}$ from a full matrix $\boldsymbol{Y}$ of multiresponse observations. The errors in the observations were assumed to be normally distributed with an unknown $m \times m$ covariance matrix $\boldsymbol{\Sigma}$. Michael Box and Norman Draper (1972) gave a corresponding function for a data matrix $\boldsymbol{Y}$ of discrete blocks of responses and applied that function to design of multiresponse experiments.

The posterior density function $p(\boldsymbol{\theta}|\boldsymbol{Y})$ found by Box and Draper (1965) is a power of the determinant $|\boldsymbol{v}(\boldsymbol{\theta})|$ whose elements appear in Eq. (7.1-3). These authors used contour plots of $p(\boldsymbol{\theta}|\boldsymbol{Y})$ to estimate highest-posterior-density (HPD) regions for a two-parameter model. This technique is useful for models with just a few parameters — say, not more than three.

Multiparameter, multiresponse models call for digital optimization. Early workers minimized $|\boldsymbol{v}(\boldsymbol{\theta})|$ by search techniques, which were tedious and gave only a point estimate of $\boldsymbol{\theta}$. Newtonlike algorithms for minimization of $|\boldsymbol{v}(\boldsymbol{\theta})|$, and for interval estimation of $\boldsymbol{\theta}$, were given by Stewart and Sørensen (1976, 1981) and by Bates and Watts (1985, 1987). Corresponding algorithms for likelihood-based estimation were developed by Bard and Lapidus (1968) and Bard (1974), extended by Klaus and Rippin (1979) and Steiner, Blau, and Agin (1986).

Several generalizations of the problem considered by Box and Draper (1965) have been analyzed in the literature. The theory has been extended to many other models and data structures, and many interesting applications have appeared. Selected results are reviewed in this chapter, along with our software for further applications.

## 7.1 PROBLEM TYPES

The initial goal of this chapter is to estimate the parameter vector $\boldsymbol{\theta}$ (and explicitly or implicitly the covariance matrix $\boldsymbol{\Sigma}$) in a model

$$Y_{iu} = F_i(\boldsymbol{x}_u, \boldsymbol{\theta}) + \mathcal{E}_{iu} \quad (i = 1, \ldots, m; \; u = 1, \ldots, n) \qquad (7.1\text{-}1)$$

for the elements $Y_{iu} = y_{iu}\sqrt{w_u}$ of a multiresponse weighted data matrix $\boldsymbol{Y}$. Each integer $u$ from 1 to $n$ denotes an independent event in which $m_u \leq m$ responses are observed. Weights $w_u$ are given to express the precision of each event relative to a standard event, just as in Chapter 6. The weighted function $F_i(\boldsymbol{x}_u, \boldsymbol{\theta}) \equiv f_i(\boldsymbol{x}_u, \boldsymbol{\theta})\sqrt{w_u}$ is an expectation model for response $i$ at the experimental point $\boldsymbol{x}_u$ and is assumed to be differentiable with respect to each parameter $\theta_r$. A notation list is given at the end of this chapter.

The weighted errors $\mathcal{E}_{iu} \equiv \varepsilon_{iu}\sqrt{w_u}$ are modeled by an $m$-dimensional normal distribution (see Eq. 4.4-3), with expected values $\mathrm{E}(\mathcal{E}_{iu})$ of zero and unknown covariances $\sigma_{ij} \equiv \mathrm{E}(\mathcal{E}_{iu}\mathcal{E}_{ju})$. Use of this distribution with Eq. (7.1-1) yields a predictive density function $p(\boldsymbol{Y}|\boldsymbol{\theta}, \boldsymbol{\Sigma})$ for observations to be taken at test conditions $\{\boldsymbol{x}_1, \ldots, \boldsymbol{x}_n\}$. This function appears in Eq. (7.1-2).

Once the weighted data $\boldsymbol{Y}$ are ready, the likelihood function $l(\boldsymbol{\theta}, \boldsymbol{\Sigma}|\boldsymbol{Y})$ can be constructed in the manner of Eq. (5.1-6) as the function $p(\boldsymbol{Y}|\boldsymbol{\theta}, \boldsymbol{\Sigma})$, with $\boldsymbol{Y}$ now given whereas $\boldsymbol{\theta}$ and $\boldsymbol{\Sigma}$ are free to vary. This function is given in Eq. (7.1-4) for full data $\boldsymbol{Y}$ and unknown covariance matrix $\boldsymbol{\Sigma}$. Multiplication of this likelihood function by a prior density $p(\boldsymbol{\theta}, \boldsymbol{\Sigma})$ in accordance with Bayes' theorem gives the posterior density function $p(\boldsymbol{\theta}, \boldsymbol{\Sigma}|\boldsymbol{Y})$, which contains all current information on $\boldsymbol{\theta}$ and $\boldsymbol{\Sigma}$. These constructions are summarized below for several problem types.

### Type 1. Full $\boldsymbol{Y}$ and Unknown $\boldsymbol{\Sigma}$

Box and Draper (1965) derived a density function for estimating the parameter vector $\boldsymbol{\theta}$ of a multiresponse model from a full data matrix $\boldsymbol{Y}$, subject to errors normally distributed in the manner of Eq. (4.4-3) with a full unknown covariance matrix $\boldsymbol{\Sigma}$. With this type of data, every event $u$ has a full set of $m$ responses, as illustrated in Table 7.1. The predictive density function for prospective data arrays $\boldsymbol{Y}$ from $n$ independent events, consistent with Eqs. (7.1-1) and (7.1-3), is

$$p(\boldsymbol{Y}|\boldsymbol{\theta}, \boldsymbol{\Sigma}) = \prod_{u=1}^{n} p(\boldsymbol{Y}_u|\boldsymbol{\theta}, \boldsymbol{\Sigma})$$

$$= |2\pi\boldsymbol{\Sigma}|^{-n/2} \exp\left\{ -\frac{1}{2} \sum_{i=1}^{m} \sum_{j=1}^{m} \sigma^{ij} \sum_{u=1}^{n} [Y_{iu} - F_i(\boldsymbol{x}_u, \boldsymbol{\theta})][Y_{ju} - F_j(\boldsymbol{x}_u, \boldsymbol{\theta})] \right\}$$

$$= |2\pi\boldsymbol{\Sigma}|^{-n/2} \exp\left\{ -\frac{1}{2} \sum_{i=1}^{m} \sum_{j=1}^{m} \sigma^{ij} v_{ji}(\boldsymbol{\theta}) \right\}$$

$$= |2\pi\boldsymbol{\Sigma}|^{-n/2} \exp\left\{ -\frac{1}{2} \sum_{i=1}^{m} [\boldsymbol{\Sigma}^{-1}v(\boldsymbol{\theta})]_{ii} \right\}$$

$$= |2\pi\boldsymbol{\Sigma}|^{-n/2} \exp\left\{ -\frac{1}{2} \operatorname{tr}\left[ \boldsymbol{\Sigma}^{-1}v(\boldsymbol{\theta}) \right] \right\} \tag{7.1-2}$$

The matrices $\boldsymbol{\Sigma}^{-1}$ and $v(\boldsymbol{\theta})$ are symmetric, with elements $\sigma^{ij}$ and

$$v_{ij}(\boldsymbol{\theta}) = \sum_{u=1}^{n} [Y_{iu} - F_i(\boldsymbol{x}_u, \boldsymbol{\theta})][Y_{ju} - F_j(\boldsymbol{x}_u, \boldsymbol{\theta})] \tag{7.1-3}$$

respectively. Here tr denotes the trace of a matrix: $\operatorname{tr} \boldsymbol{A} = \sum_i a_{ii}$.

When full data $\boldsymbol{Y}$ are available, one interprets the result of Eq. (7.1-2) as a likelihood function

$$l(\boldsymbol{\theta}, \boldsymbol{\Sigma}|\boldsymbol{Y}) \propto |\boldsymbol{\Sigma}|^{-n/2} \exp\left\{ -\tfrac{1}{2}\operatorname{tr}\left[ \boldsymbol{\Sigma}^{-1}v(\boldsymbol{\theta}) \right] \right\} \tag{7.1-4}$$

|        | $Y_{1u}$ | $Y_{2u}$ | $Y_{3u}$ | $Y_{4u}$ |
|--------|----------|----------|----------|----------|
| $u=1$  | +        | +        | +        | +        |
| $u=2$  | +        | +        | +        | +        |
| $u=3$  | +        | +        | +        | +        |
| $u=4$  | +        | +        | +        | +        |
| $u=5$  | +        | +        | +        | +        |
| $u=6$  | +        | +        | +        | +        |
| $u=7$  | +        | +        | +        | +        |
| $u=8$  | +        | +        | +        | +        |

|             | $\sigma_{j1}$ | $\sigma_{j2}$ | $\sigma_{j3}$ | $\sigma_{j4}$ |
|-------------|---------------|---------------|---------------|---------------|
| $\sigma_{1i}$ | +           | +             | +             | +             |
| $\sigma_{2i}$ | +           | +             | +             | +             |
| $\sigma_{3i}$ | +           | +             | +             | +             |
| $\sigma_{4i}$ | +           | +             | +             | +             |

**Table 7.1:** Full $Y$ and $\Sigma$ Structures.

for the postulated model and the given data, in the manner of Fisher (1922). Equation (7.1-4) is meaningful for all positive definite values of $\Sigma$, provided that the matrix $v(\theta)$ is nonsingular over the permitted range of $\theta$. In practice, $v(\theta)$ can be made nonsingular for *all* $\theta$ by using a linearly independent subset of working responses; this is done automatically in the subroutine package GREGPLUS (see Appendix C).

Box and Draper (1965) found the unprejudiced prior density function

$$p(\Sigma) \propto |\Sigma|^{-(m+1)/2} \tag{7.1-5}$$

by the method of Jeffreys, described in Section 5.5. The derivation is given on page 475 of Box and Tiao (1973, 1992). A uniform prior density may be assumed for $\theta$ over the region of appreciable likelihood, as discussed in Chapter 5; then the joint prior $p(\theta, \Sigma)$ takes the same form as $p(\Sigma)$ of Eq. (7.1-5).

Multiplying the likelihood by this joint prior, Box and Draper (1965) obtained the posterior density function

$$p(\theta, \Sigma|Y) \propto |\Sigma|^{-(n+m+1)/2} \exp\left\{-\tfrac{1}{2}\mathrm{tr}\left[\Sigma^{-1}v(\theta)\right]\right\} \tag{7.1-6}$$

for the parameters of the expectation and error models. This formula gives *all the information obtainable* from a full data array $Y$ regarding the unknown parameter vector $\theta$ and covariance matrix $\Sigma$.

One could have proceeded directly with Eq. (7.1-6) to estimate $\theta$ and $\Sigma$. This approach leads to lengthy computations if $Y$ is not full but was implemented by Stewart and Sørensen (1981). Box and Draper shortened the analysis for full data $Y$ by integrating Eq. (7.1-5) over the positive definite range of $\Sigma$, just as we shortened single-response analysis by the integration over $\sigma$ in Eq. (6.6-2). In this way, they found

$$p(\boldsymbol{\theta}|\boldsymbol{Y}) \propto |\boldsymbol{v}(\boldsymbol{\theta})|^{-n/2} \tag{7.1-7}$$

as the *marginal* posterior density for the parameter vector $\boldsymbol{\theta}$ of the expectation model. The mode of this function occurs at the minimum of the determinant $|\boldsymbol{v}(\boldsymbol{\theta})|$, thus providing a multiresponse generalization of least squares.

Another way of reducing Eq. (7.1-6) is to use the most probable value, $\widetilde{\boldsymbol{\Sigma}}(\boldsymbol{\theta})$, of the covariance matrix at each value of $\boldsymbol{\theta}$. This gives the modified posterior density function

$$\widetilde{p}(\boldsymbol{\theta}|\boldsymbol{Y}) \equiv p(\boldsymbol{\theta}, \widetilde{\boldsymbol{\Sigma}}(\boldsymbol{\theta})|\boldsymbol{Y}) \propto |\boldsymbol{v}(\boldsymbol{\theta})|^{-(n+m+1)/2} \tag{7.1-8}$$

which has the same modal $\boldsymbol{\theta}$ as Eq. (7.1-6), as shown in Section 7.11. This density function is more focused than the one in Eq. (7.1-7) and thus gives sharper estimates of $\boldsymbol{\theta}$.

Equation (7.1-6) gives full information as long as $\boldsymbol{v}(\boldsymbol{\theta})$ is nonsingular. The modal $\boldsymbol{\theta}$ occurs at the minimum of $|\boldsymbol{v}(\boldsymbol{\theta})|$, and the modal error covariance matrix is

$$\widehat{\boldsymbol{\Sigma}} = \boldsymbol{v}(\widehat{\boldsymbol{\theta}})/(n+m+1) \tag{7.1-9}$$

GREGPLUS reports this matrix near the end of the estimation from data of Type 1.

## Type 2. Block-Rectangular $Y$ and Unknown $\Sigma$

When the data form discrete rectangular blocks as in Table 7.2, we call the data structure *block-rectangular*. The lines of data in a block need not be consecutive, though for clarity they are so arranged in Table 7.2. The covariances $\sigma_{ij}$ between blocks are either irrelevant to the data (as in Table 7.2b) or are set to zero on physical grounds (as in Tables 7.2a and 7.2c), so that only the within-block elements of $\boldsymbol{\Sigma}$ have any effect. Equations (7.1-5) through (7.1-8) then take the forms

$$p(\boldsymbol{\Sigma}_b) \propto |\boldsymbol{\Sigma}_b|^{-(m_b+1)/2} \tag{7.1-10}$$

$$p(\boldsymbol{\theta}, \boldsymbol{\Sigma}_b|\boldsymbol{Y}_b) \propto |\boldsymbol{\Sigma}_b|^{-(n_b+m_b+1)/2} \exp\left\{-\tfrac{1}{2}\mathrm{tr}\left[\boldsymbol{\Sigma}_b^{-1}\boldsymbol{v}_b(\boldsymbol{\theta})\right]\right\} \tag{7.1-11}$$

$$p(\boldsymbol{\theta}|\boldsymbol{Y}_b) \propto |\boldsymbol{v}_b(\boldsymbol{\theta})|^{-n_b/2} \tag{7.1-12}$$

$$\widetilde{p}(\boldsymbol{\theta}|\boldsymbol{Y}_b) \equiv p(\boldsymbol{\theta}, \widetilde{\boldsymbol{\Sigma}}_b(\boldsymbol{\theta})|\boldsymbol{Y}_b) \propto |\boldsymbol{v}_b(\boldsymbol{\theta})|^{-(n_b+m_b+1)/2} \tag{7.1-13}$$

for the individual blocks $b = 1, \ldots, \mathrm{B}$ of responses.

These four formulas need nonsingular block matrices $\boldsymbol{v}_b(\boldsymbol{\theta})$, obtainable by proper choice of response variables for each block. GREGPLUS uses subroutines from LINPACK (Dongarra et al., 1979) to choose the working

| | Type a.<br>Full $Y$,<br>Block-Diagonal $\Sigma$ | | | | Type b.<br>Block-Rectangular $Y$,<br>Block-Diagonal $\Sigma$ | | | | Type c.<br>Irregular $Y$,<br>Diagonal $\Sigma$ | | | |
|---|---|---|---|---|---|---|---|---|---|---|---|---|
| | $Y_{1u}$ | $Y_{2u}$ | $Y_{3u}$ | $Y_{4u}$ | $Y_{1u}$ | $Y_{2u}$ | $Y_{3u}$ | $Y_{4u}$ | $Y_{1u}$ | $Y_{2u}$ | $Y_{3u}$ | $Y_{4u}$ |
| $u=1$ | + | + | + | + | + | | | | + | | + | + |
| $u=2$ | + | + | + | + | + | | | | + | + | + | + |
| $u=3$ | + | + | + | + | + | | | | + | + | | |
| $u=4$ | + | + | + | + | | + | + | | + | | | |
| $u=5$ | + | + | + | + | | + | + | | + | | | + |
| $u=6$ | + | + | + | + | | + | + | | | | | + |
| $u=7$ | + | + | + | + | | | | + | + | + | | + |
| $u=8$ | + | + | + | + | | | | + | | | + | + |
| | $\sigma_{j1}$ | $\sigma_{j2}$ | $\sigma_{j3}$ | $\sigma_{j4}$ | $\sigma_{j1}$ | $\sigma_{j2}$ | $\sigma_{j3}$ | $\sigma_{j4}$ | $\sigma_{j1}$ | $\sigma_{j2}$ | $\sigma_{j3}$ | $\sigma_{j4}$ |
| $\sigma_{1i}$ | + | 0 | 0 | 0 | + | | | | + | 0 | 0 | 0 |
| $\sigma_{2i}$ | 0 | + | + | 0 | | + | + | | 0 | + | 0 | 0 |
| $\sigma_{3i}$ | 0 | + | + | 0 | | + | + | | 0 | 0 | + | 0 |
| $\sigma_{4i}$ | 0 | 0 | 0 | + | | | | + | 0 | 0 | 0 | + |

**Table 7.2:** Block-Rectangular Data and Covariance Structures.

responses so as to obtain a nonsingular submatrix $v_b(\boldsymbol{\theta})$ for each block. A special procedure for choosing working responses proved satisfactory for Box and co-workers (1973).

Combining Eqs. (7.1-11) for all the blocks, one obtains the posterior density function

$$p(\boldsymbol{\theta}, \boldsymbol{\Sigma}|\boldsymbol{Y}) \propto \prod_{b=1}^{B} |\boldsymbol{\Sigma}_b|^{-(n_b+m_b+1)/2} \exp\left\{-\tfrac{1}{2}\mathrm{tr}\left[\boldsymbol{\Sigma}_b^{-1} v_b(\boldsymbol{\theta})\right]\right\} \qquad (7.1\text{-}14)$$

This equation could be used to estimate $\boldsymbol{\theta}$ and $\boldsymbol{\Sigma}$ jointly, but it is simpler to estimate $\boldsymbol{\theta}$ separately by one of the following methods.

Multiplication of Eqs. (7.1-12) for all the blocks gives the marginal posterior density

$$p(\boldsymbol{\theta}|\boldsymbol{Y}) \propto \prod_{b=1}^{B} |v_b(\boldsymbol{\theta})|^{-n_b/2} \qquad (7.1\text{-}15)$$

first given by Michael Box and Norman Draper (1972). Here, as in Eq. (7.1-7), information about $\boldsymbol{\Sigma}$ has been suppressed to get a simple alternative to the full posterior density function. This time the simplification is less successful, because Eq. (7.1-15) does not necessarily recover the modal parameter vector $\widehat{\boldsymbol{\theta}}$ of Eq. (7.1-14).

Multiplication of Eqs. (7.1-13) suggested the formula

$$\widetilde{p}(\boldsymbol{\theta}|\boldsymbol{Y}) \equiv p(\boldsymbol{\theta}, \widetilde{\boldsymbol{\Sigma}}(\boldsymbol{\theta})|\boldsymbol{Y}) \propto \prod_{b=1}^{\mathrm{B}} |\boldsymbol{v}_b(\boldsymbol{\theta})|^{-(n_b+m_b+1)/2} \qquad (7.1\text{-}16)$$

derived by Stewart, Caracotsios, and Sørensen (1992) and in Section 7.11. This result was obtained by using the maximum-density value of the covariance matrix $\boldsymbol{\Sigma}$ at each value of the parameter vector $\boldsymbol{\theta}$. Equation (7.1-16) and the covariance matrix estimates

$$\widehat{\boldsymbol{\Sigma}}_b = \boldsymbol{v}_b(\widehat{\boldsymbol{\theta}})/(n_b + m_b + 1) \qquad b = 1, \dots, \mathrm{B} \qquad (7.1\text{-}17)$$

give the same modal $\boldsymbol{\theta}$ as Eq. (7.1-14), and nearly the same interval estimates for the $\theta$'s. Equations (7.1-16) and (7.1-17) are used in GREGPLUS for parameter estimation from block-rectangular data structures. Other structures may be analyzable in the future by one of the following methods (not yet provided in GREGPLUS).

## Type 3. Irregular $Y$ and Unknown $\boldsymbol{\Sigma}$

If the error matrix $\mathcal{E}(\boldsymbol{\theta}) \equiv \boldsymbol{Y} - \boldsymbol{F}(\boldsymbol{x}, \boldsymbol{\theta})$ does not consist of distinct rectangular blocks, we call the data structure *irregular*. The matrix $\boldsymbol{Y}$ in Table 7.3 gives such a structure unless $\boldsymbol{\Sigma}$ is truncated to diagonal form as in Table 7.2c; this is the simplest way of dealing with irregular data.

Another way of analyzing irregular data is to treat any missing values $Y_{iu}$ as additional parameters $\theta_r$, as proposed by Box, Draper, and Hunter (1970). This method has the disadvantage that it needlessly enlarges the parameter set.

Stewart and Sørensen (1981) used Bayes' theorem and estimated $\boldsymbol{\Sigma}$ along with $\boldsymbol{\theta}$. The likelihood function is

$$l(\boldsymbol{\theta}, \boldsymbol{\Sigma}|\boldsymbol{Y}) \propto \left[ \prod_{u=1}^{n} |\boldsymbol{\Sigma}_u|^{-1/2} \right] \qquad (7.1\text{-}18)$$

$$\cdot \exp\left\{ -\tfrac{1}{2} \sum_{u=1}^{n} \sum_{i=1}^{m} \sum_{j=1}^{m} \sigma_u^{ij} \left[ Y_{iu} - F_i(\boldsymbol{x}_u, \boldsymbol{\theta}) \right] \left[ Y_{ju} - F_j(\boldsymbol{x}_u, \boldsymbol{\theta}) \right] \right\}$$

for a data matrix $\boldsymbol{Y}$ of independent events. Each submatrix $\boldsymbol{\Sigma}_u$ is constructed from the complete matrix $\boldsymbol{\Sigma}$, using the elements for working responses observed in the $u$th event. Klaus and Rippin (1979) used such a function in their algorithm for likelihood-based estimation from general data structures.

Multiplication of Eq. (7.1-18) by a locally uniform $p(\boldsymbol{\theta})$ and the prior $p(\boldsymbol{\Sigma})$ of Eq. (7.1-5) gives the posterior density function

$$p(\boldsymbol{\theta}, \boldsymbol{\Sigma}|\boldsymbol{Y}) \propto |\boldsymbol{\Sigma}|^{-(m+1)/2} \left[ \prod_{u=1}^{n} |\boldsymbol{\Sigma}_u|^{-1/2} \right] \qquad (7.1\text{-}19)$$

$$\cdot \exp\left\{ -\tfrac{1}{2} \sum_{u=1}^{n} \sum_{i=1}^{m} \sum_{j=1}^{m} \sigma_u^{ij} \left[ Y_{iu} - F_i(\boldsymbol{x}_u, \boldsymbol{\theta}) \right] \left[ Y_{ju} - F_j(\boldsymbol{x}_u, \boldsymbol{\theta}) \right] \right\}$$

|           | $Y_{1u}$ | $Y_{2u}$ | $Y_{3u}$ | $Y_{4u}$ |
|-----------|----------|----------|----------|----------|
| $u=1$     | +        |          |          |          |
| $u=2$     |          | +        |          |          |
| $u=3$     |          |          | +        |          |
| $u=4$     |          |          |          | +        |
| $u=5$     | +        | +        |          |          |
| $u=6$     | +        |          | +        |          |
| $u=7$     |          | +        | +        |          |
| $u=8$     |          | +        |          | +        |

|              | $\sigma_{j1}$ | $\sigma_{j2}$ | $\sigma_{j3}$ | $\sigma_{j4}$ |
|--------------|---------------|---------------|---------------|---------------|
| $\sigma_{1i}$ | +            | +             | +             |               |
| $\sigma_{2i}$ | +            | +             | +             | +             |
| $\sigma_{3i}$ | +            | +             | +             |               |
| $\sigma_{4i}$ |              | +             |               | +             |

**Table 7.3:** An Irregular Data Structure and Its Covariances.

for irregular $Y$ and positive definite $\Sigma$.

This function was used by Stewart and Sørensen (1981) to analyze the data of Fuguitt and Hawkins (1945, 1947). Bain (1993) found reasonable agreement, in several examples with irregular data structures, between parameter estimates based on the prior of Eq. (7.1-5) and on a detailed version (Bain 1993) of the Jeffreys prior.

### Type 4. Irregular $Y$ and Unknown Variable $\Sigma$

When one or more observed responses vary widely, it may be useful to estimate their covariances as functions of the corresponding predictors. One such model is the power function

$$\sigma_{iju} = \omega_{ij} \left| f_i(\boldsymbol{x}_u, \boldsymbol{\theta}) f_j(\boldsymbol{x}_u, \boldsymbol{\theta}) \right|^{(\gamma_i + \gamma_j)/2} = \sigma_{jiu} \qquad i, j = 1, \ldots, m \quad (7.1\text{-}20)$$

but other forms are safer when the expectation functions $f_i(\boldsymbol{x}_u, \boldsymbol{\theta})$ range through zero. Use of this expression for elements of $\Sigma$ in Eqs. (7.1-18) and (7.1-19) gives the corresponding likelihood and posterior density functions. Some results of this approach, computed by the present authors, are reported by Biegler, Damiano, and Blau (1986), as noted in Section 7.8.

### 7.2 OBJECTIVE FUNCTION

GREGPLUS treats full or block-rectangular data structures, using the objective function $S(\boldsymbol{\theta}) = -2 \ln p(\boldsymbol{\theta}|\boldsymbol{y})$. This gives

$$S(\boldsymbol{\theta}) = \sum_{b=1}^{B} \text{LPOWR}_b \ln |\boldsymbol{v}_b(\boldsymbol{\theta})| \qquad (7.2\text{-}1)$$

based on the user's arrangement of the responses into one or more blocks $b = 1, \ldots, B$. Responses $i$ and $k$ should be placed in the same block if (1) they are observed in the same events and (2) their errors of observation might be correlated. The coefficient $\text{LPOWR}_b$ takes the forms

$$\text{LPOWR}_b = \begin{cases} n_b & \text{for full data } (B = 1); \\ (n_b + m_b + 1) & \text{for block-rectangular data} \end{cases} \tag{7.2-2}$$

obtained from Eq. (7.1-7) or (7.1-13) with a simple proportionality constant. This function is safer for numerical work than the posterior density $p(\boldsymbol{\theta}|\boldsymbol{Y})$, which often exceeds the machine range when applied to large data sets. The covariance matrix $\boldsymbol{\Sigma}$ is not needed as an argument of $S$, since its maximum-density estimate is computed at the end via Eq. (7.1-9) or (7-1-17) and the elements

$$v_{ij}(\boldsymbol{\theta}) = \sum_{u=1}^{n} \mathcal{E}_{iu}(\boldsymbol{\theta})\mathcal{E}_{ju}(\boldsymbol{\theta}) \tag{7.2-3}$$

with the notation of Eq. (7.1-1) for the weighted error variables. This construction, and the working response selection procedure described below, ensure that the matrix estimate $\widehat{\boldsymbol{\Sigma}}$ will have strictly positive eigenvalues.

The permitted range of each parameter is defined as in Chapter 6:

$$\text{BNDLW}(i) \leq \theta_i \leq \text{BNDUP}(i) \qquad i = 1, \ldots, \text{NPAR} \tag{7.2-4}$$

### 7.2.1 Selection of Working Responses

GREGPLUS selects the working response variables for each application by least squares, using a modification of LINPACK Subroutine DCHDC (Dongarra et al., 1979) that forms the Cholesky decomposition of the matrix $\boldsymbol{X} = \{\sum_u \mathcal{E}_{iu}\mathcal{E}_{ju}\}$. To keep the selected responses linearly independent, a response is added to the working set only if its sum of squared errors $\sum_u \mathcal{E}_{iu}^2$ is at least $10^{-4}$ of the sum over the working responses thus far selected.

The following identities, adapted from Bard (1974) for real symmetric nonsingular matrices $\boldsymbol{X} = \{x_{ij}\}$, are used in the quadratic expansion of $S(\boldsymbol{\theta})$:

$$\frac{\partial \ln |\boldsymbol{X}|}{\partial x_{i \geq .j}} = (2 - \delta_{ij})x^{ij} \tag{7.2-5}$$

$$\frac{\partial x^{kj}}{\partial x_{k \geq .l}} = (1/2)\delta_{kl}\left[x^{ik}x^{lj} + x^{il}x^{kj}\right] \tag{7.2-6}$$

Here $\delta_{ij}$ is 1 if $i = j$ and zero otherwise; $x^{ij}$ is the corresponding element of the matrix $\boldsymbol{X}^{-1}$, and $x_{i \geq .j}$ is any element on or below the diagonal of

$X$. The needed elements of the matrix $X^{-1}$ are obtained directly from the Cholesky decomposition of $X$. The constrained minimization of $S(\boldsymbol{\theta})$ is performed by successive quadratic programming, as in Chapter 6, but with these multiresponse identities.

For inquisitive readers, we give next the detailed expressions used by GREGPLUS to compute $\boldsymbol{\theta}$-derivatives of Eqs. (7.2-1) and (7.2-3). Other readers may skip to Section 7.3.

### 7.2.2 Derivatives of Eqs. (7.2-1) and (7.2-3)

Application of Eq. (7.2-5) to the subdeterminant $|\boldsymbol{v}_b(\boldsymbol{\theta})|$ gives

$$\frac{\partial \ln |\boldsymbol{v}_b(\boldsymbol{\theta})|}{\partial v_{bi \geq j}(\boldsymbol{\theta})} = (2 - \delta_{ij})v_b^{ij}(\boldsymbol{\theta}) \tag{7.2-7}$$

for each response pair $(i, j)$ in block $b$. Hence, by the chain rule of partial differentiation,

$$\frac{\partial \ln |\boldsymbol{v}_b(\boldsymbol{\theta})|}{\partial \theta_r} = \sum_i \sum_{j \leq i} \frac{\partial \ln |\boldsymbol{v}_b(\boldsymbol{\theta})|}{\partial v_{bij}} \frac{\partial v_{bij}}{\partial \theta_r}$$

$$= \sum_i \sum_{j \leq i} (2 - \delta_{ij})v_b^{ij} \frac{\partial v_{bij}}{\partial \theta_r} \tag{7.2-8}$$

with indices $i$ and $j$ restricted to working members of response block $b$. This result, when combined with the appropriate derivative of Eq. (7.2-1), gives

$$\frac{\partial S(\boldsymbol{\theta})}{\partial \theta_r} = \sum_{b=1}^{B} \text{LPOWR}_b \sum_i \sum_{j \leq i} (2 - \delta_{ij})v_b^{ij} \frac{\partial v_{bij}}{\partial \theta_r} \tag{7.2-9}$$

Differentiation of Eq. (7.2-8) with respect to $\theta_s$ gives

$$\frac{\partial^2 \ln |\boldsymbol{v}_b(\boldsymbol{\theta})|}{\partial \theta_r \partial \theta_s} = \sum_i \sum_{j \leq i} (2 - \delta_{ij})v_b^{ij} \frac{\partial^2 v_{bij}}{\partial \theta_r \partial \theta_s}$$

$$+ \sum_i \sum_{j \leq i} (2 - \delta_{ij}) \frac{\partial v_b^{ij}}{\partial \theta_s} \frac{\partial v_{bij}}{\partial \theta_r} \tag{7.2-10}$$

The $v_b^{ij}$ derivative in Eq. (7.2-8) is evaluated via the chain rule and Eq. (7.2-5):

$$\frac{\partial v_b^{ij}}{\partial \theta_s} = \sum_k \sum_{l \leq k} (2 - \delta_{kl}) \frac{\partial v_b^{ij}}{\partial v_{bkl}} \frac{\partial v_{bkl}}{\partial \theta_s}$$

$$= -\frac{1}{2} \sum_k \sum_{l \leq k} (2 - \delta_{kl}) \left[ v_b^{ik} v_b^{lj} + v_b^{il} v_b^{kj} \right] \frac{\partial v_{bkl}}{\partial \theta_s} \tag{7.2-11}$$

Combining the last two equations with the $\theta_s$-derivative of Eq. (7.2-9), we get

$$
\begin{aligned}
\frac{\partial^2 S(\boldsymbol{\theta})}{\partial \theta_r \partial \theta_s} &= \sum_{b=1}^{B} \text{LPOWR}_b \sum_i \sum_{j \leq i} (2 - \delta_{ij}) v_b^{ij} \frac{\partial^2 v_{bij}}{\partial \theta_r \partial \theta_s} \\
&\quad - \frac{1}{2} \sum_{b=1}^{B} \text{LPOWR}_b \sum_i \sum_{j \leq i} \sum_k \sum_{l \leq k} \\
&\quad (2 - \delta_{ij})(2 - \delta_{kl}) \left[ v_b^{ik} v_b^{lj} + v_b^{il} v_b^{kj} \right] \frac{\partial v_{bkl}}{\partial \theta_s} \frac{\partial v_{bij}}{\partial \theta_r}
\end{aligned}
$$

(7.2-12)

The quadruple summation in Eq. (7.2-12) comes from the $\boldsymbol{\theta}$-derivatives of the inverse block matrices $v_b^{-1}$. In our experience, the minimization of $S$ goes best with this contribution suppressed until $\widetilde{S}$ is near a local minimum. Finally, the needed derivatives of the elements of $v_b(\boldsymbol{\theta})$ are

$$
\begin{aligned}
\frac{\partial v_{bij}}{\partial \theta_r} &= \sum_u \left[ \frac{\partial \mathcal{E}_{iu}}{\partial \theta_r} \mathcal{E}_{ju} + \mathcal{E}_{iu} \frac{\partial \mathcal{E}_{ju}}{\partial \theta_r} \right] \\
&= -\sum_u \left[ \frac{\partial F_{iu}}{\partial \theta_r} \mathcal{E}_{ju} + \mathcal{E}_{iu} \frac{\partial F_{ju}}{\partial \theta_r} \right]
\end{aligned}
$$

(7.2-13)

and

$$
\begin{aligned}
\frac{\partial^2 v_{bij}}{\partial \theta_r \partial \theta_s} &= \sum_u \left[ \frac{\partial F_{iu}}{\partial \theta_r} \frac{\partial F_{ju}}{\partial \theta_s} + \frac{\partial F_{iu}}{\partial \theta_s} \frac{\partial F_{ju}}{\partial \theta_r} \right] \\
&\quad - \sum_u \left[ \frac{\partial^2 F_{iu}}{\partial \theta_r \partial \theta_s} \mathcal{E}_{ju} + \mathcal{E}_{iu} \frac{\partial^2 F_{ju}}{\partial \theta_r \partial \theta_s} \right]
\end{aligned}
$$

(7.2-14)

The terms in the second line of Eq. (7.2-14) vanish if the model is linear in $\boldsymbol{\theta}$ and are unimportant if the data are well fitted. They also are computationally expensive. GREGPLUS omits these terms; this parallels the Gauss-Newton approximation in the normal equations of Chapter 6.

### 7.2.3 Quadratic Expansions; Normal Equations

A truncated Taylor expansion of the objective function $S(\boldsymbol{\theta})$ around the starting point $\boldsymbol{\theta}^k$ of iteration $k$ gives the approximating function

$$
\begin{aligned}
\widetilde{S}(\boldsymbol{\theta}) &= S(\boldsymbol{\theta}^k) + \sum_r \frac{\partial S}{\partial \theta_r} (\theta_r - \theta_r^k) \\
&\quad + \sum_r \sum_s \frac{1}{2!} \frac{\partial^2 S}{\partial \theta_r \partial \theta_s} (\theta_r - \theta_r^k)(\theta_s - \theta_s^k)
\end{aligned}
$$

(7.2-15)

which GREGPLUS computes from the expressions just given, using exact derivatives or difference quotients, as in Chapter 6. We express this as a symmetric matrix expansion

$$\widetilde{S}(\boldsymbol{\theta}) = S(\boldsymbol{\theta}^k) + \left[ (\boldsymbol{\theta} - \boldsymbol{\theta}^k)^T \; \vdots \; -1 \right] \begin{bmatrix} A_{\theta\theta} & \vdots & A_{\theta L} \\ \cdots\cdots\cdots \\ A_{\theta L}^T & \vdots & 0 \end{bmatrix} \begin{bmatrix} \boldsymbol{\theta} - \boldsymbol{\theta}^k \\ \cdots\cdots \\ -1 \end{bmatrix} \quad (7.2\text{-}16)$$

paralleling the Gauss-Newton form of Eq. (6.3-7). These are our multiresponse normal equations.

## 7.3 CONSTRAINED MINIMIZATION OF $S(\boldsymbol{\theta})$

The strategy described in Section 6.4 for constrained minimization of the sum of squares $S(\boldsymbol{\theta})$ is readily adapted to the multiresponse objective function $\widetilde{S}(\boldsymbol{\theta})$ of Eq. (7.2-16). The quadratic programming subroutine GRQP and the line search subroutine GRS2 are used, with the following changes:

1. The value $0.1S(\boldsymbol{\theta}^k)$ in the trust-region expression of Eq. (6.4-2) is replaced by 2.
2. If a nonpositive determinant $\boldsymbol{v}$ or $\boldsymbol{v}_b$ occurs in a line search of $S(\boldsymbol{\theta})$, subroutine GRS2 tries a new search point with $\boldsymbol{\theta} - \boldsymbol{\theta}^k$ half as large. This process is repeated until a positive determinant $\boldsymbol{v}_b$ is obtained for every response block. These precautions ensure a uniform definition of $S(\boldsymbol{\theta})$ during each iteration. GRS2 takes the same action if the user's subroutine MODEL returns a nonzero integer IERR, thus reporting an unsuccessful function evaluation.
3. In the Cholesky decomposition of $|\boldsymbol{v}_b|$ for Eq. (7.2-1) at the start of an iteration, any response $i$ whose element $v_{ii}$ is less than $10^{-4}$ of the largest $v_{kk}$ is excluded from the working set.
4. A change of the working response set is permitted only at the start of an iteration.
5. LINMOD is set to zero, since the objective $S\boldsymbol{\theta})$ in Eq. (7.2-1) cannot be linear in $\boldsymbol{\theta}$.

The removal of a response from the working set indicates one of the following conditions: linear dependence of that response on the ones in the working set, or too few observations of that response in relation to the number of unknown parameters and covariances, or too large a rounding tolerance YTOL(i) for that response. The user can control the selection of working responses by placing the desired ones early in their blocks and not using too many parameters.

After each iteration, a test is made for convergence of the parameter estimation. Convergence is declared, and the final statistics of Section 7.6 are reported, if no parameter is at a CHMAX limit and if $|\Delta\theta_i|$ for each basis parameter is less than one-tenth the half-width of its 95% HPD interval calculated from Eq. (7.5-5). If no degrees of freedom remain for the estimation, this criterion is replaced by $|\Delta\theta_i| \leq \text{RPTOL} * |\theta_i^k|$.

### 7.3.1 Final Expansions Around $\widehat{\theta}$

The final expansions of $\widetilde{S}$ parallel those given in Section 6.4.3. They give

$$\begin{aligned}
\widetilde{S}(\boldsymbol{\theta}) &= \widehat{S} + (\boldsymbol{\theta}_e - \widehat{\boldsymbol{\theta}}_e)^T \left[\widehat{\boldsymbol{A}}_{ee}\right] (\boldsymbol{\theta}_e - \widehat{\boldsymbol{\theta}}_e) \\
&= \widehat{S} + (\boldsymbol{\theta}_e - \widehat{\boldsymbol{\theta}}_e)^T \left[\widehat{\boldsymbol{A}}^{ee}\right]^{-1} (\boldsymbol{\theta}_e - \widehat{\boldsymbol{\theta}}_e)
\end{aligned} \tag{7.3-1}$$

for the full vector $\boldsymbol{\theta}_e$ of estimated parameters, and

$$\widetilde{S}(\boldsymbol{\theta}_a) = \widehat{S} + (\boldsymbol{\theta}_a - \widehat{\boldsymbol{\theta}}_a)^T \left[\widehat{\boldsymbol{A}}^{aa}\right]^{-1} (\boldsymbol{\theta}_a - \widehat{\boldsymbol{\theta}}_a) \tag{7.3-2}$$

for a test vector $\boldsymbol{\theta}_a$ of $n_a$ parameters from $\boldsymbol{\theta}_e$. The latter result gives

$$\widetilde{S}(\theta_r) = \widehat{S} + (\theta_r - \widehat{\theta}_r)^2 / \widehat{A}^{rr} \tag{7.3-3}$$

for a single test parameter $\theta_r$ from the estimated set $\boldsymbol{\theta}_e$. In Eqs. (7.3-2) and (7.3-3) the nontest members of the set $\boldsymbol{\theta}_e$ are implicitly optimized at each value of the test vector $\boldsymbol{\theta}_a$.

## 7.4 TESTING THE RESIDUALS

Once a multiresponse model is fitted, the weighted residuals

$$\widehat{\mathcal{E}}_{iu} = Y_{iu} - F_i(\boldsymbol{x}_u, \widehat{\boldsymbol{\theta}}) \tag{7.4-1}$$

should be examined for signs of weakness in the model or in the data. Plots or curve-fits of the residuals, and investigation of the largest residuals, should be done for each response. The residuals of each response should be studied as functions of experimental conditions and of the other residuals in corresponding events.

Replicate observations and/or replicate residuals provide important information on the variances and covariances of the observations and/or computed residuals. If replicates are provided in the call of GREGPLUS, then KVAR(0) is set to 1 and the input integers KVAR(1, . . . , NEVT) are signed group labels; otherwise, KVAR is set to 0. Replicate *observation* group $k$ is marked by setting KVAR(u) = k for that event, and replicate *residual* group $k$ is marked by setting KVAR(u) = −k for that event. The mean and variance for each group are computed by GREGPLUS and are used in the following tests.

Quantitative tests of multivariate residuals are available from sampling theory, as derived by Box (1949) and cited by Anderson (1984). When $\boldsymbol{\Sigma}$

is unknown but the data permit a sample estimate $\boldsymbol{v}_e$ of the experimental error contribution to the array $\boldsymbol{v}(\boldsymbol{\theta}_j)$ of Eq. (7.1-2) for a proposed model $M_j$, the goodness of fit of that model can be tested by the criterion function

$$\mathcal{M}_j \equiv \nu_j \ln |\widehat{\boldsymbol{v}}_j/\nu_j| - (\nu_j - \nu_e) \ln |(\widehat{\boldsymbol{v}}_j - \boldsymbol{v}_e)/(\nu_j - \nu_e)| - \nu_e \ln |\boldsymbol{v}_e/\nu_e| \quad (7.4\text{-}2)$$

Here $\widehat{\boldsymbol{v}}_j \equiv \boldsymbol{v}(\widehat{\boldsymbol{\theta}}_j)$ is the residual moment matrix with $\nu_j = n - p_j$ degrees of freedom, $\boldsymbol{v}_e$ is the "pure error" moment matrix with $\nu_e$ degrees of freedom, and $\widehat{\boldsymbol{v}}_j - \boldsymbol{v}_e$ is the lack-of-fit moment matrix with $\nu_j - \nu_e$ degrees of freedom. The sampling probability $\Pr(\mathcal{M} > \mathcal{M}_j)$ of $\mathcal{M}$ values larger than $\mathcal{M}_j$ is computable from Eq. 26 of Box (1949) under the hypothesis that $\boldsymbol{v}_e$ and $\widehat{\boldsymbol{v}}_j - \boldsymbol{v}_e$ are sample values from the same normal error distribution. This test, provided in the software package GREGPLUS, is the multiresponse analog of the two-tailed $F$-test provided in Box and Tiao (1973), pp. 558–565.

When $\boldsymbol{\Sigma}$ is block-diagonal, as in Table 7.2, $\widehat{\boldsymbol{\theta}}_j$ is preferably calculated according to Eq. (7.1-16). Corresponding tests of goodness of fit can then be made for the individual response blocks $b$ via the criterion functions

$$\mathcal{M}_{jb} := \nu_{jb} \ln |\widehat{\boldsymbol{v}}_{jb}/\nu_{jb}| - (\nu_{jb} - \nu_{eb}) \ln |(\widehat{\boldsymbol{v}}_{jb} - \boldsymbol{v}_{eb})/(\nu_{jb} - \nu_{eb})| -$$
$$\nu_{eb} \ln |\boldsymbol{v}_{eb}/\nu_{eb}| \qquad b = 1, \dots, B \qquad\qquad (7.4\text{-}3)$$

once the parameter count for each block is assigned. GREGPLUS automates this task, assigning to each block its number $p_{jb,\mathrm{loc}}$ of local parameters estimated, plus a fraction $(n_b - p_{jb,\mathrm{loc}})/(n - \sum_b p_{jb,\mathrm{loc}})$ of the number of parameters that appear in more than one block. An improved assignment may be provided at another time.

## 7.5 POSTERIOR PROBABILITIES AND REGIONS

Integrations of the posterior density yield regions of given posterior probability content, or probability content of given regions, in the parameter space $\boldsymbol{\theta}_e$ of the expectation model. For conciseness, we proceed as in Section 6.5 and approximate the posterior density with expansions in $\boldsymbol{\theta}_e$ around the local minimum of $\widetilde{S}$ under the given constraints.

### 7.5.1 Inferences Regarding Parameters

Equation (7.2-1), with $S(\boldsymbol{\theta})$ approximated by the expansion in Eq. (7.3-1), gives

$$p(\boldsymbol{\theta}_e|\boldsymbol{Y}) \propto \exp\left\{ -\tfrac{1}{2}(\boldsymbol{\theta}_e - \widehat{\boldsymbol{\theta}}_e)^T \left[\widehat{\boldsymbol{A}}^{ee}\right]^{-1} (\boldsymbol{\theta}_e - \widehat{\boldsymbol{\theta}}_e) \right\} \qquad (7.5\text{-}1)$$

for the joint posterior density of the estimable parameters. In this approximation $\boldsymbol{\theta}_e$ is normally distributed, with mode $\widehat{\boldsymbol{\theta}}_e$ and covariance matrix $\widehat{\boldsymbol{A}}^{ee}$.

The marginal posterior density for a $n_a$-parameter subset $\boldsymbol{\theta}_a$ of $\boldsymbol{\theta}_e$ is obtained by integrating[1] Eq. (7.5-1) from $-\infty$ to $\infty$ for the other estimated parameters. The result is

$$p(\boldsymbol{\theta}_a|\boldsymbol{Y}) \propto \exp\left\{-\tfrac{1}{2}(\boldsymbol{\theta}_a - \widehat{\boldsymbol{\theta}}_a)^T \left[\hat{\boldsymbol{A}}^{aa}\right]^{-1}(\boldsymbol{\theta}_a - \widehat{\boldsymbol{\theta}}_a)\right\} \qquad (7.5\text{-}2)$$

which reduces to

$$p(\theta_r|\boldsymbol{Y}) \propto \exp\left\{-\tfrac{1}{2}(\theta_r - \widehat{\theta}_r)^2/\hat{A}^{rr}\right\} \qquad (7.5\text{-}3)$$

for the marginal posterior densities of the individual estimated parameters. Several consequences of these results are summarized here:

1. Equation (7.5-2) is a normal density function, with mode $\widehat{\boldsymbol{\theta}}_a$ and covariance matrix

$$\operatorname{Cov}\boldsymbol{\theta}_a = \hat{\boldsymbol{A}}^{aa} \qquad (7.5\text{-}4)$$

   for the parameter subvector $\boldsymbol{\theta}_a$. This gives the variances

$$\operatorname{Var}\theta_r = \sigma_{\theta_r}^2 = \hat{A}^{rr} \qquad (7.5\text{-}5)$$

   for the individual parameter estimates. The values $\widehat{\boldsymbol{\theta}}_r$ and $\sigma_{\theta_r}$ are standard outputs from GREGPLUS; a normalized display of $\hat{\boldsymbol{A}}^{ee}$ is also provided there.

2. The probability content of the region of positive values of the estimated parameter $\theta_r$ is

$$\alpha(U) = \frac{1}{2\sqrt{\pi}} \int_U^\infty \exp(-t^2/2)\,dt \qquad (7.5\text{-}6)$$

   evaluated with the lower integration limit $U = -\widehat{\theta}_r/\sqrt{A^{rr}}$. The function $1 - \alpha(U)$ is the normal distribution, shown in Figure 4.2.

3. The interval of highest posterior density with probability content $(1 - \alpha)$ for the estimated parameter $\theta_r$ is

$$-U(\alpha/2) \le \frac{(\theta_r - \widehat{\theta}_r)}{\sigma_{\theta_r}} \le U(\alpha/2) \qquad (7.5\text{-}7)$$

   A common choice is $\alpha = 0.05$, which gives $U(\alpha/2) = 1.96$. GREGPLUS computes this interval for each estimated parameter.

---

[1] In these integrations, we assume initially that the permitted ranges of the parameters cover the region of appreciable posterior density. This assumption is ultimately removed in Section 7.5.3, as in Section 6.6.3.

4.  The region of highest joint posterior density with probability
    content $(1 - \alpha)$ for a subset $\boldsymbol{\theta}_a$ of the estimated parameters is

$$[S(\boldsymbol{\theta}_a) - S(\widehat{\boldsymbol{\theta}}_a)] \leq \chi^2(n_a, \alpha) \qquad (7.5\text{-}8)$$

This region can be further approximated as

$$(\boldsymbol{\theta}_a - \widehat{\boldsymbol{\theta}}_a)^T \left[\widehat{\boldsymbol{A}}^{aa}\right]^{-1} (\boldsymbol{\theta}_a - \widehat{\boldsymbol{\theta}}_a) \leq \chi^2(n_a, \alpha) \qquad (7.5\text{-}8a)$$

by use of Eq. (7.3-2). The resulting boundary is ellipsoidal when
$n_a \geq 2$ and encloses a volume proportional to $|\widehat{\boldsymbol{A}}^{aa}|^{1/2}$. The
latter feature is relevant to experimental design, as discussed in
Section 7.6. When $n_a = 1$, this region corresponds to that of
Eq. (7.5-7). The function $\chi^2(n_a, \alpha)$ is the large-sample limit of
the $F$-distribution encountered in Sections 6.5 and 6.6.

For full data structures (see Table 7.1), one can estimate posterior densities and regions of $\boldsymbol{\theta}_e$ by expanding $|\boldsymbol{v}(\boldsymbol{\theta})|$ quadratically instead of $S(\boldsymbol{\theta})$. The results parallel those for least-squares estimation from single-response data in the sense that $\boldsymbol{\Sigma}$ is integrated out. This approach, however, is not appropriate for multiresponse problems, since the determinant $|\boldsymbol{v}(\boldsymbol{\theta})|$ (a polynomial of order $m$ in the elements $v_{ij}(\boldsymbol{\theta})$), is then of order 4 or greater in $\boldsymbol{\theta}$. The resulting posterior density expressions become very difficult to integrate over $\Sigma$ when either $\boldsymbol{Y}$ or $\boldsymbol{\Sigma}$ is not full. For full $\boldsymbol{Y}$ and $\boldsymbol{\Sigma}$, interval estimates of $\boldsymbol{\theta}$ by this approach are available (Stewart and Sørensen 1976, 1981; Bates and Watts 1988); the resulting intervals are somewhat wider than those given here.

### 7.5.2 Inferences Regarding Functions

The results just given can be transformed in the manner of Section 6.6.2 to get multiresponse inferences for auxiliary functions. One defines a vector $\boldsymbol{\phi}_a(\boldsymbol{\theta})$ of such functions and computes the matrix $\widehat{\boldsymbol{A}}^{aa}$ as in Eqs. (6.6-19) and (6.6-20). The results of Section 7.5.1 can then be applied to the chosen functions by substituting $\widehat{\boldsymbol{A}}$ for $\widehat{\boldsymbol{A}}$ and $\boldsymbol{\phi}$ for $\boldsymbol{\theta}$ in those equations. GREGPLUS does these calculations automatically on request to obtain the interval estimates and covariance matrix for a linearly independent subset of the auxiliary functions.

### 7.5.3 Discrimination Among Rival Models

Rival models $M_1, M_2, \ldots, M_J$ are to be compared on a common data matrix $\boldsymbol{Y}$ of weighted observations from $n$ independent multiresponse events. Prior probabilities $p(M_1), \ldots, p(M_J)$, here considered equal, are assigned to the models without looking at the data. The relative posterior probabilities of the models are to be calculated in the manner of Section 6.6.3. The main points of the analysis are summarized here from the work of Stewart, Shon, and Box (1998).

*Case 1: Full Data, Known Covariance Matrix*

For a given experimental design and expectation model, let $p(\boldsymbol{Y}|M_j, \boldsymbol{\Sigma})$ denote the probability density of prospective data in $\boldsymbol{Y}$-space, predicted via Eq. (7.1-2). According to Bayes' theorem, the *posterior* probability of Model $M_j$ conditional on *given* arrays $\boldsymbol{Y}$ and $\boldsymbol{\Sigma}$ is then

$$p(M_j|\boldsymbol{Y}, \boldsymbol{\Sigma}) = p(M_j)p(\boldsymbol{Y}|M_j, \boldsymbol{\Sigma})/C \qquad (7.5\text{-}9)$$

in which $C$ is a normalization constant, equal for all the models.

If Model $M_j$ contains an unknown parameter vector $\boldsymbol{\theta}_j$, then integration of Eq. (7.5-9) over the permitted range of $\boldsymbol{\theta}_j$ gives the total posterior probability for this model:

$$p(M_j|\boldsymbol{Y}, \boldsymbol{\Sigma}) = p(M_j)\int p(\boldsymbol{Y}|\boldsymbol{\theta}_j, M_j, \boldsymbol{\Sigma})p(\boldsymbol{\theta}_j|M_j)d\boldsymbol{\theta}_j/C \qquad (7.5\text{-}10)$$

This integration includes only the parameters in the estimated set $\boldsymbol{\theta}_{je}$, since any other parameters are held constant. Equation (7.5-10) can be expressed more naturally as

$$p(M_j|\boldsymbol{Y}, \boldsymbol{\Sigma}) = p(M_j)\int \ell(\boldsymbol{\theta}_j|\boldsymbol{Y}, M_j, \boldsymbol{\Sigma})p(\boldsymbol{\theta}_j|M_j)d\boldsymbol{\theta}_j/C \qquad (7.5\text{-}11)$$

by rewriting $p(\boldsymbol{Y}|\boldsymbol{\theta}_j, M_j, \boldsymbol{\Sigma})$ as a likelihood function in the manner of Eq. (5.1-6).

We treat the prior density $p(M_j|\boldsymbol{\theta}_j)$ as uniform in $\boldsymbol{\theta}_j$ over the region of appreciable likelihood $\ell(\boldsymbol{\theta}_j|\boldsymbol{Y}, M_j, \boldsymbol{\Sigma})$. With this approximation, Eq. (7.5-11) gives

$$p(M_j|\boldsymbol{Y}) = p(M_j)p(\boldsymbol{\theta}_j|M_j)\int \ell(\boldsymbol{\theta}_j|\boldsymbol{Y}, M_j, \boldsymbol{\Sigma})d\boldsymbol{\theta}_j/C \qquad (7.5\text{-}12)$$

as the posterior probability of Model $M_j$.

The probability expression just obtained, being based on data, has a sampling distribution over conceptual replications of the experiments. We evaluate the ratio $p(\boldsymbol{\theta}_j|M_j)/C$ by requiring that $p(M_j|\boldsymbol{Y}, \boldsymbol{\Sigma})$ have expectation $p(M_j)$ over such replications when Model $M_j$ is treated as true. With this sensible requirement, Eq. (7.5-12) yields a posterior probability

$$p(M_j|\boldsymbol{Y}, \boldsymbol{\Sigma}) = p(M_j)\frac{\int \ell(\boldsymbol{\theta}_j|\boldsymbol{Y}, M_j, \boldsymbol{\Sigma})d\boldsymbol{\theta}_j}{\mathrm{E}[\int \ell(\boldsymbol{\theta}_j|\boldsymbol{Y}, M_j, \boldsymbol{\Sigma})d\boldsymbol{\theta}_j]} \qquad (7.5\text{-}13)$$

for each candidate model, with the denominator calculated as if that model were true.

The likelihood function here has the form (see Eq. (7.1-4) with $\boldsymbol{\Sigma}$ given)

$$\ell(\boldsymbol{\theta}_j|\boldsymbol{Y}, \boldsymbol{\Sigma}) \propto \exp\left\{-\frac{1}{2}\operatorname{tr}\left[\boldsymbol{\Sigma}^{-1}\boldsymbol{v}(\boldsymbol{\theta})\right]\right\}$$

$$\propto \exp\left\{-S(\boldsymbol{\theta}_j)/2\right\} \qquad (7.5\text{-}14)$$

in which $S(\boldsymbol{\theta}_j)$ denotes the generalized sum of squares $\operatorname{tr}\boldsymbol{\Sigma}^{-1}\boldsymbol{v}(\boldsymbol{\theta})$ of Eq. (7.1-6) for model $j$. Expressing $S(\boldsymbol{\theta}_j)$ as its minimum value $\widehat{S}_j$ plus a nonnegative function $G_j(\boldsymbol{\theta}_j)$ and inserting the result into Eq. (7.5-13), we obtain

$$p(M_j|\boldsymbol{Y}, \boldsymbol{\Sigma}) = p(M_j)\frac{\exp\{-\widehat{S}_j/2\}\int \exp\{-G(\boldsymbol{\theta}_j)/2\}d\boldsymbol{\theta}_j}{\mathrm{E}[\exp\{-\widehat{S}_j/2\}\int \exp\{-G(\boldsymbol{\theta}_j)/2\}d\boldsymbol{\theta}_j]}$$

$$= p(M_j)\frac{\exp\{-\widehat{S}_j/2\}}{\mathrm{E}[\exp\{-\widehat{S}_j/2\}]} \qquad (7.5-15)$$

Here we have taken $\exp\{-G(\boldsymbol{\theta}_j)/2\}$ at its expected value for each point in the range of $\boldsymbol{\theta}_j$; consequently, the integrals over $\boldsymbol{\theta}_j$ have canceled out.

The denominator of Eq. (7.5-15) can be evaluated in the manner of Eq. (6.6-29), noting that the sum of squares here has $\nu_j = nm - p_j$ degrees of freedom. The expectation $\mathrm{E}[]$ is again proportional to $2^{-\nu_j/2}$ and thus to $2^{p_j/2}$. Hence, Eq. (7.5-15) gives the posterior probabilities

$$p(M_j|\boldsymbol{Y}, \boldsymbol{\Sigma}) \propto p(M_j)2^{-p_j/2}\exp(-\widehat{S}_j/2) \qquad j = 1, \ldots, J \qquad (7.5\text{-}16)$$

This result parallels Eq. (6.6.-30); the quotient $\widehat{S}_j/\sigma^2$ there corresponds, for $m = 1$, to the minimum of $S(\boldsymbol{\theta}_j)$ in Eq. (7.5-14). The parameterization penalty factor $2^{-p_j/2}$ holds again here, and holds also for block-diagonal or irregular data structures, as shown by Stewart, Shon, and Box (1998).

*Case 2: Block-Diagonal, Unknown Covariance Matrix*

Suppose that $\boldsymbol{v}(\boldsymbol{\theta})$ is block-diagonal and that $\boldsymbol{\Sigma}$ is unknown (though its relevant elements will form a block-diagonal array, as in Table 7.2b). Suppose that $\boldsymbol{Y}$ yields sample estimates $\boldsymbol{v}_{eb}$ with $\nu_{be}$ degrees of freedom for the experimental error contributions to one or more of the residual block matrices $\boldsymbol{v}_b(\boldsymbol{\theta}_j)$. Then the posterior probabilities based on the combined data take the form

$$p(M_j|\boldsymbol{Y}) \propto p(M_j)2^{-p_j/2}\prod_b |\widehat{\boldsymbol{v}}_{bj}|^{-\nu_{eb}/2} \qquad (7.5\text{-}17)$$

for any number of sampled blocks, by analogy with Eq. (6.6-34).

## 7.6 SEQUENTIAL PLANNING OF EXPERIMENTS

Sequential planning is particularly helpful in multiresponse experiments to select informative sets of responses for supplemental events. Replicate experiments at the outset would make this selection easier. The discussion in Section 6.7 is also applicable here, with the matrices $\hat{A}$ and $\hat{\mathcal{A}}$ calculated by the methods of this chapter. Sequential planning for multiresponse model discrimination, and for discrimination combined with estimation, may be implemented in GREGPLUS and Athena at another time.

## 7.7 EXAMPLES

Two examples are treated here: one on multicomponent diffusion and one on reaction kinetics. Parameter estimation and model discrimination are demonstrated, along with goodness-of-fit testing when replicates are available.

### Example 7.1. Kinetics of a Three-Component System

Consider the chemical conversion of pure species 1 to species 2 and 3 in a batch isothermal reactor. Simulated yields $y_{iu}$ for various times $t_u$ are as follows:

| $t_u$, min | $w_u$ | $y_{1u}$ | $y_{2u}$ | $y_{3u}$ |
|:----------:|:-----:|:--------:|:--------:|:--------:|
| 0.5 | 2 | 0.936 | 0.043 | 0.014 |
| 1.0 | 1 | 0.855 | 0.152 | 0.068 |
| 1.0 | 1 | 0.785 | 0.197 | 0.096 |
| 2.0 | 1 | 0.628 | 0.130 | 0.090 |
| 2.0 | 1 | 0.617 | 0.249 | 0.118 |
| 4.0 | 1 | 0.480 | 0.184 | 0.374 |
| 4.0 | 1 | 0.423 | 0.298 | 0.358 |
| 8.0 | 1 | 0.166 | 0.147 | 0.651 |
| 8.0 | 1 | 0.205 | 0.050 | 0.684 |
| 16.0 | 2 | 0.044 | 0.024 | 0.945 |

**Table 7.4:** Simulated Data, Adapted from Box and Draper (1965).

The expected values of the yields are modeled by the differential equations

$$\frac{df_1}{dt} = -k_1 f_1$$

$$\frac{df_2}{dt} = k_1 f_1 - k_2 f_2$$

$$\frac{df_3}{dt} = k_2 f_2$$

with coefficients $k_1$ and $k_2$ min$^{-1}$. These equations have the solution

$$f_1 = \exp -k_1 t$$
$$f_2 = [\exp(-k_1 t) - \exp(-k_2 t)]\, k_1/(k_2 - k_1)$$
$$f_3 = 1 - f_1 - f_2$$

under the given initial condition. The parameters for this model are chosen as $\theta_1 = \ln k_1$ and $\theta_2 = \ln k_2$ so that the assumption of a locally uniform $p(\boldsymbol{\theta})$ will exclude negative rate coefficients in accord with chemical theory.

The parameters are readily estimated by GREGPLUS, as illustrated in Appendix C. Here we describe some calculations based on a modified data set, in which the first and last events are given doubled weights $w_u = 2$.

Let us first assume that the errors in $f_1$ and $f_2$ are uncorrelated; then $\boldsymbol{\Sigma}$ is a diagonal matrix, as in Table 7.2c. The resulting parameter estimates (95% HPD intervals) are

$$\theta_1 = -1.569 \pm 0.067 \quad \theta_2 = -0.684 \pm 0.252 \quad \text{for diagonal } \boldsymbol{\Sigma}$$

and correspond to a least-squares solution.

To analyze these data with a full covariance structure, we put the responses in one block. The resulting 95% HPD interval estimates,

$$\theta_1 = -1.559 \pm 0.053 \quad \theta_2 = -0.698 \pm 0.226 \quad \text{for full } \boldsymbol{\Sigma}$$

are tighter than the previous set, so use of the full covariance structure is preferable.

The estimation can be greatly improved by strategic addition of data. Runs starting with component 2 would be most useful. Trial simulations indicate that an added run at $t_u = 2.0$ starting with component 2 would give a 2.46-fold increase in the normal-equation determinant $|\hat{\boldsymbol{A}}|$.

### Example 7.2. Multicomponent Diffusion in a Porous Catalyst

Feng, Kostrov and Stewart (1974) reported multicomponent diffusion data for gaseous mixtures of helium (He), nitrogen (N$_2$), and methane (CH$_4$) through an extruded platinum-alumina catalyst as functions of pressure (1 to 70 atm), temperature (300 to 390 K), and terminal compositions. The experiments were designed to test several models of diffusion in porous media over the range between Knudsen and continuum diffusion in a commercial catalyst (Sinclair-Engelhard RD-150) with a wide pore-size distribution.

The results of the study are summarized in Table 7.5, along with a brief account of the features of each $p_j$-parameter model. Each model was fitted by least squares to 283 observations $y_{iu}$ of the functions $\ln N_{iuz}$, where $N_{iuz}$ is the measured axial flux of species $i$ in the $u$th event, in g-moles per second per cm$^2$ of particle cross section. This corresponds to using the same variance $\sigma^2$ for each response function $\ln N_{iuz}$. Lacking replicates, we compare the models according to Eq. (7.5-16) with a variance estimate $s^2 = 0.128/(283-6)$, the residual mean-square deviation of the observations from Model 3. The weighted sum in Eq, 7.5-16 is then computed for Model $j$ as $\sum_{i,u} \epsilon_{iuj}^2/s^2 = 2,164\widehat{S}_j$.

The results thus found are striking. Model 3 is the clear winner on these data, with a posterior probability share above 0.999. This model is also preferable on physical grounds. The variance estimate was obtained from the residuals for Model 3; therefore, the goodness of fit of this model could not be tested.

| Model $j$ | Features, Authors | $p_j$ | $\widehat{S}_j$ | $\chi_j^2$ | Posterior Probability Share, $\Pi(M_j|\boldsymbol{Y})$ |
|---|---|---|---|---|---|
| 1g | Bulk and Knudsen diffusion, one pore size parameter Mason and Evans (1969) Gunn and King (1960); | 2 | 1.655 | 3582. | $10^{-717}$ |
| 2g | Bulk and Knudsen diffusion, full pore size distribution $\epsilon(r)$ Johnson and Stewart (1965) | 1 | 1.431 | 3097. | $10^{-612}$ |
| 3g | Bulk and Knudsen diffusion Feng and Stewart (1974) | 4 | 0.441 | 954. | $10^{-147}$ |
| 3Ag | Bulk and Knudsen diffusion, two pore sizes inferred from $\epsilon(r)$ Feng and Stewart (1974) | 2 | 0.988 | 2138. | $10^{-404}$ |
| 1 | Model 1g plus two surface diffusivities | 4 | 0.776 | 1679. | $10^{-304}$ |
| 2 | Model 2g plus two surface diffusivities | 3 | 0.717 | 1552. | $10^{-276}$ |
| 3 | Model 3g plus two surface diffusivities | 6 | **0.128** | **277** | **0.999** |
| 3A | Model 3Ag plus two surface diffusivities | 4 | 0.204 | 441. | $10^{-35}$ |

**Table 7.5:** Diffusion Model Discrimination; RD-150 Catalyst.

## 7.8 PROCESS INVESTIGATIONS

Investigations of multiresponse modeling are summarized in Table 7.6 for various chemical process systems. All data types except that of Table 7.2a are represented. Investigations 1–3 and 5–7 used data of the type in Table 7.1, though the values chosen for Investigations 2 and 3 constituted lines 1–8 of the irregular data array of Fuguitt and Hawkins (1945). Investigations 9–11 involved independent blocks of data from different laboratories. Investigations 1, 2, 6, and 7 gave point estimates $\hat{\boldsymbol{\theta}}$ only; the others included interval estimates of the parameters.

The first real-life application of Eq. (7.1-7) was made by Mezaki and Butt (1968) to a four-component chemical reaction system. They estimated the six kinetic parameters at each temperature by minimizing $|\boldsymbol{v}(\boldsymbol{\theta})|$, and also by minimizing the total sum of squares, $\mathrm{tr}\,[\boldsymbol{v}(\boldsymbol{\theta})]$. The parameter estimates by the two methods differed significantly, and the minimum-$|\boldsymbol{v}|$ estimates were more consistent with the literature. The greater steepness of the $|\boldsymbol{v}(\boldsymbol{\theta})|$ function gave faster convergence toward $\hat{\boldsymbol{\theta}}$. Computation of a grid of values around the endpoint confirmed the local minimization of $|\boldsymbol{v}(\boldsymbol{\theta})|$.

Investigation 2, by Box et al. (1973), is a classic study of causes and cures for singularities of $\boldsymbol{v}(\boldsymbol{\theta})$. For the data of Fuguitt and Hawkins (1945), use of the five reported responses gave a singular $\boldsymbol{v}(\boldsymbol{\theta})$ within rounding error for all parameter values, making the minimum-$|\boldsymbol{v}|$ criterion useless. To correct this situation the matrix $\boldsymbol{Y}^T\boldsymbol{Y}$ was analyzed, along with the original description of the data, to find a linearly independent subset of three responses; the minimization then went smoothly with a determinant $|\boldsymbol{v}(\boldsymbol{\theta})|$ of order 3. McLean et al. (1979) recommended testing the rank of $\boldsymbol{v}(\boldsymbol{\theta})$ rather than $\boldsymbol{Y}^T\boldsymbol{Y}$; this is done in GREGPLUS and Athena.

Investigation 3 gave a more detailed treatment of the data considered by Box et al. (1973), using an early version of Subroutine GREG. Local Taylor expansions of $|\boldsymbol{v}(\boldsymbol{\theta})|$ were used to compute constrained Newton steps toward the minimum and to obtain interval estimates of $\boldsymbol{\theta}$. Local rank testing of $\boldsymbol{v}$ was included in the selection of working responses. A collocation algorithm was used to compute the parametric sensitivities $\partial F_i(\boldsymbol{x}_u)/\partial\theta_r$ directly.

Investigation 4 treated the full data of Fuguitt and Hawkins (1945,1947) with modal and interval estimation of $\boldsymbol{\theta}$ and $\boldsymbol{\Sigma}$ via Eq. (7.1-19) for this irregular data structure.

Investigation 5 is a pioneering example of sequential design and analysis of multiresponse experiments. The vanadia-catalyzed oxidation of $o$-xylene was investigated in a recycle reactor at a temperature of 320°C. A comprehensive 15-parameter kinetic model based on earlier studies was postulated. The parameters were estimated after 15 preliminary runs, and every fourth run thereafter, by minimizing the determinant $|\boldsymbol{v}(\boldsymbol{\theta})|$ with the search algorithm of Powell (1964). Each designed set of four runs was selected from a grid of candidates to minimize the predicted volume of the 15-parameter joint confidence region by the method of Draper and Hunter (1966). The joint confidence region shrank considerably over the early iterations of this process, and the improvement continued less rapidly through four design cycles (16 runs). Five final runs were then performed and a reduced model

| INVESTIGATION | | PROBLEM TYPE | $m$, or $\{m_b\}$[1] VALUES | MODEL EQS.[2] | DESCENT ALGORITHM | NO. of $\theta$'s ESTIMATED |
|---|---|---|---|---|---|---|
| 1. Ethanol dehydration | Mezaki and Butt (1968) | 1 | 4 | ODE | Sobol (1963) | 6 |
| 2. α-Pinene decomposition | Box et al. (1973) | 1 | 3 | ODE | Hooke and Jeeves (1962) | 5 |
| 3. α-Pinene decomposition | Stewart and Sørensen (1976) | 1 | 3 | ODE | GREG | 6 |
| 4. α-Pinene decomposition | Stewart and Sørensen (1981) | 3 | 4 | ODE | GREG | 12–13 |
| 5. o-Xylene oxidation | Boag, Bacon and Downie (1978) | 1 | 5 | AE | Powell (1964) | 15 |
| 6. Oil shale decomposition | Ziegel and Gorman (1980) | 1 | 1–3 | ODE | Rosenbrock (1960) | 3–6 |
| 7. Petrochemical process | Ziegel and Gorman (1980) | 1 | 4 | AE | Rosenbrock (1960) | |
| | Stewart and Sørensen, Consultants | 1 | 4 | AE | GREG | 28–35 |
| 8. Dow test problem | Biegler, Damiano and Blau (1986): | | | | | |
| | Stewart, Caracotsios, and Sørensen (1992) | 3 | 2 | ODAE | GREG | 7 |
| | Other solutions | 2c | 2–4 | ODE, ODAE | Various | 7–9 |
| 9. Propylene hydrogenation: | Stewart, Shabaker and Lu (1988); Stewart, Shon and Box (1998) | 2b | {1, 1} | AE | GREG | 58–70 |
| 10. Catalytic reforming | Weidman and Stewart (1990) | 2b | {5, 1} | PDAE | GREG | 11–12 |
| 11. Distillation tray models | Young and Stewart (1992, 1995) | 2b | {1, 1, 1} | AE | GREG | 152–157 |

[1] Bracketed lists show multiblock data structures.
[2] Equation types are denoted by acronyms:
AE for algebraic equations, ODE for ordinary differential equations, ODAE for ordinary differential and algebraic equations, and PDAE for partial differential and algebraic equations.

**Table 7.6:** Multiresponse Investigations with Unknown Covariance Matrix.

was fitted, retaining eight significant parameters.

Investigations 6 and 7 were reported by Ziegel and Gorman (1980). The oil shale study was based on data from the literature, and gave a considerably improved representation by means of a multiresponse model that was fitted by minimizing $|v(\theta)|$. The petrochemical process is unnamed for proprietary reasons, but the study is notable as an early application of multiresponse, multiparameter modeling to a complex practical process. Models of the latter process were also investigated successfully with the subroutine GREG of that time, as indicated in Table 7.6.

Investigation 8 is a set of contributed solutions to a maximum-likelihood estimation problem posed by Blau, Kirkby and Marks (1981) of the Dow Chemical Company. Data on four responses were provided, along with a nine-parameter reaction model and a four-parameter model for the variances as functions of the responses. Five contributed solutions were reported by Biegler, Damiano, and Blau (1986) including ours, which is described briefly here. The four responses satisfy two linear equality constraints (material balances); thus, the covariance matrix has rank 2 at most, and 2 in our solution. There were occasional missing observations, as in Tables 7.3 and 7.2c. We treated $\Sigma$ as a nondiagonal matrix; consequently, the variance model provided by Dow needed to be generalized to the form in Eq. (7.1-20), which corresponds to Table 7.3. The resulting parameter estimates showed $\Sigma$ to be nondiagonal and showed moderate heteroscedasticity (nonconstant covariances), with $\gamma_i$ values of 0.0 and $0.190 \pm 0.118$, respectively, for the responses code-named [HABM] and [AB]. The five contributed solutions represent the data tolerably, but the parameter estimates differ appreciably because of differences in model forms.

Investigation 9 dealt with reaction rate models for the catalytic hydrogenation of propylene over Pt-alumina. Computations via Eq. (7.1-15) were given for 15 reaction models, the best of which were constructed from evidence on multiple surface species along with the reactor data.

The two responses were observed separately in different laboratories by Rogers (1961) and Shabaker (1965). Corrections for catalyst deactivation during each of the 28 days of Shabaker's experiments added 55 ($= 2 \times 28 - 1$) parameters to each model; earlier workers with these data used comparable numbers of parameters implicitly in their graphical activity corrections. The models were ranked approximately in 1988 by a heuristic criterion and more clearly in 1998 by posterior probabilities computed from Eq. (7.5-17).

Investigation 10 was a study of fixed-bed reactor models and their application to the data of Hettinger et al. (1955) on catalytic reforming of $C_7$ hydrocarbons. The heuristic posterior density function $p(\theta|Y)$ proposed by Stewart (1987) was used to estimate the rate and equilibrium parameters of various reaction schemes, two of which were reported in the article. The data were analyzed with and without models for the intraparticle and boundary-layer transport. The detailed transport model led to a two-dimensional differential-algebraic equation system, which was solved via finite-element discretization in the reactor radial coordinate and adaptive stepwise integration downstream. (Weidman 1990). Parametric sensitivities were calculated during the integrations via the algorithm of Caracotsios and Stewart (1985), updated as DDAPLUS of Appendix B. The detailed transport model was able to describe the data with a simpler reaction scheme. Models capable of exactly fitting the smaller data block ($n_b = 2$, $m_b = 1$) were avoided to keep aromatic yields as working responses.

Investigation 11 dealt with transport models for the fractionation performance of sieve trays. Data from three laboratories were analyzed. Two to six variables were observed in each experiment. These observations were treated in Eq. (7.1-1) as independent events $u$, with equal weights $w_u$ for all values reported from a given laboratory. Thus, the observations from Laboratory $i$ were reduced to a column of values $y_{iu}$ with a common variance $\sigma_{ii}$. The resulting covariance matrix was diagonal, as in Table 7.2c, and the $Y$ array was block-rectangular with block widths $m_b = 1$.

For each model studied in Investigation 11, 3 to 9 predictive parameters were estimated, plus 149 parameters representing adjusted conditions of the experiments. The estimation was done according to Eq. (7.1-15), since Eq. (7.1-16) was not available at that time. A promising model was selected from among several candidates and was reduced to a concise form by Young and Stewart (1995).

Since the covariance matrix was diagonal in Investigation 11, the minimization of $-2\ln p(\boldsymbol{\theta}|\boldsymbol{Y})$ gave optimally weighted least-squares solutions. The same was true in Investigation 9.

## 7.9 CONCLUSION

In this chapter, Bayesian and likelihood-based approaches have been described for parameter estimation from multiresponse data with unknown covariance matrix $\boldsymbol{\Sigma}$. The Bayesian approaches permit objective estimates of $\boldsymbol{\theta}$ and $\boldsymbol{\Sigma}$ by use of the noninformative prior of Jeffreys (1961). Explicit estimation of unknown covariance elements is optional for problems of Types 1 and 2 but mandatory for Types 3 and 4.

The posterior density function is the key to Bayesian parameter estimation, both for single-response and multiresponse data. Its mode gives point estimates of the parameters, and its spread can be used to calculate intervals of given probability content. These intervals indicate how well the parameters have been estimated; they should always be reported.

For simplicity, the usual notational distinction between independent variables $x_{iu}$ and observations $y_{iu}$ has been maintained in this chapter, with the experimental settings $x_{iu}$ regarded as perfectly known. If the settings are imprecise, however, it is more natural to regard them as part of the observations $y_{iu}$; this leads to various error-in-variables estimation methods. Full estimation of $\boldsymbol{\Sigma}$ is then not possible (Solari 1969), and some assumptions about its elements are necessary to analyze the data. Conventional "error-in-variables" treatments use least squares, with a scalar variance $\sigma^2$.

Multiresponse analysis can provide fuller information, as illustrated in Investigation 11. There is a growing literature on error-in-variables methods.

Many multiresponse investigations have used procedures based on Eq. (7.1-7) or (7.1-15), which exclude $\Sigma$ from the parameter set but implicitly estimate $\widehat{\Sigma}$ nonetheless. Additional formulas of this type are provided in Eqs. (7.1-8,9) and (7.1-16,17), which give estimates of $\boldsymbol{\theta}$ that are more consistent with the full posterior density functions for those problem structures; see Eqs. (7.1-6) and (7.1-14).

Explicit estimation of $\Sigma$ requires enough data to include each independent element $\sigma_{ij}$ as a parameter. Replicate experiments are particularly effective for this purpose, provided that they are reported precisely enough to give error information [unlike the repeated values produced by rounding in the data of Fuguitt and Hawkins (1945, 1947)].

## 7.10 NOTATION

| | |
|---|---|
| $\boldsymbol{A}_{\theta\theta}$ | submatrix of elements $\partial^2 S(\boldsymbol{\theta})/\partial\theta_r\partial\theta_s$ in Eq. (7.2-16) |
| $\hat{\boldsymbol{A}}$ | matrix of Eq. (7.2-16) at $\widehat{\theta}$ |
| $\hat{\boldsymbol{A}}_{ee}$ | modal value of $\boldsymbol{A}_{\theta\theta}$ for estimated parameters |
| $\hat{\boldsymbol{A}}^{ee}$ | inverse of submatrix $\hat{\boldsymbol{A}}_{ee}$ for estimated parameters |
| $\hat{\boldsymbol{\mathcal{A}}}_{ee}$ | modal value of $\hat{\boldsymbol{\mathcal{A}}}$ for estimated functions |
| $\hat{\boldsymbol{\mathcal{A}}}^{ee}$ | inverse of submatrix $\hat{\boldsymbol{\mathcal{A}}}_{ee}$ for estimated functions |
| $\hat{\boldsymbol{A}}^{aa}$ | submatrix of $\hat{\boldsymbol{A}}^{ee}$ for parameter subvector $\boldsymbol{\theta}_a$ |
| $\hat{\boldsymbol{A}}_{rs}$ and $\hat{\boldsymbol{A}}^{rs}$ | elements of $\hat{\boldsymbol{A}}_{ee}$ and $\hat{\boldsymbol{A}}^{ee}$ |
| $b$ | response block number, with range from 1 to $B$ |
| $i$ | response index in Eq. (7.1-1) |
| $\mathcal{E}_{iu}$ | error term in Eq. (7.1-1) |
| $f_i(\boldsymbol{x}_u, \boldsymbol{\theta})$ | expectation model for response $i$ in event $u$ |
| $F_i(\boldsymbol{x}_u, \boldsymbol{\theta})$ | $\equiv \sqrt{w_u} f_i(\boldsymbol{x}_u, \boldsymbol{\theta})$, weighted expectation value for response $i$ in event $u$ |
| $\boldsymbol{F}(\boldsymbol{x}, \boldsymbol{\theta})$ | array of weighted expectation model values $F_i(\boldsymbol{x}_u, \boldsymbol{\theta})$ |
| $\mathcal{E}_{iu}$ | $= Y_{iu} - F_{iu}(\boldsymbol{x}_u, \boldsymbol{\theta})$, error in weighted observation $Y_{iu}$ |
| $l(\boldsymbol{\theta}, \boldsymbol{\Sigma}|\boldsymbol{Y})$ | likelihood function in Eq. (7.1-4) or (7.1-18) |
| LPOWR$_b$ | coefficient defined in Eq. (7.2-2) |
| $m$ | number of working responses in data array |
| $m_b$ | number of working responses in block $b$ |
| $m_u$ | number of working responses in event $u$ |
| $n$ | number of events in data collection |
| $n_a$ | number of parameters in subvector $\boldsymbol{\theta}_a$ of $\boldsymbol{\theta}_e$ |
| $n_b$ | number of events in response block $b$ |
| $p_j$ | number of parameters estimated in Model M$_j$ |
| $p(x)$ | prior probability density of a continuous random variable at $x$, or prior probability of value $x$ for a discrete variable |
| $p(\boldsymbol{\theta}, \boldsymbol{\Sigma}|\boldsymbol{Y})$ | posterior density function in Eq. (7.1-6) or (7.1-19) |
| $p(\boldsymbol{\theta}|\boldsymbol{Y})$ | marginal posterior density function in Eq. (7.1-7) |

| | |
|---|---|
| $\widetilde{p}(\boldsymbol{\theta}|\boldsymbol{Y})$ | modified posterior density function in Eq. (7.1-8) |
| $S(\boldsymbol{\theta})$ | objective function in Eq. (7.2-1) |
| $\widetilde{S}(\boldsymbol{\theta})$ | local quadratic approximation of $S(\boldsymbol{\theta})$, Eq. (7.2-15) |
| $\Delta\widetilde{S}^k$ | increment of $\widetilde{S}$ computed by GRQP in iteration $k$ |
| $t$ | time |
| $u$ | event index in Eq. (7.1-1) |
| $\boldsymbol{v}(\boldsymbol{\theta})$ | error product matrix in Eq. (7.1-3) |
| $\boldsymbol{v}_b(\boldsymbol{\theta})$ | subvector of $\boldsymbol{v}(\boldsymbol{\theta})$ for block $b$ |
| $\boldsymbol{x}_u$ | vector of experimental settings for event $u$ |
| $w_u$ | weight assigned to observations in event $u$ |
| $\boldsymbol{Y}$ | matrix of weighted observations $y_{iu}\sqrt{w_u}$ |
| $y_{iu}$ | unweighted observation of response $i$ in event $u$ |

*Greek letters*

| | |
|---|---|
| $\alpha$ | probability content of region of positive $\theta_r$ in Eq. (7.5-6), or of region $S(\boldsymbol{\theta}) > S(\boldsymbol{\theta}_a)$ in Eq. (7.5-8) |
| $\gamma_i$ | heteroscedasticity exponent for response $i$ in Eq. (7.1-20) |
| $\delta_{ij}$ | Kronecker symbol, unity when $i = j$ and zero otherwise |
| $\varepsilon_{iu}$ | $=(y_{iu} - f_i(\boldsymbol{x}_u, \boldsymbol{\theta})$ error term in observation $y_{iu}$; depends on $\boldsymbol{\theta}$ when model and observation $y_{iu}$ are given |
| $\boldsymbol{\theta}$ | vector of parameters in expectation term of Eq. (7.1-1) |
| $\widehat{\boldsymbol{\theta}}$ | modal estimate of $\boldsymbol{\theta}$ |
| $\theta_r$ | element $r$ of parameter vector $\boldsymbol{\theta}$ |
| $\nu_{bj}$ | residual degrees of freedom for Model M$_j$ in block $b$ |
| $\sum$ | summation sign |
| $\boldsymbol{\Sigma}$ | error covariance matrix; see Eq. (4.4-1) and Problem **4.C**. |
| $\boldsymbol{\Sigma}_u$ | error covariance matrix for event $u$ |
| $\boldsymbol{\Sigma}_b$ | error covariance matrix for response block $b$ |
| $\widehat{\boldsymbol{\Sigma}}$ | modal estimate of $\boldsymbol{\Sigma}$ |
| $\widehat{\boldsymbol{\Sigma}}(\boldsymbol{\theta})|\boldsymbol{Y}$ | maximum-density estimate of $\boldsymbol{\Sigma}$ at $\boldsymbol{\theta}$, conditional on $\boldsymbol{Y}$ |
| $\sigma_{ij}$ | element $ij$ of covariance matrix $\boldsymbol{\Sigma}$ |
| $\sigma_u^{ij}$ | element $ij$ of precision matrix $\boldsymbol{\Sigma}_u^{-1}$ |
| $\sigma_{\theta_r}^2$ | variance of marginal posterior distribution of $\theta_r$ |
| $\boldsymbol{\theta}$ | parameter vector in Eq. (7.1-1) |
| $\widehat{\boldsymbol{\theta}}$ | modal (maximum-density) value of $\boldsymbol{\theta}$ |
| $\theta_r$ | element $r$ of $\theta$ |
| $\omega_{ij}$ | reference value of $\sigma_{ij}$ in Eq. (7.1-20) |

*Subscripts*

| | |
|---|---|
| $a$ | of parameter subvector $\boldsymbol{\theta}_a$ |
| $b$ | of response block $b$ |
| $e$ | of estimated parameters |
| $u$ | of event $u$ |

*Superscripts*

| | |
|---|---|
| $ij$ | element $ij$ of inverse matrix |
| $-1$ | inverse |
| $T$ | transpose |

|   | modal (maximum-density) value |
|---|---|

*Functions and operations*

| $\equiv$ | preceding notation is defined by next expression |
|---|---|
| $E(z)$ | expectation of random variable $z$ |
| Pr | cumulative probability |
| $\text{tr}(\boldsymbol{X})$ | $=\sum_i x_{ii}$, trace of matrix $\boldsymbol{X}$ |
| $|\boldsymbol{X}|$ | determinant of matrix $\boldsymbol{X}$ |
| $\propto$ | proportionality sign |
| $\prod$ | product |

## ADDENDUM: PROOF OF EQS. (7.1-16) AND (7.1-17)

The following identities for real nonsingular matrices $\boldsymbol{X}$ and $\boldsymbol{B}$ are derived by Bard (1974):

$$\frac{\partial \ln |\boldsymbol{X}|}{\partial \boldsymbol{X}} = \left(\boldsymbol{X}^{-1}\right)^T \tag{A.1}$$

$$\frac{\partial x^{kl}}{\partial x_{ij}} = -x^{ki} x^{jl} \tag{A.2}$$

$$\frac{\partial}{\partial \boldsymbol{X}} \text{tr}(\boldsymbol{X}^T \boldsymbol{B}) = \boldsymbol{B} \tag{A.3}$$

Here the elements of $d\boldsymbol{X}$ are treated as independent. The matrix $\boldsymbol{X}$ of interest here is $\boldsymbol{\Sigma}_b^{-1}$, the inverse covariance matrix of response block $b$. Elements of $\boldsymbol{X}^{-1}$ are marked with superscript indices, and $(\boldsymbol{X})^T$ is the transpose of $\boldsymbol{X}$.

For analysis of Eq. (7.1-14), define the function

$$G(\boldsymbol{\theta}, \boldsymbol{\Sigma}) = -2 \ln p(\boldsymbol{\theta}, \boldsymbol{\Sigma}) + \text{const.}$$

$$= \sum_{b=1}^{B}(n_b + m_b + 1) \ln |\boldsymbol{\Sigma}_b| + \sum_{b=1}^{B} \text{tr}\left[\boldsymbol{\Sigma}_b^{-1} \boldsymbol{v}_b(\boldsymbol{\theta})\right] \tag{A.4}$$

for nonsingular $\boldsymbol{\Sigma}_b$ and $\boldsymbol{v}_b(\boldsymbol{\theta})$ in each block $b$. Differentiation with respect to $\boldsymbol{\Sigma}_b^{-1}$ for a particular block $b$ gives

$$\frac{\partial G}{\partial \boldsymbol{\Sigma}_b^{-1}} = -(n_b + m_b + 1)\boldsymbol{\Sigma}_b + \boldsymbol{v}_b(\boldsymbol{\theta}) \tag{A.5}$$

after use of Eqs. (A.1) and (A.3). When this derivative matrix is zero, $\partial G/\partial \boldsymbol{\Sigma}_b$ is also zero in view of Eq. (A.2). The resulting stationary value of $G$ is a local minimum with respect to $\boldsymbol{\Sigma}_b$, and the corresponding estimate of $\boldsymbol{\Sigma}_b$ is

$$[\widetilde{\boldsymbol{\Sigma}}_b(\boldsymbol{\theta})|\boldsymbol{Y}] = \boldsymbol{v}_b(\boldsymbol{\theta})/(n_b + m_b + 1) \tag{A.6}$$

Hence,

$$[\widetilde{\boldsymbol{\Sigma}}_b^{-1}(\boldsymbol{\theta})|\boldsymbol{Y}] = (n_b + m_b + 1)\boldsymbol{v}_b^{-1}(\boldsymbol{\theta}) \tag{A.7}$$

and substitution into Eq. (A.4) gives

$$G(\boldsymbol{\theta}, \widetilde{\boldsymbol{\Sigma}}(\boldsymbol{\theta})|\boldsymbol{Y}) = -2 \ln p(\boldsymbol{\theta}, \widetilde{\boldsymbol{\Sigma}}(\boldsymbol{\theta})|\boldsymbol{Y}) + \text{const.}$$

$$= \sum_{b=1}^{B} (n_b + m_b + 1) \ln \left| \frac{\boldsymbol{v}_b(\boldsymbol{\theta})}{(n_b + m_b + 1)} \right|$$

$$+ \sum_{b=1}^{B} (n_b + m_b + 1) \text{tr} \left[ \boldsymbol{v}_b^{-1}(\boldsymbol{\theta}) \boldsymbol{v}_b(\boldsymbol{\theta}) \right] \tag{A.8}$$

$$= \sum_{b=1}^{B} (n_b + m_b + 1) \ln |\boldsymbol{v}_b(\boldsymbol{\theta})| + \text{const.}$$

Solving for $p(\boldsymbol{\theta}, \widetilde{\boldsymbol{\Sigma}}(\boldsymbol{\theta})|\boldsymbol{Y})$ then gives Eq. (7.1-16), and maximization of the latter function over $\boldsymbol{\theta}$ gives the modal parameter vector $\widehat{\boldsymbol{\theta}}$, valid equally for Eqs. (7.1-16) and (7.1-14). Application of Eq. (A.6) at $\widehat{\boldsymbol{\theta}}$ yields Eq. (7.1-17) as the modal estimate of the covariance matrix.

## PROBLEMS

### 7.A Testing Independence of Error Distributions

Modify Example C.4 of Appendix C to do the parameter estimation on the assumption of a diagonal $\boldsymbol{\Sigma}$ matrix. To do this, set the number of blocks NBLK equal to the number of responses and the block widths IBLK(K,1) equal to 1. Then do the following calculations:

(a) Estimate $\boldsymbol{\theta}$ with GREGPLUS and the modified program at LEVEL=22.

(b) Rerun Example C.4, using the modal parameter values from (a) as initial guesses.

(c) Compare the closeness of the interval estimates found in your two solutions.

### 7.B Effect of Mass-Balancing in Multiresponse Estimation

Modify Example C.4 to use the array OBS directly, ignoring the mass-balance condition $\sum_{i=1}^{3} y_{iu,\text{adj}} = 1$. The full set of experimental data is given in the table below:

| $t_u$ | $y_{1u}$ | $y_{2u}$ | $y_{3u}$ |
|-------|----------|----------|----------|
| 0.5   | 0.959    | 0.025    | 0.028    |
| 0.5   | 0.914    | 0.061    | 0.000    |
| 1.0   | 0.855    | 0.152    | 0.068    |
| 1.0   | 0.785    | 0.197    | 0.096    |
| 2.0   | 0.628    | 0.130    | 0.090    |
| 2.0   | 0.617    | 0.249    | 0.118    |
| 4.0   | 0.480    | 0.184    | 0.374    |
| 4.0   | 0.423    | 0.298    | 0.358    |
| 8.0   | 0.166    | 0.147    | 0.651    |
| 8.0   | 0.205    | 0.050    | 0.684    |
| 16.0  | 0.034    | 0.000    | 0.899    |
| 16.0  | 0.054    | 0.047    | 0.991    |

Then run GREGPLUS with the modified program. Compare the modal and 95% HPD interval estimates of the parameters with those obtained in Example C.4.

## 7.C Sequential Design Including Feed Composition

Extend Example C.4 to include candidate experiments with pure B as feed at each of the five $t_u = \{0.25, 3.0, 6.0, 12.0, 32.0\}$. Run GREGPLUS with the modified program to select the best candidate experiment according to each of the following criteria:

(a) Maximum determinant $|\hat{\boldsymbol{A}}^{ee}|$.

(b) Minimum trace of $\hat{\boldsymbol{A}}^{ee}$.

## 7.D Thermal Decomposition of Oil Shale

Hubbard and Robinson (1950) did extensive experiments on thermal decomposition of Colorado oil shales. The following lumped components were reported.

1. Gas: the noncondensed vapors formed during the heating of the shale.
2. Oil: the organic matter that was vaporized during heating of the shale and subsequently condensed.
3. Bitumen: the benzene-soluble organic matter found in the shale after the heating period.
4. Kerogen: the benzene-insoluble organic portion of the shale.

In addition, we define gasifiable inorganic matter other than water as a fifth component. The following data were obtained by Hubbard and Robinson in batch heating experiments at 400°C with a shale containing 7.9% kerogen.

| $u$ | $t_u$, min | $y_{1u}$ | $y_{2u}$ | $y_{3u}$ |
|-----|-----------|----------|----------|----------|
| 1   | 5         | 4.3      | 0.0      | 0.0      |
| 2   | 7         | 9.4      | 0.0      | 2.2      |
| 3   | 10        | 7.9      | 0.7      | 11.5     |
| 4   | 15        | 8.6      | 7.2      | 7.7      |
| 5   | 20        | 10.1     | 11.5     | 15.1     |
| 6   | 25        | 12.2     | 15.8     | 17.3     |
| 7   | 30        | 10.1     | 20.9     | 17.3     |
| 8   | 40        | 12.2     | 26.6     | 20.1     |
| 9   | 50        | 12.2     | 32.4     | 20.1     |
| 10  | 60        | 11.5     | 38.1     | 22.3     |
| 11  | 80        | 7.7      | 43.2     | 20.9     |
| 12  | 100       | 14.4     | 49.6     | 11.5     |
| 13  | 120       | 15.1     | 51.8     | 6.5      |
| 14  | 150       | 15.8     | 54.7     | 3.6      |

These data are to be modeled by Eq. (7.1-1), with weights $w_u = 1$ and the differential equation system

$$\frac{d}{dt}\begin{pmatrix} F_1 \\ F_2 \\ F_3 \\ F_4 \\ F_5 \end{pmatrix} = \begin{bmatrix} 0 & 0 & k_4 & 0 & k_5 \\ 0 & 0 & k_2 & k_3 & 0 \\ 0 & 0 & -(k_2 + k_4) & k_1 & 0 \\ 0 & 0 & 0 & -(k_1 + k_3) & 0 \\ 0 & 0 & 0 & 0 & -k_5 \end{bmatrix}\begin{pmatrix} F_1 \\ F_2 \\ F_3 \\ F_4 \\ F_5 \end{pmatrix} \quad (7.\text{D-}1)$$

with initial conditions

$$\text{at } t = t_0 \equiv \theta_6, \qquad \mathbf{F} = (0, 0, 0, 100, \theta_7)^T \qquad (7.\text{D-}2)$$

for the expectation functions. This model adds a fifth reaction and a fifth response to the scheme of Ziegel and Gorman (1980).

(a) Write a subroutine MODEL to simulate the experiments with Eqs. (7.D-1,2) at the current parameter vector $\boldsymbol{\theta} \equiv (k_1, k_2, k_3, k_4, k_5, f_5^0, \theta_6, \theta_7)^T$. Analytic solutions exist for various special cases, but numerical integration with DDAPLUS is recommended.

(b) Use GREGPLUS at LEVEL 22 with NBLK=3 (full $\mathbf{Y}$ and diagonal $\boldsymbol{\Sigma}$) to estimate $\boldsymbol{\theta}$ from all the data, starting at

$$\boldsymbol{\theta}^0 = (0.3, 0.5, 0.2, 0.3, 0.5, 0.5, 5.0)^T$$

This method gives an optimally weighted least-squares solution. Constraints $0 \le \theta_6 < 1$ and $\theta_7 > 0$ are recommended.

(c) Change NBLK to 1 (full $\mathbf{Y}$ and full $\boldsymbol{\Sigma}$) and estimate the parameters again, starting from the parameters found in (a). Report the initial and final $S$ values, modal parameter values, 95% HPD intervals, and normalized covariance matrix for the parameters.

(d) Compare the closeness of the HPD intervals found in (b) and (c).

## REFERENCES and FURTHER READING

Aitken, A. C., On least squares and linear combination of observations, *Proc. Roy. Soc. Edinburgh*, **55** 42–47 (1935).

Anderson, T. W., *An Introduction to Multivariate Statistical Analysis*, 2nd edition, Wiley, New York (1984).

Bain, R. S., Solution of nonlinear algebraic equation systems and single and multiresponse nonlinear parameter estimation problems, Ph. D. thesis, University of Wisconsin–Madison (1993).

Bard, Y., *Nonlinear Parameter Estimation*, Academic Press, New York (1974).

Bard, Y., and L. Lapidus, Kinetics analysis by digital parameter estimation, *Catal. Rev.*, **2**, 67–112 (1968).

Bates, D. M., and D. G. Watts, A multi-response Gauss-Newton algorithm, *Commun. Stat. – Simul. Comput.*, **13**, 705–715 (1984).

Bates, D. M., and D. G. Watts, Multiresponse estimation with special application to systems of linear differential equations, *Technometrics*, **27**, 329–339 (1985).

Bates, D. M., and D. G. Watts, A generalized Gauss-Newton procedure for multi-response parameter estimation, *SIAM J. Sci. Statist. Comput.*, **7**, 49–55 (1987).

Bates, D. M., and D. G. Watts, *Nonlinear Regression Analysis and Its Applications*, Wiley, New York (1988).

Bayes, T. R., An essay towards solving a problem in the doctrine of chances, *Phil. Trans. Roy. Soc. London*, **53**, 370–418 (1763). Reprinted in *Biometrika*, **45**, 293–315 (1958).

Biegler, L. T., J. J. Damiano, and G. E. Blau, Nonlinear parameter estimation: a case study comparison, *AIChE J.*, **32**, 29–45 (1986).

Blau, G. E., L. Kirkby, and M. Marks, An industrial kinetics problem for testing noninear parameter estimation algorithms, Dow Chemical Company (1981).

Boag, I. F., D. W. Bacon, and J. Downie, Using a statistical multiresponse method of experimental design in a reaction network study, *Can. J. Chem. Eng.*, **56**, 389–395 (1978).

Box, G. E. P., A general distribution theory for a class of likelihood criteria, *Biometrika*, **36**, 317–346 (1949).

Box, G. E. P., and N. R. Draper, The Bayesian estimation of common parameters from several responses, *Biometrika*, **52**, 355–365 (1965).

Box, M. J., and N. R. Draper, Estimation and design criteria for multiresponse non-linear models with non-homogeneous variance, *J. Roy. Statist. Soc.*, Series C *(Appl. Statist.,)* **21**, 13–24 (1972).

Box, M. J., N. R. Draper, and W. G. Hunter, Missing values in multiresponse nonlinear data fitting, *Technometrics*, **12**, 613–620 (1970).

Box, G. E. P., and W. J. Hill, Discrimination among mechanistic models, *Technometrics*, **9**, 57–71 (1967).

Box, G. E. P., W. G. Hunter, J. F. MacGregor, and J. Erjavec, Some problems associated with the analysis of multiresponse data, *Technometrics*, **15**, 33–51 (1973).

Box, G. E. P., and H. L. Lucas, Design of experiments in non-linear situations, *Biometrika*, **46**, 77–90 (1959).

Box, G. E. P., and G. C. Tiao, *Bayesian Inference in Statistical Analysis*, Addison-Wesley, Reading, MA (1973). Reprinted by Wiley, New York (1992).

Buzzi-Ferraris, G., P. Forzatti, and P. Canu, An improved version of a sequential design criterion of discriminating among rival response models, *Chem. Eng. Sci.*, **45**, 477 (1990).

Caracotsios, M., *Model Parametric Sensitivity Analysis and Nonlinear Parameter Estimation. Theory and Applications.* Ph. D. thesis, University of Wisconsin–Madison (1986).

Caracotsios, M., and W. E. Stewart, Sensitivity analysis of initial value problems with mixed ODEs and algebraic equations, *Comput. Chem. Eng.*, **9**, 359–365 (1985).

Caracotsios, M., and W. E. Stewart, Sensitivity analysis of initial-boundary-value problems with mixed PDEs and algebraic equations. Applications to chemical and biochemical systems, *Comput. Chem. Eng.*, **19**, 1019–1030 (1995).

Chow, G. C., A comparison of the information and posterior probability criteria for model selection, *J. Econometrics*, **14**, 21 (1981).

Dongarra, J. J., J. R. Bunch, C. B. Moler, and G. W. Stewart, *LINPACK Users' Guide*, SIAM, Philadelphia (1979).

Draper, N. R., and W. G. Hunter, Design of experiments for parameter estimation in multiresponse situations, *Biometrika*, **53**, 525–533 (1966).

Feng, C. F., V. V. Kostrov, and W. E. Stewart, Multicomponent diffusion of gases in porous solids. Models and experiments, *Ind. Eng. Chem. Fundam.*, **5**, 5–9 (1974).

Fisher, R. A., On the mathematical foundations of theoretical statistics, *Phil. Trans. Roy. Soc. London, A*, **222**, 309–368 (1922).

Fuguitt, R. E., and J. E. Hawkins, The liquid-phase thermal isomerization of $\alpha$-pinene, *J. Am. Chem. Soc.*, **67**, 242–245 (1945).

Fuguitt, R. E., and J. E. Hawkins, Rate of the thermal isomerization of $\alpha$-pinene in the liquid phase, *J. Am. Chem. Soc.*, **69**, 319–322 (1947).

Gauss, C. F., *Theoria Motus Corporum Coelestium in Sectionibus Conicis Solem Ambientium*, Perthes et Besser, Hamburg (1809). Reprinted as *Theory of Motion of the Heavenly Bodies Moving about the Sun in Conic Sections,* Translated by C. H. Davis, Dover, New York (1963).

Gauss, C. F., Theoria combinationis observationum erroribus minimis obnoxiae: Pars prior; Pars posterior; Supplementum, *Commentatines societatis regiae scientiarum Gottingensis recentiores*, **5**, (1823); **6**, (1828). Translated by G. W. Stewart as *Theory of the Combination of Observations Least Subject to Errors, Part One, Part Two, and Supplement*, Society for Industrial and Applied Mathematics, Philadelphia, (1995). A valuable Afterword by the translator is included.

Guay, M., and D. D. McLean, Optimization and sensitivity aanalysis for multiresponse parameter estimation in systema of ordinary differential equations, *Comput. Chem. Eng.*, **19**, 1271–1285 (1995).

Hettinger, W. P., Jr., C. D. Keith, J. L. Gring, and J. W. Teter, Hydroforming reactions. Effect of certain catalyst properties and poisons, *Ind. Eng. Chem.*, **47**, 719–730 (1955).

Hooke, R., and T. A. Jeeves, Direct search solution of numerical and statistical problems, *J. Assoc. Comput. Mach.*, **8**, 212–229 (1962).

Hubbard, A. B., and W. E. Robinson, A thermal decomposition study of Colorado oil shale, *Report of Investigations* **4744**, U.S. Department of the Interior, Bureau of Mines (1950).

Jeffreys, H., *Theory of Probability*, 3rd edition, Clarendon Press, Oxford (1961).

Klaus, R., and D. W. T. Rippin, A new flexible and easy-to-use general-purpose regression program, *Comput. Chem. Eng.*, **3**, 105–115 (1979).

Legendre, A. M., *Nouvelles Méthodes pour la Determination des Orbites de Comètes*, Paris (1805).

Lu, Y. T., Reaction modeling of propylene hydrogenation over alumina-supported platinum, MS Thesis, University of Wisconsin–Madison (1988).

Mason, E. A., and R. B. Evans III, *J. Chem. Educ.*, **46**, 359 (1969).

McLean, D. D., D. J. Pritchard, D. W. Bacon, and J. Downie, Singularities in multiresponse modeling, *Technometrics*, **21**, 291–298 (1979).

Mezaki, R., Modification of a reaction model involving compettive-noncompetitive adsorption, *J. Catal.*, **10**, 238 (1968).

Mezaki, R., and J. B. Butt, Estimation of rate constants from multiresponse kinetic data, *Ind. Eng. Chem. Fundam.*, **7**, 120–125 (1968).

Powell, M. J. D., An efficient method for finding the minimum of a function of several variables without calculating derivatives, *Computer J.*, **7**, 155–162 (1964).

Rogers, G. B., Kinetics of the catalytic hydrogenation of propylene, PhD thesis, University of Wisconsin–Madison (1961).

Rosenbrock, H. H., An automatic method for finding the greatest or least value of a function, *Computer J.*, **3**, 175–184 (1960).

Shabaker, R. H., Kinetics and effectiveness factors for the hydrogenation of propylene on a platinum-a;umina catalyst, PhD thesis, University of Wisconsin–Madison (1965).

Sobol, M., NU Scoop Linear Surface Minimization Routine, *IBM Share Program* **1551** (1963).

Solari, M. E., The "maximum likelihood solution" of the problem of estimating a linear functional relationship, *J. Roy. Statist. Soc.*, B, **31**, 372–375 (1969).

Sørensen, J. P., *Simulation, Regression and Control of Chemical Reactors by Collocation Techniques*, Doctoral thesis, Danmarks tekniske Højskole, Lyngby (1982).

Steiner, E. C., G. E. Blau, and G. L. Agin, *Introductory Guide to SIMUSOLVE*, Mitchell & Gauthier, Concord, MA (1986).

Stewart, W. E., Multiresponse parameter estimation with a new and noninformative prior, *Biometrika*, **74**, 557–562 (1987). This prior is superseded by Eq. (7.1-5).

Stewart, W. E., M. Caracotsios, and J. P. Sørensen, Parameter estimation from multiresponse data, *AIChE J.*, **38**, 641–650 (1992); Errata, **38**, 1302 (1992).

Stewart, W. E., R. H. Shabaker, and Y. T. Lu, Kinetics of propylene hydrogenation on platinum-alumina, *Chem. Eng. Sci.*, **43**, 2257–2262 (1988); Errata, **44**, 205 (1989).

Stewart, W. E., Y. Shon, and G. E. P. Box, Discrimination and goodness of fit of multiresponse mechanistic models, *AIChE J.*, **44**, 1404–1412 (1998).

Stewart, W. E., and J. P. Sørensen, Sensitivity and regression of multicomponent reactor models, *4th Int. Symp. Chem. Reaction Eng.*, DECHEMA, Frankfurt, **I**, 12–20 (1976).

Stewart, W. E., and J. P. Sørensen, Bayesian estimation of common parameters from multiresponse data with missing observations, *Technometrics*, **23**, 131–141 (1981); Errata, **24**, 91 (1982).

Wei, J., and C. D. Prater, The structure and analysis of complex reaction systems, *Adv. Catal.*, **13**, 203–392 (1962).

Weidman, D. L. Catalyst particle modeling in fixed-bed reactors, Ph. D. thesis, University of Wisconsin–Madison (1990).

Weidman, D. L., and W. E. Stewart, Catalyst particle modelling in fixed-bed reactors, *Chem. Eng. Sci.*, **45**, 2155-2160 (1990).

Yang, K. H., and O. A. Hougen, Determination of mechanisms of catalyzed gaseous reactions, *Chem. Eng. Prog.*, **46**, 146 (1950).

Young, T. C., and W. E. Stewart, Correlation of fractionation tray performance via a cross-flow boundary-layer model, *AIChE J.*, **38**, 592–602 (1992); Errata, **38**, 1302 (1992).

Young, T. C., and W. E. Stewart, Concise correlation of sieve-tray heat and mass transfer, *AIChE J. (R&D Note)*, **41**, 1319-1320 (1995).

Ziegel, E. R., and J. W. Gorman, Kinetic modeling with multiresponse data, *Technometrics*, **22**, 139–151 (1980).

# Appendix A

# Solution of Linear Algebraic Equations

Linear systems of algebraic equations occur often in science and engineering as exact or approximate formulations of various problems. Such systems have appeared many times in this book. Exact linear problems appear in Chapter 2, and iterative linearization schemes are used in Chapters 6 and 7.

This chapter gives a brief summary of properties of linear algebraic equation systems, in elementary and partitioned form, and of certain elimination methods for their solution. Gauss-Jordan elimination, Gaussian elimination, LU factorization, and their use on partitioned arrays are described. Some software for computational linear algebra is pointed out, and references for further reading are given.

## A.1 INTRODUCTORY CONCEPTS AND OPERATIONS

A system of $m$ linear algebraic equations in $n$ unknowns,

$$
\begin{array}{rrrrr}
a_{11}x_1 + & a_{12}x_2 + & \ldots & + a_{1n}x_n = & b_1 \\
a_{21}x_1 + & a_{22}x_2 + & \ldots & + a_{2n}x_n = & b_2 \\
\vdots & \vdots & \vdots & \vdots & \vdots \\
a_{m1}x_1 + & a_{m2}x_2 + & \ldots & + a_{mn}x_n = & b_m
\end{array}
\tag{A.1-1}
$$

can be written more concisely as

$$
\sum_{j=1}^{n} a_{ij}x_j = b_i \qquad i = 1, \ldots, m
\tag{A.1-2}
$$

Here $a_{ij}$ denotes the coefficient of the $j$th unknown $x_j$ in the $i$th equation, and the numbers $a_{ij}$ and $b_i$ (hence $x_j$) all are real. Often the number $m$ of equations is equal to the number $n$ of unknowns, but exceptions are common in optimization and modeling and will be noted in Chapters 6 and 7.

Array notations are convenient for such problems. Let the bold letters

$$
A \equiv \begin{pmatrix} a_{11} & a_{12} & \ldots & a_{1n} \\ a_{21} & a_{22} & \ldots & a_{2n} \\ \vdots & \vdots & \vdots & \vdots \\ a_{m1} & a_{m2} & \ldots & a_{mn} \end{pmatrix} \qquad x \equiv \begin{pmatrix} x_1 \\ x_2 \\ \vdots \\ x_n \end{pmatrix} \qquad b \equiv \begin{pmatrix} b_1 \\ b_2 \\ \vdots \\ b_m \end{pmatrix}
\tag{A.1-3}
$$

denote, respectively, the coefficient matrix $\boldsymbol{A}$, the unknown column vector $\boldsymbol{x}$, and the right-hand column vector $\boldsymbol{b}$. Another way of representing an array is to enclose the expression for its elements in curly brackets: $\boldsymbol{A} = \{a_{ij}\}$, $\boldsymbol{x} = \{x_j\}$, $\boldsymbol{b} = \{b_i\}$. Defining the matrix-vector product

$$\boldsymbol{Ax} \equiv \left\{ \sum_{j=1}^{n} a_{ij}x_j \right\} \tag{A.1-4}$$

in accordance with Eq. (A.1-2), one can summarize Eq. (A.1-1) concisely as

$$\boldsymbol{Ax} = \boldsymbol{b} \tag{A.1-5}$$

Methods for solving Eq. (A.1-1) can then be described in terms of operations on the arrays $\boldsymbol{A}$ and $\boldsymbol{b}$. Before doing so, we pause to summarize a few essentials of linear algebra; fuller treatments are given in Stewart (1973) and in Noble and Daniel (1988).

Matrices can be added if their corresponding dimensions are equal. The addition is done element by element:

$$\boldsymbol{A} + \boldsymbol{B} = \{a_{ij} + b_{ij}\} \tag{A.1-6}$$

The difference $\boldsymbol{A} - \boldsymbol{B}$ is formed analogously. It follows that two matrices are equal if their corresponding elements are equal; their difference is then a zero matrix $\boldsymbol{0}$, with every element equal to zero.

Multiplication of a matrix $\boldsymbol{A}$ by a scalar $s$ is done element by element,

$$s\boldsymbol{A} = s\{a_{ij}\} = \{sa_{ij}\} = \{a_{ij}s\} = \boldsymbol{A}s \tag{A.1-7}$$

and is commutative and associative as shown.

Matrices $\boldsymbol{A}$ and $\boldsymbol{B}$ can be multiplied provided that they are *conformable* in the order written, that is, that the number of columns (*column order*) of the first matrix equals the number of rows (*row order*) of the second. The product $\boldsymbol{C}$ of a $m \times n$ matrix $\boldsymbol{A}$ by a $n \times p$ matrix $\boldsymbol{B}$ has the elements

$$c_{ik} = [\boldsymbol{AB}]_{ik} = \sum_{j=1}^{n} a_{ij}b_{jk} \quad \begin{cases} i = 1, \ldots, m \\ k = 1, \ldots, p \end{cases} \tag{A.1-8}$$

Thus, the element in row $i$ and column $k$ of the product matrix is the sum over $j$ of the products (element $j$ in row $i$ of the first matrix) times (element $j$ in column $k$ of the second matrix). When $\boldsymbol{AB}$ equals $\boldsymbol{BA}$, we say that the multiplication is *commutative;* this property is limited to special pairs of square matrices of equal order.

The *transpose* of a matrix $\boldsymbol{A}$ is

$$\boldsymbol{A}^T = \{a_{ij}\}^T = \{a_{ji}\} \tag{A.1-9}$$

Thus, the rows of $\boldsymbol{A}^T$ are the columns of $\boldsymbol{A}$, and the columns of $\boldsymbol{A}^T$ are the rows of $\boldsymbol{A}$. A matrix that equals its transpose is called *symmetric*. The transpose of a product of square or rectangular matrices is given by

$$(\boldsymbol{AB})^T = \boldsymbol{B}^T \boldsymbol{A}^T \qquad \text{(A.1-10)}$$

that is, by the product of the transposed matrices in reverse order.

A vector $\boldsymbol{x}$, by convention, denotes a column of elements, hence a one-column matrix $\{x_{i1}\}$. The transpose $\boldsymbol{x}^T$ is a matrix $\{x_{1i}\}$ by Eq. (A.1-9) and is a row of elements, ordered from left to right. The multiplication rule (A.1-8) and its corollary (A.1-10) thus hold directly for conforming products involving vectors. For example, Eq. (A.1-5) takes the alternate form

$$\boldsymbol{x}^T \boldsymbol{A}^T = \boldsymbol{b}^T \qquad \text{(A.1-11)}$$

when transposed according to Eq. (A.1-10). Such transpositions arise naturally in certain computations with stoichiometric matrices, as noted in Chapter 2.

A set of row or column vectors, $\boldsymbol{v}_1, \ldots, \boldsymbol{v}_p$, is called *linearly independent* if its only vanishing linear combination $\sum_i c_i \boldsymbol{v}_i$ is the trivial one, with coefficients $c_i$ all zero.[1] Such a set provides the *basis vectors* $\boldsymbol{v}_1, \ldots, \boldsymbol{v}_p$ of a $p$-dimensional *linear space* of vectors $\sum_i c_i \boldsymbol{v}_i$, with the *basis variables* $c_1, \ldots, c_p$ as coordinates.

Constructions of basis vectors for the rows or columns of the coefficient matrix arise naturally in solving linear algebraic equations. The number, $r$, of row vectors in any basis for the rows is called the *rank* of the matrix, and is also the number of column vectors in any basis for the columns.

Equation (A.1-1) has solutions if and only if $\boldsymbol{b}$ is a linear combination of nonzero column vectors of $\boldsymbol{A}$; then we say that $\boldsymbol{b}$ is in the *column space* of $\boldsymbol{A}$. In such cases, if the rank, $r$, of $\boldsymbol{A}$ equals $n$ there is a unique solution, whereas if $r < n$ there is an infinity of solutions characterized by $(n - r)$ free parameters. The possible outcomes are identified further in Section A.3.

The situation is simplest if the matrix $\boldsymbol{A}$ is square $(m = n)$; then a unique solution vector $\boldsymbol{x}$ exists if and only if the rank of $\boldsymbol{A}$ equals its order $n$. Such a matrix $\boldsymbol{A}$ has a unique *inverse*, $\boldsymbol{A}^{-1}$, defined by

$$\boldsymbol{A}^{-1}\boldsymbol{A} = \boldsymbol{A}\boldsymbol{A}^{-1} = \boldsymbol{I} \qquad \text{(A.1-12)}$$

and is said to be *nonsingular*, or *regular*. Here $\boldsymbol{I}$ is the *unit matrix*

$$\boldsymbol{I} = \mathrm{diag}\{1, \ldots, 1\} \equiv \begin{pmatrix} 1 & & \\ & \ddots & \\ & & 1 \end{pmatrix} \qquad \text{(A.1-13)}$$

---

[1] Otherwise, the set is called *linearly dependent*.

of the same order as $\boldsymbol{A}$. The solution of Eq. (A.1-5) is then expressible as

$$x = A^{-1}b \qquad (A.1\text{-}14)$$

whatever the value of the $n$-vector $\boldsymbol{b}$, though for computations there are better schemes, as described in the following sections.

## A.2 OPERATIONS WITH PARTITIONED MATRICES

It is frequently useful to classify the equations and/or the variables of a problem into subsets; then Eq. (A.1-1) can be represented in partitioned forms such as

$$\begin{pmatrix} A_{11} & A_{12} \\ A_{21} & A_{22} \\ A_{31} & A_{32} \end{pmatrix} \begin{pmatrix} x_1 \\ x_2 \end{pmatrix} = \begin{pmatrix} b_1 \\ b_2 \\ b_3 \end{pmatrix} \qquad (A.2\text{-}1)$$

Here each entry denotes a subarray and is shown accordingly in bold type. Thus, the vector $\boldsymbol{x}$ is partitioned here into two subvectors, and the matrix $\boldsymbol{A}$ is partitioned into six submatrices. The submatrix $\boldsymbol{A}_{hk}$ contains the coefficients in the row subset $h$ for the elements of the subvector $\boldsymbol{x}_k$; thus, these arrays conform for multiplication in the order written.

Equations (A.1-4,6,7,8, and 10) lead to analogous rules for addition and multiplication of partitioned arrays; thus, Eq. (A.1-4) yields the formula

$$\boldsymbol{Ax} = \{\boldsymbol{A}_{hk}\}\{\boldsymbol{x}_k\} = \left\{ \sum_k \boldsymbol{A}_{hk}\boldsymbol{x}_k \right\} = \{[\boldsymbol{Ax}]_h\} \qquad (A.2\text{-}2)$$

for postmultiplication of a partitioned matrix $\{\boldsymbol{A}_{hk}\}$ by a conformally partitioned vector $\{\boldsymbol{x}_k\}$. Equation (A.1-8) yields the formula

$$\boldsymbol{C}_{hq} = [\boldsymbol{AB}]_{hq} = \sum_k \boldsymbol{A}_{hk}\boldsymbol{B}_{kq} \qquad (A.2\text{-}3)$$

for the submatrices $\boldsymbol{C}_{hq}$ of a product of conformally partitioned matrices $\{\boldsymbol{A}_{hk}\}$ and $\{\boldsymbol{B}_{kq}\}$.

Application of Eq. (A.2-2) to Eq. (A.2-1) yields the partitioned equations

$$\boldsymbol{A}_{11}\boldsymbol{x}_1 + \boldsymbol{A}_{12}\boldsymbol{x}_2 = \boldsymbol{b}_1$$
$$\boldsymbol{A}_{21}\boldsymbol{x}_1 + \boldsymbol{A}_{22}\boldsymbol{x}_2 = \boldsymbol{b}_2 \qquad (A.2\text{-}4)$$
$$\boldsymbol{A}_{31}\boldsymbol{x}_1 + \boldsymbol{A}_{32}\boldsymbol{x}_2 = \boldsymbol{b}_3$$

when the resulting row groups are written out. The following example makes use of Eq. (A.2-3).

## Example A.1. Symbolic Inversion of a Partitioned Matrix

A formal inverse is desired for nonsingular matrices of the form

$$A = \begin{pmatrix} A_{11} & A_{12} \\ A_{21} & A_{22} \end{pmatrix} \qquad (A.2\text{-}5)$$

in which $A_{11}$ and $A_{22}$ are nonsingular (hence square). This result is used in Sections 6.5 and 7.5 for statistical estimation of parameter subsets.

Let us denote the inverse by

$$A^{-1} = \begin{pmatrix} \alpha & \beta \\ \gamma & \delta \end{pmatrix} \qquad (A.2\text{-}6)$$

Then Eqs. (A.1-12) and (A.2-3) require

$$AA^{-1} = \begin{pmatrix} A_{11} & A_{12} \\ A_{21} & A_{22} \end{pmatrix} \begin{pmatrix} \alpha & \beta \\ \gamma & \delta \end{pmatrix} = \begin{pmatrix} I_{11} & 0_{12} \\ 0_{21} & I_{22} \end{pmatrix} \qquad (A.2\text{-}7)$$

Interpreting this matrix equality as described in Eq. (A.1-6), we obtain four equations for the four unknown submatrices $\alpha$, $\beta$, $\gamma$, and $\delta$:

$$A_{11}\alpha + A_{12}\gamma = I_{11} \qquad (A.2\text{-}8)$$
$$A_{11}\beta + A_{12}\delta = 0_{12} \qquad (A.2\text{-}9)$$
$$A_{21}\alpha + A_{22}\gamma = 0_{21} \qquad (A.2\text{-}10)$$
$$A_{21}\beta + A_{22}\delta = I_{22} \qquad (A.2\text{-}11)$$

Premultiplication of Eq. (A.2-9) by $A_{11}^{-1}$ gives

$$\beta = -A_{11}^{-1}A_{12}\delta \qquad (A.2\text{-}12)$$

Premultiplication of Eq. (A.2-10) by $A_{22}^{-1}$ gives

$$\gamma = -A_{22}^{-1}A_{21}\alpha \qquad (A.2\text{-}13)$$

Insertion of the last result into Eq. (A.2-8) gives

$$[A_{11} - A_{12}A_{22}^{-1}A_{21}]\alpha = I_{11} \qquad (A.2\text{-}14)$$

whence the submatrix $\alpha$ is

$$\alpha = [A_{11} - A_{12}A_{22}^{-1}A_{21}]^{-1} \qquad (A.2\text{-}15)$$

Insertion of Eq. (A.2-12) into Eq. (A.2-11) gives

$$[-A_{21}A_{11}^{-1}A_{12} + A_{22}]\delta = I_{22} \qquad (A.2\text{-}16)$$

whence the submatrix $\boldsymbol{\delta}$ is

$$\boldsymbol{\delta} = [\boldsymbol{A}_{22} - \boldsymbol{A}_{21}\boldsymbol{A}_{11}^{-1}\boldsymbol{A}_{12}]^{-1} \tag{A.2-17}$$

Assembling these results, we obtain the formula

$$\boldsymbol{A}^{-1} =$$
$$\begin{pmatrix} [\boldsymbol{A}_{11} - \boldsymbol{A}_{12}\boldsymbol{A}_{22}^{-1}\boldsymbol{A}_{21}]^{-1} & -\boldsymbol{A}_{11}^{-1}\boldsymbol{A}_{12}[\boldsymbol{A}_{22} - \boldsymbol{A}_{21}\boldsymbol{A}_{11}^{-1}\boldsymbol{A}_{12}]^{-1} \\ -\boldsymbol{A}_{22}^{-1}\boldsymbol{A}_{21}[\boldsymbol{A}_{11} - \boldsymbol{A}_{12}\boldsymbol{A}_{22}^{-1}\boldsymbol{A}_{21}]^{-1} & [\boldsymbol{A}_{22} - \boldsymbol{A}_{21}\boldsymbol{A}_{11}^{-1}\boldsymbol{A}_{12}]^{-1} \end{pmatrix}$$
$$\tag{A.2-18}$$

for the inverse of a matrix of any order, in terms of operations on its sub-arrays.

## A.3 GAUSS-JORDAN REDUCTION

Gauss-Jordan reduction is a straightforward elimination method that solves for an additional unknown $x_k$ at each stage. An augmented array

$$\boldsymbol{\alpha} = \begin{pmatrix} a_{11} & a_{12} & \cdots & a_{1n} & b_1 \\ a_{21} & a_{22} & \cdots & a_{2n} & b_2 \\ \vdots & \vdots & \vdots\vdots\vdots & \vdots & \vdots \\ a_{m1} & a_{m2} & \cdots & a_{mn} & b_m \end{pmatrix} \tag{A.3-1}$$

is initialized with the elements of Eq. (A.1-1) and is transformed by stages to the solution. Stage $k$ goes as follows:

1. Find $\alpha_{pk}$, the absolutely largest current element in column $k$ that lies in a row not yet selected for pivoting. If $\alpha_{pk}$ is negligible, skip to step 4; otherwise, accept it as the pivot for stage $k$ and go to step 2.

2. Eliminate $x_k$ from all rows but $p$ by transforming the array as follows:

$$\alpha_{ij}^{(\text{new})} = [\alpha_{ij} - (\alpha_{ik}/\alpha_{pk})\alpha_{pj}]^{(\text{old})} \begin{cases} j = k, \ldots, n+1 \\ i \neq p \end{cases} \tag{A.3-2}$$

3. Normalize the current pivotal row as follows:

$$\alpha_{pj}^{(\text{new})} = [\alpha_{pj}/\alpha_{pk}]^{(\text{old})} \qquad j = k \ldots, n \tag{A .3-3}$$

thus obtaining a coefficient of unity for $x_k$ in the solution.

4. If $k < \min(m,n)$, increase $k$ by 1 and go to 1. Otherwise, the elimination is completed.

The use of the largest available pivot element, $\alpha_{pk}$, in column $k$ limits the growth of rounding error by ensuring that the multipliers $(\alpha_{ik}/\alpha_{pk})$ in Eq. (A.3-2) all have magnitude 1 or less.

This algorithm takes $\min(m, n)$ stages. The total number of pivots accepted is the rank, $r$, of the equation system. The final array can be decoded by associating each coefficient column with the corresponding variable. Rearranging the array to place the pivotal rows and columns in the order of selection, one can write the results in the partitioned form

$$\begin{pmatrix} I_r & A'_{12} \\ 0 & 0 \end{pmatrix} \begin{pmatrix} x_1 \\ x_2 \end{pmatrix} = \begin{pmatrix} b'_1 \\ b'_2 \end{pmatrix} \tag{A.3-4}$$

Here $I_r$ is the unit matrix of order $r$, $x_1$ is the vector of $r$ *pivotal variables* (those whose columns were selected for pivoting), $x_2$ is the vector of $(n-r)$ nonpivotal variables, and $A'_{12}$ is the $r \times (n-r)$ matrix of transformed coefficients of the $x_2$ variables. Expanding this partitioned equation, we obtain the pivotal equation set

$$x_1 + A'_{12}x_2 = b'_1 \tag{A.3-5}$$

and the nonpivotal equation set

$$0 = b'_2 \tag{A.3-6}$$

The possible outcomes of the reduction can now be summarized:

1. If $r = m = n$, the solution is unique. Equation (A.3-4) reduces to

$$I_n x = x = b' \tag{A.3-7}$$

   and all the equations and variables are pivotal.

2. If $r = n < m$, the solution is overdetermined. Equation (A.3-4) reduces to

$$I_n x = x = b'_1 \tag{A.3-8}$$
$$0 = b'_2 \tag{A.3-9}$$

   If the vector $b'_2$ is zero, the problem is consistent and the solution is given by Eq. (A.3-8); otherwise, no $n$-vector $x$ can satisfy all the equations. Depending on the causes of the discrepancies, one may need to reject or revise certain data, or else seek a compromise solution that minimizes the discrepancies in some sense; see Chapters 2, 6, and 7.

3. If $r = m < n$, the solution is underdetermined. Equation (A.3-4) reduces to the form in Eq. (A.3-5), with $b'_1 = b'$ since all rows are pivotal when $r = m$. Thus, the pivotal variables (the set $x_1$) are obtained as explicit functions of the nonpivotal variables

(the set $x_2$). The $x_2$ variables are called the *parameters* of the solution.

4. If $r$ is less than both $m$ and $n$, the reduction yields Eqs. (A.3-5) and (A.3-6). The vector $b_2'$ provides a numerical consistency test, and the $x_2$ variables serve as parameters of the solution.

The algorithm just described can also compute the inverse of a nonsingular matrix. For simplicity, consider the equations to be ordered originally so that each diagonal element $\alpha_{kk}$ becomes a pivot. Then the foregoing reduction transforms the coefficient matrix $A$ into a unit matrix. It follows that the algorithm is equivalent to premultiplication of the initial array by $A^{-1}$. A simple way of computing $A^{-1}$, therefore, is to include a unit matrix in the initial $\alpha$ array. The final array then takes the form

$$A^{-1}\,(\,A \quad b \quad I\,) = (\,I \quad x \quad A^{-1}\,) \tag{A.3-10}$$

giving $A^{-1}$ in the space where $I$ was initially inserted.

The inversion can be done more compactly by computing the inverse in the space originally occupied by $A$. Simple changes in steps 2 and 3 will accomplish this:

1. See above.
2. Extend the range of $j$ in Eq. (A.3-2) to all $j \neq p$. After the computation of $(\alpha_{ik}/\alpha_{pk})^{(old)}$, insert that quotient as the new element $\alpha_{ik}$.
3. Replace $\alpha_{pk}$ by 1, and then divide every element of row $p$ by $\alpha_{pk}^{(old)}$.
4. See above.

Should $A$ be singular, this method still gives a *basis inverse*, namely, the inverse of the submatrix of $A$ in which the pivots were found. A version of this algorithm using strictly positive, diagonal pivots is used for constrained parameter estimation in Subroutine GREGPLUS, as described in Sections 6.4 and 7.3; there the determinant of the basis submatrix is obtained directly as the product of the pivotal divisors. Basis inverse matrices also prove useful in Sections 6.6 and 7.5 for interval estimation of parameters and related functions.

## A.4 GAUSSIAN ELIMINATION

Gauss gave a shorter algorithm that introduces fewer roundings. The method has two parts: elimination and backsubstitution. We describe it here for the array $\alpha$ of Eq. (A.3-1) with $m = n$.

Stage $k$ of the Gaussian elimination goes as follows:

1. Find $\alpha_{pk}$, the absolutely largest of the elements $\alpha_{kk}, \ldots, \alpha_{nk}$. If $\alpha_{pk}$ is negligible, stop and declare the matrix singular. Otherwise, accept $\alpha_{pk}$ as the pivot for stage $k$, and (if $p > k$)

interchange rows $p$ and $k$, thus bringing the pivot element to the $(k, k)$ position.

2. If $k < n$, eliminate $x_k$ from rows $k + 1, \ldots, n$ by transforming the array as follows [here $^{(\text{old})}$ denotes element values just before this calculation]:

$$\alpha_{ij}^{(\text{new})} = [\alpha_{ij} - (\alpha_{ik}/\alpha_{kk})\alpha_{kj}]^{((\text{old}))} \begin{cases} j = k, \ldots, n + 1 \\ i = k + 1, \ldots, n \end{cases} \quad \text{(A.4-1)}$$

3. If $k < n$, increase $k$ by 1 and go to 1. Otherwise, the elimination is completed.

At the end of the elimination, the array takes the upper triangular form

$$\alpha' = \begin{pmatrix} u_{11} & u_{12} & \cdots & u_{1n} & b_1'' \\ & u_{22} & \cdots & u_{2n} & b_2'' \\ & & \ddots & \vdots & \vdots \\ & & & u_{nn} & b_n'' \end{pmatrix} \quad \text{(A.4-2)}$$

Associating each coefficient $u_{ij}$ with the corresponding element of $x$, we get the transformed equation system

$$\begin{aligned} u_{11}x_1 + u_{12}x_2 + \cdots + u_{1n}x_n &= b_1'' \\ u_{22}x_2 + \cdots + u_{2n}x_n &= b_2'' \\ \ddots \qquad \vdots \qquad &\vdots \\ u_{nn}x_n &= b_n'' \end{aligned} \quad \text{(A.4-3)}$$

which is readily solved, beginning with the last equation:

$$\begin{aligned} x_n &= b_n''/u_{nn} \\ x_{n-1} &= (b_{n-1}'' - u_n x_n)/u_{n-1,n-1} \\ &\vdots \\ x_i &= \left( b_i'' - \sum_{j=i+1}^{n} u_{ij}x_j \right) /u_{ii} \end{aligned} \quad \text{(A.4-4)}$$

The solution is direct when Eqs. (A.4-4) are applied in the order given, since each right-hand member then contains only quantities already found.

For computation of a particular solution vector $x$, this method requires $\frac{1}{3}(n^3 - n) + n^2$ operations of multiplication or division, versus $\frac{1}{2}(n^3 - n) + n^2$ for the Gauss-Jordan method. Thus, the Gauss method takes about two-thirds as many operations. The computation of an inverse matrix takes $O(n^3)$ operations of multiplication or division for either method.

## A.5 LU FACTORIZATION

LU factorization is an elegant scheme for computing $x$ for various values of the right-hand vector $b$.

The Gaussian algorithm described in Section A.4 transforms the matrix $A$ into an upper triangular matrix $U$ by operations equivalent to premultiplication of $A$ by a nonsingular matrix. Denoting the latter matrix by $L^{-1}$, one obtains the representation

$$L^{-1}A = U \qquad\qquad (A.5\text{-}1)$$

for the transformation process. Premultiplication of this formula by $L$ then gives

$$A = LU \qquad\qquad (A.5\text{-}2)$$

as an LU factorization of $A$.

In applications one never computes the matrix $L$, but works with the lower triangular matrix $L^{-1}$ defined by the row interchange list $p(k)$ and the lower triangular array of multipliers $m_{ik} = (\alpha_{ik}/\alpha_{kk})$ for stages $k = 1, \dots, n$ of the elimination. For efficient use of computer memory the diagonal multipliers $m_{kk} \equiv 1$ are not stored, while the other multipliers are saved as the subdiagonal elements of the transformed $A$ matrix [in the region that is empty in Eq. (A.4-2).]

Premultiplication of Eq. (A.1-5) by $L^{-1}$, followed by use of (A.5-1), gives

$$Ux = L^{-1}b = b'' \qquad\qquad (A.5\text{-}3)$$

as a matrix statement of Eq. (A.4-3). Premultiplication of Eq. (A.5-3) by $U^{-1}$ gives

$$x = U^{-1}b'' \qquad\qquad (A.5\text{-}4)$$

as a formal statement of the back-solution algorithm in Eq. (A.4-4). This two-step processing of $b$ can be summarized as

$$x = U^{-1}(L^{-1}b) \qquad\qquad (A.5\text{-}5)$$

which is faster and gives less rounding error than the alternative of computing the inverse matrix

$$A^{-1} = U^{-1}L^{-1} \qquad\qquad (A.5\text{-}6)$$

and then using Eq. (A.1-14).

The pivot selection procedure used here is known as *partial pivoting*, because only one column is searched for the current pivot. A safer but slower procedure, known as *full pivoting*, is to select the current pivot as the absolutely largest remaining eligible element in the current transformed matrix.

## A.6 SOFTWARE

Excellent implementations of linear algebra algorithms are available from several sources, including MATLAB, IMSL, NAG, LINPACK and LA-PACK. LINPACK and LAPACK are available from www.netlib.org, along with codes for many other numerical algorithms. The LINPACK codes were written for serial machines, whereas LAPACK is designed for efficient utilization of parallel machines as well. LAPACK codes are used where applicable in the software described in this book.

## PROBLEMS

### A.A Matrix and Matrix-Vector Multiplication

Prove Eq. (A.1-8) by evaluating the contribution of element $x_k$ to the $i$th element of the product vector $(AB)x = A(Bx)$. Here $A$ is $m \times n$, $B$ is $n \times p$, and $x$ is an arbitrary conforming vector.

### A.B Transposition Rule for Matrix Products

Prove Eq. (A.1-10) by use of Eqs. (A.1-8) and (A.1-9). Again, take $A$ to be $m \times n$ and $B$ to be $n \times p$.

### A.C Inversion Formula

Verify Eq. (A.2-18) by premultiplication with the matrix of Eq. (A.2-5) according to Eq. (A.2-3).

### A.D Symmetry and Matrix Inversion

Show, by use of Eqs. (A.1-12) and (A.1-10), that every symmetric invertible matrix $A$ has a symmetric inverse.

### A.E Inverse of a Matrix Product

Consider a product

$$P = ABC \tag{A.E-1}$$

of nonsingular matrices of equal order. Show that

$$P^{-1} = C^{-1}B^{-1}A^{-1} \tag{A.E-2}$$

by forming $P^{-1}P$ and simplifying. This result is used in Chapters 6 and 7 to obtain probability distributions for auxiliary functions of model parameters.

### A.F Solution of a Linear System

Use Gaussian elimination and back-substitution to solve the linear system of Eq. (A.1-2), with $m = n = 3$, $a_{ij} = 1/(i+j-1)$, and $b_i^T = \{1, 0, 0\}$. The coefficient matrix $A$ given here is called the *Hilbert matrix*; it arises in a nonrecommended brute-force construction of orthogonal polynomials from monomials $t^j$.

## REFERENCES and FURTHER READING

Anderson, E., Z. Bai, C. Bischof, J. Demmel, J. J. Dongarra, J. Du Croz, A. Greenbaum, S. Hammarling, A. McKenney, S. Ostrouchov, and D. Sorensen, *LAPACK Users' Guide,* Society for Industrial and Applied Mathematics, Philadelphia (1992).

Dongarra, J. J., J. R. Bunch, C. B. Moler, and G. W. Stewart, *LINPACK Users' Guide,* SIAM, Philadelphia (1979).

Golub, G. H., and C. F. Van Loan, *Matrix Computations,* Johns Hopkins University Press, Baltimore (1983).

Noble, B., and J. W. Daniel, *Applied Linear Algebra,* 3rd edition, Prentice-Hall, Englewood Cliffs, NJ (1988).

Press, W. H., B. P. Flannery, S. A. Teukolsky, and W. T. Vetterling, *Numerical Recipes (FORTRAN Version),* Cambridge University Press (1989).

Stewart, G. W., *Introduction to Matrix Computations,* Academic Press, New York (1973).

# Appendix B
# DDAPLUS Documentation

This appendix explains how to use DDAPLUS to solve nonlinear initial-value problems containing ordinary differential equations with or without algebraic equations, or to solve purely algebraic nonlinear equation systems by a damped Newton method. Three detailed examples are given.

## B.1 WHAT DDAPLUS DOES

The DDAPLUS algorithm (Caracotsios and Stewart, 1985), updated here, is an extension of the DDASSL (Petzold, 1982) implicit integrator. DDAPLUS solves differential-algebraic equation systems of the form

$$\boldsymbol{E}(t, \boldsymbol{u}; \boldsymbol{\theta})\boldsymbol{u}' = \boldsymbol{f}(t, \boldsymbol{u}; \boldsymbol{\theta}) \quad \text{for } t \in [t_0, T] \qquad \text{(B.1-1a)}$$

for the state vector $\boldsymbol{u}(t; \boldsymbol{\theta})$ as a function of time or distance, $t$, at a given value of a parameter vector $\boldsymbol{\theta}$. Here $\boldsymbol{u}' = d\boldsymbol{u}/dt \equiv (\partial \boldsymbol{u}/\partial t)_{\boldsymbol{\theta}}$, and $\boldsymbol{f}(t, \boldsymbol{u}; \boldsymbol{\theta})$ is a user-defined function. The limit $T$ may be equal to $t_0$, greater, or less.

The matrix function $\boldsymbol{E}(t, \boldsymbol{u}; \boldsymbol{\theta})$ indicates the form of the equation system. The user specifies this function by use of the argument **Info(13)** described in Section B.4, and a subroutine **Esub** described in Section B.7. Algebraic equations give vanishing rows in $\boldsymbol{E}$, and differential equations in $t$ give nonzero rows. Any pattern of zero and/or nonzero elements $E_{ij}$ is permitted; a system with $\boldsymbol{E} = \boldsymbol{0}$ can be solved with $T = t_0$, given a good enough initial guess. The functions $\boldsymbol{E}(t, \boldsymbol{u}; \boldsymbol{\theta})$ and $\boldsymbol{f}(t, \boldsymbol{u}; \boldsymbol{\theta})$ are assumed differentiable with respect to $t$, $\boldsymbol{u}$, and $\boldsymbol{\theta}$ during the computation from $t_0$ to $T$. Finite jumps of $\boldsymbol{E}$ and $\boldsymbol{f}$ are permitted across $t_0$ to introduce new equation forms; see **Info(1)**, **Info(4)**, and **Info(17)** in Section B.4. Multiple jumps can be handled by restarts from successive new $t_0$ values.

The initial state is determined by the differentiability condition

$$\boldsymbol{E}(t, \boldsymbol{u}; \boldsymbol{\theta})[\boldsymbol{u} - \boldsymbol{u}_{0_-}(\boldsymbol{\theta})] = \boldsymbol{0} \quad \text{at} \quad t = t_0 \qquad \text{(B.1-1b)}$$

along with any algebraic equations in the system (B.1-1a). Here $\boldsymbol{u}_{0_-}$ is the user's input state, applicable just before $t_0$ at the chosen value of the parameter vector $\boldsymbol{\theta}$. More explicitly, we can write the full set of initial conditions as

$$\boldsymbol{E}_D(t, \boldsymbol{u}; \boldsymbol{\theta})[\boldsymbol{u} - \boldsymbol{u}_{0_-}(\boldsymbol{\theta})] - \boldsymbol{f}_A(t, \boldsymbol{u}; \boldsymbol{\theta}) = \boldsymbol{0} \quad \text{at} \quad t = t_0 \qquad \text{(B.1-2)}$$

189

with subscripts $D$ and $A$ denoting, respectively, differential and algebraic equations. DDAPLUS gives an error stop if the rank of $\boldsymbol{E}$ does not equal the number of differential equations, as required for solvability of the system.

When $\boldsymbol{\theta}$ is variable, its influence on the solution in the neighborhood of a given $\boldsymbol{\theta}$ value is described by the matrix function

$$W(t) \equiv \frac{\partial \boldsymbol{u}(t; \boldsymbol{\theta})}{\partial \boldsymbol{\theta}^T} \tag{B.1-3}$$

whose elements $\partial u_i / \partial \theta_j$ are the first-order *parametric sensitivities* of the problem. DDAPLUS computes, on request, the function $\boldsymbol{W}(t)$ at the same $\boldsymbol{\theta}$, by solving the sensitivity equations

$$E\frac{d\boldsymbol{W}}{dt} + \left[\frac{\partial}{\partial \boldsymbol{u}^T}[\boldsymbol{E}\boldsymbol{u}' - \boldsymbol{f}]_{t,\,\boldsymbol{u}',\,\boldsymbol{\theta}}\right]\boldsymbol{W} = -\frac{\partial}{\partial \boldsymbol{\theta}^T}[\boldsymbol{E}\boldsymbol{u}' - \boldsymbol{f}]_{t,\,\boldsymbol{u},\,\boldsymbol{u}'} \quad \text{for } t \in [t_0, T]$$
$$\equiv \boldsymbol{B}(t \neq t_0) \tag{B.1-4a}$$

and their initial conditions at $t = t_0$,

$$\left[\frac{\partial}{\partial \boldsymbol{u}^T}\{\boldsymbol{E}_D[\boldsymbol{u} - \boldsymbol{u}_{0-}] - \boldsymbol{f}_A\}_{t,\,\boldsymbol{\theta}}\right]\boldsymbol{W} = -\frac{\partial}{\partial \boldsymbol{\theta}^T}\{\boldsymbol{E}_D[\boldsymbol{u} - \boldsymbol{u}_{0-}] - \boldsymbol{f}_A\}_{t,\,\boldsymbol{u}}$$
$$\equiv \boldsymbol{B}(t_0), \tag{B.1-4b}$$

which are the total $\boldsymbol{\theta}$-derivatives of equations (B.1-1a) and (B.1-2). The columns of the current $\boldsymbol{B}$ can be approximated by DDAPLUS as difference quotients. Columns $\boldsymbol{B}_j$ in which $\partial \boldsymbol{E}/\partial \theta_j$ is zero can be computed analytically if desired, via a user-provided subroutine **Bsub** as outlined in Section B.7.

The state trajectory $\boldsymbol{u}(t)$ is computed by the implicit integrator DDASSL (Petzold 1982; Brenan, Campbell, and Petzold 1989), updated here to handle the initial condition of Eq. (B.1-2). The DDASSL integrator is especially designed to handle stiff, coupled systems of ordinary differential and algebraic equations. It employs a variable-order, variable-step predictor-corrector approach initiated by Gear (1971). The derivative vector $\boldsymbol{u}'_{n+1}$, applicable at $t_{n+1}$, is approximated in the corrector stage by a backward difference formula, and the resulting system of algebraic equations is solved for the state vector $\boldsymbol{u}_{n+1}$ by a modified Newton method. The integration step size and order are automatically adjusted to expedite progress while meeting the user's tolerances on error per step.

The coefficient matrix $\boldsymbol{G}$ for the Newton iterations has the form

$$G(t \neq t_0) \equiv \left[C_J\boldsymbol{E} + \frac{\partial}{\partial \boldsymbol{u}^T}[\boldsymbol{E}\boldsymbol{u}' - \boldsymbol{f}]_{t,\,\boldsymbol{u}',\,\boldsymbol{\theta}}\right] \quad \text{for } t \in [t_0, T] \tag{B.1-5a}$$

for Eq. (B.1-1a) and

$$G(t_0) \equiv \left[\frac{\partial}{\partial \boldsymbol{u}^T}\{\boldsymbol{E}_D[\boldsymbol{u} - \boldsymbol{u}_{0-}(\boldsymbol{\theta})] - \boldsymbol{f}_A\}_{t,\,\boldsymbol{\theta}}\right] \quad \text{at } t = t_0 \tag{B.1-5b}$$

for Eq. (B.1-2). Here $C_J$ is the coefficient in the backward-difference formula for $u'$. The elements of $G$ can be updated by DDAPLUS with difference approximations, as difference quotients, or (if $\partial E/\partial u^T$ is zero) they can be computed via a user-provided subroutine **Jac**, as outlined in Section B.8.

The sensitivity trajectory $W(t)$, when requested, is computed by the method of Caracotsios and Stewart (1985), extended here to include Eq. (B.1-4b). The sensitivities are computed at $t_0$ and after each step, via a direct linear algorithm that utilizes the current matrix $G$ of equation (B.1-5b) or (B.1-5a) to solve equation (B.1-4b) or a backward-difference form of (B.1-4a).

## B.2 OBJECT CODE

The object code for DDAPLUS and its dependencies is provided in the software package **Athena Visual Studio**, which is available at www.AthenaVisual.com.

To complete the codes required for a given problem, the user provides a driver program (here called MAIN) and four subroutines: **fsub**, **Esub**, **Jac**, and **Bsub**. Each of these subroutines must be declared EXTERNAL in the MAIN program. Table B.1 summarizes the roles of these user-provided subroutines; all are needed for proper linking, but a dummy form suffices for any subroutine that will not be executed. Further information on these subroutines is provided in Sections B.3 through B.6, and detailed examples are given in Section B.10.

All the subroutines described in this appendix can be created automatically by Athena Visual Studio in ANSI standard FORTRAN-77. Since FORTRAN-90/95 is compatible with this standard, there should be no difficulty in using the package under FORTRAN-90/95. The user needs only to supply the (1) model equations, (2) the initial conditions, and (3) any pertinent data. The user input is decribed in more detail in the *Help* menu of Athena Visual Studio, by clicking *Book Examples* and then *Appendix B* and finally selecting the appropriate example.

## B.3 CALLING DDAPLUS

The code DDAPLUS is normally called via Athena Visual Studio. For very detailed control of the available options, DDAPLUS may be called as follows from a user-provided MAIN program for the given problem:

```
CALL DDAPLUS(t,tout,Nstvar,U,UPRIME,Rtol,Atol,Info,
A Rwork,Lrw,Iwork,Liw,Rpar,Ipar,Idid,LUNREP,LUNERR,Irange,
B Imodel,Iaux,fsub,Esub,Jac,Bsub,Ipass,Filepath,Lenpath)
```

## B.4 DESCRIPTION OF THE CALLING ARGUMENTS

In this section we describe the arguments used in calls to DDAPLUS.

| Name | When executed | Purpose |
|------|---------------|---------|
| MAIN | *For each problem* | Driver; calls DDAPLUS |
| **fsub** | *For each problem* (See Section B.4) | Evaluates the vector $\boldsymbol{f}(t, \boldsymbol{u}; \boldsymbol{\theta})$ of Eq. (B.1-1a) |
| **Esub** | *When Info(13)$\neq$ 0* (See Sections B.4, B.5) | Evaluates the nonzero elements of the current matrix $\boldsymbol{E}$ |
| **Jac** | *When Info(5)=1* (See Sections B.4, B.6) | Evaluates the nonzero elements of the current matrix $\partial \boldsymbol{f}/\partial \boldsymbol{u}^T$ |
| **Bsub** | *When Info(12)$>$ 0 and Info(14)=1* (See Sections B.4, B.7) | Evaluates the nonzero elements of the current vector $\partial \boldsymbol{f}/\partial \theta_j$ |

**Table B.1:** Codes the User Provides to Go with DDAPLUS.

### t, tout

**t** is the current value of the independent variable $t$, and **tout** is the next $t$ value where the solution is desired. **tout** must be different from **t** except on entry to a new solution region; equality is permitted then for display of entry conditions. The direction of integration is determined by the sign of the user's input value **Rwork(3)** and cannot be changed once a solution is started. Thus, the **tout** values for a given problem must form a strictly increasing sequence if **Rwork(3)**$\geq$ 0, or a strictly decreasing sequence if **Rwork(3)** is negative.

### Nstvar

**Nstvar** is the number of state variables, and is the dimension of the state vector $\boldsymbol{u}$ in Eqs. (B.1-1a) and (B.1-1b).

### U(Nstvar,Nspar+1), UPRIME(Nstvar,Nspar+1)

These arrays are used for storage of the state and sensitivity functions and their $t$-derivatives. Their leading dimension must be **Nstvar** whenever sensitivity functions are included. The state vector $\boldsymbol{u}$ and its $t$-derivative are stored in the first columns of these arrays:

$$\mathbf{U}(i, 1) = u_i(t) \qquad i = 1, 2, \ldots, \textbf{Nstvar}$$

$$\mathbf{UPRIME}(i, 1) = u_i'(t) \qquad i = 1, 2, \ldots, \textbf{Nstvar}$$

The sensitivity functions, when requested, are stored in the remaining columns:

$$\mathbf{U}(i, 1 + j) = \frac{\partial u_i(t)}{\partial \theta_j} \qquad j = 1, 2, \ldots, \textbf{Nspar}$$

$$\mathbf{UPRIME}(i, 1 + j) = \frac{\partial u_i'(t)}{\partial \theta_j} \qquad j = 1, 2, \ldots, \textbf{Nspar}$$

Here **Nspar** is the number of parameters for which the sensitivity functions are requested; it is provided by the user as **Info(12)**.

On the initial entry to DDAPLUS, the initial values $u_{i0\_}$ just before $t_0$ should be provided as the first **Nstvar** elements of **U**. An initial guess is needed, as in Example B.4, for any element $u_i$ whose $t$-derivative does not appear. If **Info(12)**$> 0$, as in Examples B.2 and B.3, the values $\partial u_{i0\_}/\partial \theta_j$ should be provided as the remaining elements **U**(i,1+j) of the starting vector **U**, as in Examples B.2 and B.3. No input value of **UPRIME** is required, since DDAPLUS computes this vector. On return from DDAPLUS, **U** and **UPRIME** will contain the solution and its $t$-derivative at the current **t**.

## Rtol(•,•), Atol(•,•)

These are relative and absolute error tolerances set by the user. They can be either scalars or arrays, depending on the user's specification of **Info(2)** as described below. DDAPLUS uses these inputs to construct a vector of mixed local error tolerances

$$\mathbf{Wt} = \mathbf{Atol} + \mathbf{Rtol}*|\mathbf{U}|$$

for the changes in elements of **U** across the current integration step. If the norm (see **Info(15)** of the vector $\{\varepsilon_M U_i/\mathrm{Wt}_i\}$ at the start of any integration step exceeds 0.01 (here $\varepsilon_M U_i$ is the rounding adjustment to $U_i$), DDAPLUS will pause and increase **Atol** and **Rtol** to allow the computation to continue.

The true (global) error of a computed solution is its deviation from the true solution of the initial value problem. Practically all present-day integrators, including this one, control the local error for each step and do not attempt to control the global error directly.

## Info(1)

This parameter must be set to zero whenever beginning a problem or when changing any equations. It is reserved for use by DDAPLUS at all other times, except as noted under **Idid**$=-1$.

*Set* **Info(1)**$=0$ *to indicate that this is an initial call of DDAPLUS, or the start of a new region with changed equations,* with $t_0$ reset to the current **t**.

## Info(2)

This parameter enables the user to specify the desired local accuracy of the computed solution via the error tolerances **Rtol** and **Atol**. The simplest way is to take **Rtol** and **Atol** to be scalars. For greater flexibility (different tolerances for different components of **U**), **Rtol** and **Atol** can be specified as arrays. The following options are available:

*Set* **Info(2)**$=0$ *if you specify* **Rtol** *and* **Atol** *as scalars.* These tolerances will be applied to each state variable if **Info(15)**$=0$ or to each element of **U** if **Info(15)**$=1$.

*Set* **Info(2)=1** *if you specify* **Rtol** *and* **Atol** *as arrays.* Tolerances on the state variables suffice if **Info(15)=0**; tolerances are needed for the full vector **U** if **Info(15)=1**.

### Info(3)

The code integrates from **t** toward **tout** by steps. If you wish, it will return the computed solution and derivative at the next integration point or at **tout**, whichever comes first. This is a good way to proceed if you want to monitor the solution. Results at **tout** are computed by interpolation; thus, you can use small increments of **tout** without significantly slowing down the computation.

*Set* **Info(3)=0** *if you want the solution only at* **tout**.

*Set* **Info(3)=1** *if you want the solution at the next integration point or at* **tout**, *whichever comes first.*

### Info(4)

To handle closely spaced **tout** points efficiently, the code may integrate past one or more **tout** values and then interpolate the solution to those points. However, sometimes it is not possible to integrate beyond some point **tstop** because the equations change or become undefined. Then the user must tell the code not to go past **tstop**.

*Set* **Info(4)=0** *if the integration can be carried out with no restrictions on the independent variable.*

*Set* **Info(4)=1** *if the equations change or are not defined past a stopping point* **tstop**. *Specify this point by setting* **Rwork(1)=tstop**.

Another stopping criterion is described under **Info(17)**: a stopping value of a designated element of **U** can be specified. The two criteria can also be used together as in Example B.3.

### Info(5)

DDAPLUS solves the corrector equations for each $t$-step by Newton's method, using an iteration matrix $G(t \neq t_0)$, of order **Nstvar**, defined by Eq. (B.1-5a). On request, Eq. (B.1-2) is solved similarly with the matrix $G(t_0)$ defined by Eq. (B.1-5b). If the user does not provide a detailed subroutine **Jac** (see Section B.6 below), DDAPLUS will approximate $G(t)$ by finite differences.

*Set* **Info(5)=0** *if* $G(t)$ *is to be computed by DDAPLUS via finite differences.* This method is very efficient when $G(t)$ is banded and the option **Info(6)=1** is used.

*Set* **Info(5)=1** *if the nonzero elements of* $G(t)$ *are to be computed with the aid of a detailed subroutine* **Jac**; *see Section B.6.*

### Info(6)

DDAPLUS will perform better if it is told to use band-matrix algorithms whenever the lower and upper bandwidths of $G$ satisfy **2ml + mu** < **Nstvar**. The matrix $G$ (and $E$ if nondiagonal) will then be stored more compactly, numerical differencing will be done more efficiently, and the matrix algorithms will run much faster. If you do not indicate that the iteration matrix $G$ is banded, DDAPLUS will treat $G$ as a full square matrix.

Set **Info(6)=0** *if* $G(t)$ *is to be treated as a full square (dense) matrix.*

Set **Info(6)=1** *if* $G(t)$ *is to be treated as a band matrix. Set the lower and upper bandwidths,* **Iwork(1)=ml** *and* **Iwork(2)=mu**. These bandwidths are nonnegative integers such that the nonzero elements of the matrix lie in a band extending from the **ml***th* subdiagonal to the **mu***th* superdiagonal. The efficiency will be further enhanced if the equations and unknowns are arranged so that **ml≤mu** (Anderson et al. 1992).

## Info(7)

The user can specify a maximum absolute stepsize to prevent the integrator from jumping over very large regions.

Set **Info(7)=0** *if you want the code to set its own maximum stepsize.*

Set **Info(7)=1** *if you want to define a maximum absolute stepsize.* Provide this value by setting **Rwork(2)=hmax**.

## Info(8)

The initial stepsize can be chosen by DDAPLUS or by the user. In either case, the sign of the initial step is taken from **Rwork(3)**, which DDAPLUS sets to +0.001 whenever the initial **Rwork(3)** is zero.

Set **Info(8)=0** *if you want the code to choose the initial stepsize.* DDAPLUS will compute a near-optimal stepsize using the algorithm of Shampine (1987). If the vector **UPRIME** is initially zero, **Rwork(3)** should be set to a safe stepsize large enough to give a nonzero value of the right-hand vector $f$; this is better than using **hmax** to limit the range searched by the Shampine algorithm.

Set **Info(8)=1** *if you want to choose the initial step; then insert the chosen value* **h0** *into* **Rwork(3)**. Differences among neighboring solutions, as in line searches, are obtained more accurately when the solutions compared have the same **h0**.

## Info(9)

If storage is a severe problem, you can save space by reducing the maximum integrator order **maxord**. Each unit decrease of **maxord** below the default value of 5 saves **Nstvar** locations, or **Nstvar∗(Nspar+1)** when sensitivity calculations are requested; however, the progress is likely to be slower. In any case, **maxord** must be at least 1 and no greater than 5.

Set **Info(9)=0** *if you want the maximum integrator order to be 5.*

Set **Info(9)=1** *if you want the maximum integrator order to be less than 5; specify its value by setting* **Iwork(3)=maxord**.

**Info(10)**
This input can be used to activate special procedures for maintaining non-negative solution elements $u_i$. DDAPLUS will report if it is unable to do this; in that event the problem probably needs to be reformulated (for example, by restoring some neglected reverse reaction steps).

*Set* **Info(10)=0** *if positive and negative solution elements are permitted.*

*Set* **Info(10)=1** *to request nonnegativity of all $u_i$ at $t = t_0$.*

*Set* **Info(10)=2** *to request nonnegativity of all $u_i$ for all $t$.*

**Info(11)**
This input value is now ignored, since DDAPLUS provides consistent initial conditions by solving any algebraic rows present in Eq. (B.1-1a). The initial derivative vector **UPRIME($t_0$)** is calculated automatically, so its input value need not be given.

The calculation of **UPRIME** will involve the derivative $(\partial f_A/\partial t)_{u,\theta}$ if any algebraic equation in the set (B.1-1a) has a term depending explicitly on $t$. In this case, provide a nonzero $\Delta t$ in **Rwork(44)**; DDAPLUS will then evaluate $(\partial f_A/\partial t)_{u,\theta}$ as a divided difference. Use **Rwork(44)=0** otherwise.

**Info(12)**
This parameter controls the option of calculating the first-order sensitivity functions.

*Set* **Info(12)=0** *if you want the code to calculate the state functions only.*

*Set* **Info(12)=Nspar** *if you want the computation to include the sensitivity functions.*

**Nspar** is the number of parameters (independent of $t$) whose sensitivity functions are to be calculated. These parameters must be stored in the array **Rpar**, in locations specified via the array **Ipar** as described at the end of this section.

**Info(13)**
This parameter is used to select the form of the $t$-derivative terms.
*Set* **Info(13)=0** *if $E$ is a unit matrix. Then Eq. (B.1-1a) takes the form*

$$u' = f(t, u; \theta)$$

*Set* **Info(13)=±1** *if $E$ is diagonal but not a unit matrix.* Use the $-$ sign if $E$ will be constant between $t_0$ and the next restart; use the $+$ sign if $E$ will vary with $t$ or $u$, or with $\theta$ in a sensitivity calculation. Provide the nonzero elements $E_{ii}(t, u; \theta)$ in a subroutine **Esub** as described in Section B.5.

*Set* **Info(13)=±2** *if $E$ is a nondiagonal matrix, to be stored in the same form as $G$; see* **Info(6)**, **Iwork(1)**, *and* **Iwork(2)**. Use the $-$ sign if $E$ will not change between $t_0$ and the next restart; use the $+$ sign if $E$ will vary with $t$ or $\boldsymbol{u}$, or with $\boldsymbol{\theta}$ in a sensitivity calculation. Provide the nonzero elements $E_{ij}(t, \boldsymbol{u}; \boldsymbol{\theta})$ in a subroutine **Esub** as described in Section B.5.

## Info(14)

The DDAPLUS sensitivity algorithm uses the matrix function $\boldsymbol{B}(t)$ defined in Eqs. (B.1-4a) and (B.1-4b). This function will be approximated by divided differences in DDAPLUS, unless the user sets **Info(14)=1** and provides the derivatives $\partial f_i / \partial \theta_j$ with respect to one or more parameters in Subroutine **Bsub**. *Set* **Info(14)=0** *if you want DDAPLUS to approximate* $\boldsymbol{B}(t)$ *by differencing.*

*Set* **Info(14)=1** *if you want to evaluate derivatives* $\partial f_i / \partial \theta_j$ *for one or more parameters with a detailed Subroutine* **Bsub***; see Section B.7.* This method is preferable for parameters that are very small or zero.

## Info(15)

This integer is used to specify which elements of **U** are to be included in the local error tests described under **Info(2)**.

*Set* **Info(15)=0** *if only the state variables* $u_i = U(i, 1)$ *are to be included.* This option is normally preferred, especially for parameter estimation where it gives better consistency between solutions with and without sensitivity analysis.

*Set* **Info(15)=1** *if all elements of* **U** *are to be included.* This option slows down the integrator quite a bit.

## Info(16)

This integer specifies the kind of norm to be used in testing error vectors.

*Set* **Info(16)=0** *to request the infinity-norm (maximum absolute element).*

*Set* **Info(16)=1** *to request the Euclidean norm (root-mean-square value).*

## Info(17)

This integer, when positive, is called **IYstop** and indicates that the state or sensitivity variable **U(IYstop)** must not cross the value **Cstop** in the current integration. If the index **Info(17)** is outside the set of variables indicated in **Info(15)**, DDAPLUS will write a diagnostic message and return control to the calling program.

Before calling DDAPLUS, the user's program MAIN must insert the target value **Cstop** into **Rwork(42)**, and may insert a multiplier for the stopping tolerance as described under **Rwork(43)**. This stopping criterion can be used concurrently with the **tstop** criterion (see **Info(4)**); DDAPLUS will return with **Idid = 4** if **U(IYstop)** attains its stopping value within the specified tolerance before **tstop** is reached.

## Info(18)

This integer controls the optional reporting of the index $i$ for the controlling variable $U_i$ in tests of rounding error, truncation error or iteration convergence. The index is written to LUNERR for any failed test if **Info(18)=2**; for any test causing termination if **Info(18)=1**, and never if **Info(18)=0**.

The index is also returned in **Iwork(4)** when such a test causes termination.

## Info(29)

This integer controls the amount of printout during the integration by DDAPLUS. Set **Info(29)=-1** to suppress any printout; set **Info(29)** to a zero or positive value to request various levels of solution information from DDAPLUS.

## Rwork(•), Lrw

**Rwork** is a double precision work array of length **Lrw** used by DDAPLUS. A value **Lrw=5000** suffices for many medium-sized problems. If the input value **Lrw** is too small, (for example, if **Lrw=1**), DDAPLUS will promptly terminate and state the additional length required. Some of the contents of **Rwork** are summarized here; the user may access these via the program MAIN.

**Rwork(1)** is an input/output register. It must be initialized to the value **tstop** if the input option **Info(4)=1** is used.

**Rwork(2)** is an input register. It must be set to the desired steplength limit **hmax** if the option **Info(7)=1** is used.

**Rwork(3)** is an input/output register. The $t$-steps are taken forward if the initial value here is zero or positive, and backward otherwise. This register is initialized by DDAPLUS to 0.001 unless MAIN sets **Info(8)=1** and $|\mathbf{Rwork(3)}| > 0$. On return, this register contains the signed step **h** to be attempted next.

**Rwork(4)** contains the endpoint of the latest completed integration step. The returned value of **t** may differ when **Idid=3, 4,** or **5** because of interpolations then performed.

**Rwork(7)** contains the signed step **h** used in the last successful integration step.

**Rwork(41)** contains the relative perturbation step size for finite-difference computation of $\mathbf{G}(t)$. The default stepsize is the square root of the machine precision; DDAPLUS uses this unless MAIN provides a larger value.

**Rwork(42)** must be set to the desired stopping value **Cstop** of the state or sensitivity variable **U(IYstop)**, whenever the user inserts an index **IYstop** into **Info(17)**.

**Rwork(43)** is an optional input, available to loosen the stopping criterion

$$|\mathbf{U(IYstop)} - \mathbf{Cstop}| \leq \max[1.0, \mathbf{Rwork(43)}] * (\mathbf{Wt(IYstop}$$

used whenever a positive index **IYstop** is provided in **Info(17)**. If **U(IYstop)** begins within this range, stopping is deferred until reentry of this range.

**Rwork(44)** is reserved for a user-specified $\Delta t$ value for $t$-differencing of $\boldsymbol{f}_A$ as described under **Info(11)**.

## Iwork(•), Liw

**Iwork** is an integer work array of length **Liw** used by DDAPLUS. A value **Liw=500** suffices for many medium-sized problems. If the input value **Liw** is too small (for example, if **Liw=1**), DDAPLUS will promptly terminate and state the additional length required. Some of the contents of **Iwork** are summarized here; the user may access these via his/her program MAIN.

**Iwork(1)** is accessed only when the user has requested band matrix algorithms by specifying **Info(6)=1**. This register must then be set to the lower bandwidth **ml**, defined under **Info(6)**, of the iteration matrix $G(t)$.

**Iwork(2)** likewise is accessed only when the user has requested band matrix algorithms by specifying **Info(6)=1**. This register must then be set to the upper bandwidth **mu**, defined under **Info(6)**, of the iteration matrix $G(t)$.

**Iwork(3)** contains the maximum permitted integrator order; see **Info(9)**.

**Iwork(4)**, when nonzero, contains the index $i$ of the dominant variable in the test failure that stopped the DDAPLUS computation. **Info(18)** can be used to get related information.

**Iwork(7)** contains the integrator order to be attempted in the next step.

**Iwork(8)** contains the integrator order used in the latest step.

**Iwork(10)** contains the total number of integration steps completed so far in the current problem.

**Iwork(11)** contains the number of integration steps completed since the setting of the current $t_0$; see **Info(1)**.

**Iwork(12)** contains the number of residual vector evaluations for Eq. (B.1-1a), plus any needed for Eq. (B.1-2), since the setting of the current $t_0$.

**Iwork(13)** contains the number of evaluations of the iteration matrix $G(t)$ since the setting of the current $t_0$.

**Iwork(14)** contains the number of error test failures since the setting of the current $t_0$.

**Iwork(15)** contains the number of convergence test failures since the setting of the current $t_0$.

**Iwork(16)** contains the maximum number of iterations permitted for the damped Newton method in an integration step. The default value of 4 is used if the input value is anything less.

**Iwork(17)** contains the maximum number of damped Newton iterations permitted for the completion of the initial state; see **Info(11)**. If the input value of **Iwork(17)** is zero, the code resets it to 10.

**Iwork(18)** indicates the number of damped Newton iterations allowed, with updated matrix $G(t_0)$ each time (see Eq. (B.1-2)), before changing the update interval to **Iwork(18)** iterations in the completion of the initial state. If the input value of **Iwork(18)** is zero, the code resets it to 5.

**Iwork(19)** contains the number of additional locations needed in IWORK.

**Iwork(20)** contains the number of additional locations needed in RWORK.

**Iwork(21)** contains the maximum number of steps before reaching **tout**.

## Rpar(0:•), Ipar(0:•)

**Rpar** is a double precision work array, and **Ipar** is an integer work array. These are provided so that DDAPLUS can pass special information to the user-provided subroutines. The parameters whose sensitivity coefficients are desired must be stored in the **Rpar** array in the locations specified by the elements $1, \ldots,$ Nspar of the array **Ipar**. For example, the input **Ipar(2)=3** would indicate that the second sensitivity parameter was stored in **Rpar(3)**. The location **Ipar(0)** is reserved for a user-defined constant, such as the address (later in **Ipar**) of a list of pointers to other information. The location **Rpar(0)** may be used to store values of a user-defined continuation parameter.

## B.5 EXIT CONDITIONS AND CONTINUATION CALLS

The main task of DDAPLUS is to compute the state vector $u$ (and the sensitivity matrix $W$ if requested) at the next output point. The status of the solution upon return to the calling program is described by the parameter **Idid**, whose possible values are explained below.

### Idid=0

The initial **UPRIME** and **h** are ready at $t = t_0$, along with **U** updated from $t_{0-}$ whenever algebraic equations are present. This is the normal return when DDAPLUS is called with **tout=t**. To commence integration, MAIN must call DDAPLUS again with **tout** moved from **t** in the direction indicated by the sign of **h** in **Rwork(3)**.

### Idid=1

A successful integration step partway to **tout** is being reported under the option **Info(3)=1**.

### Idid=2

A successful integration step to **tout** is being reported. If you want to continue the integration, MAIN must define a new **tout** and call the code again. You cannot change the direction of integration without restarting.

### Idid=3

The integration has proceeded successfully across **tout**. The values of **U** and **UPRIME** (including sensitivity functions when requested) have been computed at **tout** by interpolation. If you want to continue the integration, MAIN must define a new **tout** and call the code again. You cannot change the direction of integration without restarting.

### Idid=4

The integration has gone successfully to a point where **U(IYstop)** is within the stopping tolerance of **Cstop**; see **Info(17)**. If you want to continue the integration with the same state equations, MAIN must define a new **tout** and call the code again with **Info(17)** and/or **Rwork(42, 43)** reset. If the state equations change at this point, MAIN must set the index **Irange** according to the new equation forms and call DDAPLUS with **Info(1)=0**. You cannot change the direction of integration without restarting.

## Idid=5

The integration has gone successfully to a point that matches **tstop** within 100 times the machine precision. If the solution is to be continued, MAIN must reset **Irange**, **Info(4)** and **Rwork(1)** according to the new equation forms and call DDAPLUS with **Info(1)=0**. You cannot change the direction of the integration without restarting.

## Idid=−1

The code has taken about **Iwork(21)** steps (counted as **Iwork(11)** − **Iwork(10)**) on this call. If you want to continue, set **Info(1)=1** in MAIN and call DDAPLUS again; another **Iwork(21)** steps will be allowed.

## Idid=−2

The error tolerances have been reset (increased) to values the code estimates to be appropriate for continuing. If you are sure you want to proceed with these adjusted tolerances, MAIN should leave **Info(1)=1** and call the code again. Otherwise, start over with **Info(1)=0** and new tolerances of your choosing.

## Idid=−3

A solution component is zero and you have set the corresponding **Atol** component equal to zero. If you are sure you want to continue, MAIN must insert positive values of those components of **Atol** that correspond to zero solution components; then set **Info(1)=0** and call the code again.

## Idid=−6

Repeated error test failures occurred on the last attempted step in DDAPLUS. A singularity may be present. If you are certain you want to continue, MAIN should restart the solution from the last **U** value with **Info(1)=0**.

## Idid=−7

Repeated convergence test failures occurred on the last attempted step. An inaccurate or ill-conditioned matrix $G(t)$ may be the cause. If you are absolutely sure you want to continue, MAIN should restart the solution with **Info(1)=0**.

## Idid=−8

The latest iteration matrix is singular. Some of the equations or variables may be redundant. If so, you need to review the problem formulation.

## Idid=−9

There were multiple convergence test failures preceded by multiple error test failures on the last attempted step. This may be an indication that your problem is ill-posed, or not solvable with this code, or that there is a discontinuity or singularity in the equations. A discontinuity requires a restart from that point $(\mathbf{t}, \mathbf{Y})$, using the new equation forms.

## Idid=−10
DDAPLUS had multiple convergence test failures because **Ires** was set to −1 by one of your subroutines, indicating difficulty there. If you are absolutely certain you want to continue, restart the solution; see **Info(1)**. You may want to include WRITE statements in your subroutines to explain whenever **Ires** is set to −1.

## Idid=−11
Control is being returned to the calling program because **Ires** (see description in subroutine **fsub**) has been set to −2. You cannot continue the integration when this condition occurs. An explanation will appear in the output if the value Ires=−2 was set by a module of DDAPLUS; if not, rerun the problem after inserting WRITE statements in your subroutines to explain whenever one of them sets **Ires** to −2.

## Idid=−12
DDAPLUS failed to complete the initialization of **UPRIME** or **U** or **h**. This could happen (in the presence of algebraic state equations) because of a poor initial guess or because a solution does not exist. The difficulty might also have been caused by an inaccurate or ill-conditioned iteration matrix $G(t_0)$.

## Idid=−33
The code has encountered trouble from which it cannot recover. In such cases a message is written explaining the trouble, and control is returned to the calling program. This **Idid** value usually indicates an invalid input.

## LUNERR, LUNREP
**LUNERR** and **LUNREP** are the logical unit numbers to which messages and results from DDAPLUS are to be written, respectively.

## Irange
This integer argument is for use in problems whose equation forms change at one or more stopping values of **t** along the solution path. When such a point is reached, program MAIN must select the equation forms for the next range of **t** by setting **Irange** to the appropriate value. Since such a change of equation forms generally causes a jump of **UPRIME**, program MAIN must set **Info(1)=0** before calling DDAPLUS to enter the new region.

## Imodel
This integer argument indicates the model that is currently being processed by the DDAPLUS solver.

## Ipass
This integer argument must be set equal to zero. It is used by the DDAPLUS license manager.

## Filepath
This character argument contains the file path of the user's model.

## Filepath

This integer argument contains the length of the **Filepath.**

## B.6 THE SUBROUTINE fsub

This subroutine defines the function vector $\boldsymbol{f}(t, \boldsymbol{u}; \boldsymbol{\theta})$ in Eq. (B.1-1a). It may be organized as in the following pseudocode.

> Subroutine fsub ( t, Nstvar, U, fval, Rpar, Ipar, Irange, Imodel, Ires )
> - 
> - This subroutine calculates the right-hand terms
> - $f_i(t, \boldsymbol{u}; \boldsymbol{\theta})$ of the differential and algebraic equations
> - at the current t, U and Rpar.
> - 
> - Declare required arrays and variables
> - 
>   Double precision t, U(∗), fval(∗), Rpar(0:∗), · · · · ·
>   Integer Nstvar, Ipar(0:∗), Irange, Ires, · · · · ·
> - 
> - Calculate right-hand function values
> - 
>   fval(•) = · · · · ·       (Use form indicated by Irange)
> - 
> - End of the fsub subroutine
> - 
>   End

Here the vector **fval** is constructed according to Eq. (B.1-1a):

$$\mathbf{fval}(i) = f_i(t, \boldsymbol{u}(t); \boldsymbol{\theta}) \qquad i = 1, \ldots, \text{Nstvar}$$

The argument **Ires** is zero on entry and should not be altered unless **fsub** encounters trouble, such as an illegal or undefined value of an argument or calculated quantity. To attempt recovery in DDAPLUS by taking a smaller step, return with **Ires=−1**. To terminate the execution, return with **Ires=−2**.

## B.7 THE SUBROUTINE Esub

The form of this subroutine depends on the value of **Info(13)**. If **Info(13)=0**, a dummy form of **Esub** suffices:

> Subroutine Esub
> End

When **Info(13)** ≠ **0**, **Esub** may be organized as in the following pseudocode

Subroutine Esub ( t, Nstvar, U, Ework, Rpar, Ipar, Irange, Imodel, Ires )

- •
- • This subroutine calculates the nonzero elements
- • of the current matrix $E(t, u; \theta)$. It is executed only
- • if MAIN sets Info(13) to 1, −1, 2, or −2 before calling DDAPLUS.
- •
- • Declare arrays and variables
- •

    Double precision t, U(*), Ework( ), Rpar(0:*), · · · · ·
    Integer Nstvar, Ipar(0:*), Irange, Ires, · · · · ·

- •
- • Calculate the nonzero elements of Ework
- •

    Ework(•) = · · · · ·        (Use form indicated by Info(13) and Irange)

- •
- • End of Subroutine Esub
- •

    End

If **Info(13)** $= \pm 1$, then $E$ must be a diagonal matrix. Its work space is dimensioned as **Ework(*)** and its elements are inserted as follows:

```
Do 10 I=1,Nstvar
    Ework(I)=E(I,I)
10 Continue
```

If **Info(13)** $= \pm 2$ and **Info(6)** $= 0$, then $E$ is treated as a dense matrix. It is dimensioned by **Esub** as **Ework(Nstvar,Nstvar)**, and its elements are inserted as follows:

```
K=0
Do 20 J=1,Nstvar
    Do 10 I=1,Nstvar
        Ework(I,J)=E(I,J)
10   Continue
20 Continue
```

If **Info(13)** $= \pm 2$ and **Info(6)** $= 1$, then $E$ must be a banded matrix. Its work space is dimensioned as **Ework(*)** and its elements are inserted as follows:

```
Mband=ML+MU+1
LDband=2*ML+MU+1
Do 20 J=1,Nstvar
    I1=MAX(1,J−MU)
    I2=MIN(Nstvar,J+ML)
    K=I1−J+Mband + (J−1)*LDband
    Do 10 I=I1,I2
        Ework(K)=E(I,J)
        K=K+1
10   Continue
```

20 Continue

A negative **Info(13)** signifies that $E$ is constant throughout the problem, or until such time as DDAPLUS may be restarted to change equation forms. In parametric sensitivity calculations this requires that, over the same time interval, $E$ be independent of the sensitivity parameters as well as of $t$ and $u$.

The argument **Ires** is for use as in **fsub**, to notify DDAPLUS of any difficulty encountered here.

## B.8 THE SUBROUTINE Jac

The form of this subroutine depends on the choice of **Info(5)** and **Info(6)**. If **Info(5)=0**, a dummy form of **Jac** suffices:

```
Subroutine Jac
End
```

When **Info(5)=1**, **Jac** may be organized as in the following pseudocode:

```
Subroutine Jac ( t, Nstvar, U, Pdwork, Rpar, Ipar, Irange, Imodel,
Ires )
```
- •
- • This subroutine calculates the nonzero elements of $\partial f/\partial u^T$,
- • for use by DDAPLUS in computing $G(t \neq t_0)$ or $G(t_0)$ for
- • systems with $\partial E/\partial u^T = 0$. Jac is executed only
- • if MAIN sets Info(5) to 1 before calling DDAPLUS.
- • If $\partial E/\partial u^T$ is nonzero, set Info(5)=0 and use the
- • dummy form of Jac given above.
- •
- • Declare arrays and variables
- •
```
      Double precision t, U(∗), Pdwork(  ), Rpar(0:∗), · · · · ·
      Integer Nstvar, Ipar(0:∗), Irange, Ires, · · · · ·
```
- •
- • Insert the nonzero elements $\partial f_i/\partial u_j$ into Pdwork:
- •
```
      Pdwork(•) = · · · · ·       (Use form indicated by Info(6) and Irange)
      IF(unsuccessful) THEN
         Ires = −1
      END IF
```
- •
- • End of the Jac subroutine
- •
```
      End
```

If **Info(5)=1** and **Info(6)=0**, then $\partial f/\partial u^T$ is treated as a dense matrix. **Esub** can then declare **Pdwork** as a matrix with leading dimension **Nstvar** and insert its nonzero elements as follows:

```
Do 20 J=1,Nstvar
   Do 10 I=1,Nstvar
      Pdwork(I,J)=∂fᵢ/∂uⱼ
10  Continue
20 Continue
```

If **Info(5)=1** and **Info(6)=1**, then $\partial \boldsymbol{f}/\partial \boldsymbol{u}^T$ is treated as a banded matrix. Then **Esub** must declare **Pdwork** as a vector and insert its nonzero elements as follows:

```
Mband=ML+MU+1
LDband=2*MI+MU+1
Do 20 J=1,Nstvar
   I1=MAX(1,J−MU)
   I2=MIN(Nstvar,J+ML)
   K=I1−J+Mband + (J−1)*LDband
   Do 10 I=I1,I2
      Pdwork(K)=∂fᵢ/∂uⱼ
      K=K+1
10  Continue
20 Continue
```

The argument **Ires** is zero on entry, and should not be altered unless the current call of **Jac** is unsuccessful. In the latter event, set **Ires** to a negative value before return; DDAPLUS will then set **Ires=−2** and terminate.

After each successful call of **Jac**, DDAPLUS completes the iteration matrix **G** in Pdwork with the following formulas, which follow from Eqs. (B.1-5a) and (B.1-5b) when $\partial \boldsymbol{E}/\partial \boldsymbol{u}^T$ is zero:

$$G_{ij}(t \neq t_0) = \left[ C_J\, E_{ij} - \frac{\partial f_i}{\partial u_j} \right]_{t,\,\boldsymbol{\theta}} \tag{B.8-1a}$$

$$G_{ij}(t_0) = \begin{cases} E_{ij}(t_0, \boldsymbol{\theta}) & \text{if row } i \text{ of } \boldsymbol{E} \text{ has a nonzero element} \\[2ex] -\partial f_i/\partial u_j \Big|_{t_0, \boldsymbol{\theta}} & \text{otherwise} \end{cases}$$

$$\tag{B.8-1b}$$

The elements are stored by DDAPLUS in accordance with **Info(6)**, taking the bandwidths from **Iwork(1)** and **Iwork(2)** when needed.

## B.9 THE SUBROUTINE Bsub

The form of this subroutine depends on the choices of **Info(12)** and **Info(14)**. When **Info(12)=0** or **Info(14)=0**, a dummy form of **Bsub** suffices:

```
Subroutine Bsub
End
```

When **Info(12)>0** and **Info(14)=1**, **Bsub** will be called by DDAPLUS, and may be organized as in the following pseudocode.

Subroutine Bsub ( t, Nstvar, U, fpj, Jspar, Rpar, Ipar, Irange, Imodel, Ires )

- 
- This subroutine calculates the nonzero elements of fpj≡ $\partial \boldsymbol{f}/\partial \theta_j$,
- for use by DDAPLUS in constructing a column of the matrix $\boldsymbol{B}(t)$.
- Bsub is executed only if MAIN sets Info(12)>0 and Info(14)=1
- before calling DDAPLUS. If $\partial \boldsymbol{E}/\partial \theta_j$ is nonzero, leave Ires=0
- to activate DDAPLUS's finite-difference algorithm for this column of

$\boldsymbol{B}(t)$.

- 
- Declare arrays and variables
- 

        Double precision t, U(∗), fpj(∗), Rpar(0:∗), ·····
        Integer Nstvar, Jspar, Ipar(0:∗), Irange, Ires, ·····

- 

        Go to ( 10, 20, ·····) Jspar

- 
- Calculate nonzero elements of column vector $\partial \boldsymbol{f}/\partial \theta_j$
- 

        IF(formulas for the current column are being provided) THEN
            fpj(•) = ·····       (Use form indicated by Irange)
            IF(Column successfully evaluated) THEN
                Ires = 1
            ELSE
                Ires = −1
            END IF
        END IF

- 
- End of the Bsub subroutine
- 

        End

**Ires** is defined differently in this subroutine. Its input value is zero, and its output value is chosen as follows. An output value **Ires=1** indicates a successful calculation; an unaltered **Ires** value (zero) indicates that **Bsub** does not provide an expression for the current column $\partial \boldsymbol{f}/\partial \theta_j$, and that DDAPLUS should calculate this column by finite differences. Any other output value of **Ires** indicates an error condition; DDAPLUS will then set **Ires=−2** and terminate.

If **Bsub** returns **Ires=1**, then DDAPLUS evaluates the current column of $\boldsymbol{B}$ from the returned vector fpj as follows:

$$B_{ij}(t \neq t_0) = \left[\frac{\partial f_i}{\partial \theta_j}\right]_{t,\,\boldsymbol{u},\,\boldsymbol{u}'} \tag{B.9-1a}$$

and

$$
B_{ij}(t_0) =
\begin{cases}
0 & \text{if row } i \text{ of } \boldsymbol{E}(t_0) \text{ is not zero} \\
\left.\partial f_i/\partial\theta_j\right|_{t_0,\,\boldsymbol{u}(t_0)} & \text{otherwise}
\end{cases}
\tag{B.9-1b}
$$

These formulas follow from Eqs. (B.1-4a,b) when $\partial\boldsymbol{E}/\partial\theta_j$ is zero.

The argument **Jspar** of this subroutine is set by DDAPLUS and should not be altered by the user. This integer (numbered over the sensitivity parameters only) identifies the parameter for which the sensitivity functions are currently being computed.

## B.10 NUMERICAL EXAMPLES

Three examples are given here to demonstrate various capabilities of DDAPLUS. In the first example, DDAPLUS is used to solve a system of ordinary differential equations for the concentrations in an isothermal batch reactor.In the second example, the same state equations are to be integrated to a given time limit, or until one of the state variables reaches a given limit. The last example demonstrates the use of DDAPLUS to solve a differential-algebraic reactor problem with constraints of electroneutrality and ionization equilibria.

### Example B.1. Integration of State Equations

Consider an isothermal batch reactor, in which the concentrations $u_i(t)$ are described by the following system of ordinary differential equations:

$$
\begin{aligned}
u_1'(t) &= -\theta_1 u_1(t)u_2(t) + \theta_2 u_3(t) \\
u_2'(t) &= -\theta_1 u_1(t)u_2(t) + \theta_2 u_3(t) - 2\theta_3 u_2^2(t) + 2\theta_4 u_4(t) \\
u_3'(t) &= \theta_1 u_1(t)u_2(t) - \theta_2 u_3(t) \\
u_4'(t) &= \theta_3 u_2^2(t) - \theta_4 u_4(t)
\end{aligned}
$$

This system is to be integrated numerically from $t=0$ to $t=10$ for the following initial conditions and parameter values:

$$
u_1(0) = \theta_5 \quad u_2(0) = \theta_6 \quad u_3(0) = \theta_7 \quad u_4(0) = \theta_8
$$

$$
\theta_1 = 100.0 \quad \theta_2 = 2.0 \quad \theta_3 = 10^4 \quad \theta_4 = 10^{-1}
$$

$$
\theta_5 = 1.0 \quad \theta_6 = 1.0 \quad \theta_7 = 0 \quad \theta_8 = 0
$$

The following Athena Visual Studio code solves this test example by calls to DDAPLUS and tabulates the four state variables as functions of time.

## Athena Visual Studio Code for Example B.1

```
! Appendix B: Example B.1
! Computer-Aided Modeling of Reactive Systems
!--------------------------------------------
 Global Theta1,Theta2,Theta3,Theta4 As Real

! Enter the reaction rate constants
!-------------------------------
 Theta1=100.0
 Theta2=2.0
 Theta3=10000.0
 Theta4=0.1

@Initial Conditions
 U(1)=1.0
 U(2)=1.0
 U(3)=0.0
 U(4)=0.0

@Model Equations
 F(1)=-Theta1*U(1)*U(2)+Theta2*U(3)
 F(2)=-Theta1*U(1)*U(2)+Theta2*U(3) -
      2.0*Theta3*U(2)*U(2)+2.0*Theta4*U(4)
 F(3)= Theta1*U(1)*U(2)-Theta2*U(3)
 F(4)= Theta3*U(2)*U(2)-Theta4*U(4)

@Solver Options
 Neq=4                ! Number of Equations
 Tend=10.0            ! End of Integration
 Npts=0.01;0.1;1.0    ! Points of Intermediate Output
```

## Numerical Results for Example B.1

```
Number of State Equations......................    4
Number of Sensitivity Parameters...............    0
Number of Integration Output Points............    5
```

| Time | U(1) | U(2) | U(3) | U(4) |
|------|------|------|------|------|
| 0.00000E+00 | 1.00000E+00 | 1.00000E+00 | 0.00000E+00 | 0.00000E+00 |
| 1.00000E-02 | 9.76198E-01 | 3.37956E-03 | 2.38021E-02 | 4.86409E-01 |
| 1.00000E-01 | 9.69063E-01 | 1.28664E-03 | 3.09374E-02 | 4.83888E-01 |
| 1.00000E+00 | 9.29980E-01 | 1.80190E-03 | 7.00204E-02 | 4.64089E-01 |
| 1.00000E+01 | 9.04015E-01 | 2.12353E-03 | 9.59846E-02 | 4.50946E-01 |

```
EXIT DDAPLUS: SOLUTION FOUND

Number of Steps Taken Thus Far...................    231
Number of Function Evaluations...................    563
Number of Jacobian Evaluations...................    23
Number of Jacobian Factorizations...............    23
```

The deferred updating of the Jacobian matrix PD evidently works very well; 23 updates sufficed for the 231 integration steps. Here 563 evaluations of the vector $f$ were needed besides those used in updating PD; this corresponds to about two Newton iterations per integration step.

## Example B.2. Integration with Two Stopping Criteria

Consider again the batch reactor of Example B.1, in which the species concentrations $u_i(t)$ are described by four ordinary differential equations:

$$u_1'(t) = -\theta_1 u_1(t) u_2(t) + \theta_2 u_3(t)$$
$$u_2'(t) = -\theta_1 u_1(t) u_2(t) + \theta_2 u_3(t) - 2\theta_3 u_2^2(t) + 2\theta_4 u_4(t)$$
$$u_3'(t) = \theta_1 u_1(t) u_2(t) - \theta_2 u_3(t)$$
$$u_4'(t) = \theta_3 u_2^2(t) - \theta_4 u_4(t)$$

This system is to be integrated numerically for the following initial conditions and parameter values:

$$u_1(0) = \theta_5 \quad u_2(0) = \theta_6 \quad u_3(0) = \theta_7 \quad u_4(0) = \theta_8$$

$$\theta_1 = 100.0 \quad \theta_2 = 2.0 \quad \theta_3 = 10^4 \quad \theta_4 = 10^{-1}$$

$$\theta_5 = 1.0 \quad \theta_6 = 1.0 \quad \theta_7 = 0 \quad \theta_8 = 0$$

The solution is to be continued until $t = 10$ or until $u_1 = 0.92$, whichever happens first.

The following Athena Visual Studio code solves this problem.

### Athena Visual Studio Code for Example B.2

```
! Appendix B: Example B.2
! Computer-Aided Modeling of Reactive Systems
!-------------------------------------------
 Global Theta1,Theta2,Theta3,Theta4 As Real

! Enter the reaction rate constants
!-------------------------------------------
 Theta1=100.0
 Theta2=2.0
 Theta3=10000.0
 Theta4=0.1

@Initial Conditions
 U(1)=1.0
 U(2)=1.0
 U(3)=0.0
 U(4)=0.0

@Model Equations
 F(1)=-Theta1*U(1)*U(2)+Theta2*U(3)
 F(2)=-Theta1*U(1)*U(2)+Theta2*U(3) -
      2.0*Theta3*U(2)*U(2)+2.0*Theta4*U(4)
 F(3)= Theta1*U(1)*U(2)-Theta2*U(3)
 F(4)= Theta3*U(2)*U(2)-Theta4*U(4)

@Solver Options
 Neq=4              ! Number of Equations
 Tend=10.0          ! End of Integration
 Ustop=0.92         ! Stop Control on State Variable U(i)
 Uindex=1           ! Index of State Variable U(i)
 Npts=30            ! Number of Points for Intermediate Output
```

## Numerical Results for Example B.2

```
Number of State Equations......................     4
Number of Sensitivity Parameters...............     0
Number of Integration Output Points............    32
```

| Time | U(1) | U(2) | U(3) | U(4) |
|------|------|------|------|------|
| 0.00000E+00 | 1.00000E+00 | 1.00000E+00 | 0.00000E+00 | 0.00000E+00 |
| 3.22581E-01 | 9.56286E-01 | 1.45976E-03 | 4.37136E-02 | 4.77413E-01 |
| 6.45161E-01 | 9.41671E-01 | 1.65206E-03 | 5.83292E-02 | 4.70009E-01 |
| 9.67742E-01 | 9.30882E-01 | 1.79045E-03 | 6.91178E-02 | 4.64546E-01 |
| 1.29032E+00 | 9.23055E-01 | 1.88912E-03 | 7.69455E-02 | 4.60583E-01 |
| 1.45233E+00 | 9.20000E-01 | 1.92724E-03 | 8.00000E-02 | 4.59036E-01 |

```
EXIT DDAPLUS: SOLUTION FOUND

Number of Steps Taken Thus Far...................   202
Number of Function Evaluations...................   499
Number of Jacobian Evaluations...................    22
Number of Jacobian Factorizations...............     22
```

## Example B.3. Integration of a Differential-Algebraic Reactor Model

Biegler, Damiano, and Blau (1986) presented a 10-component reactor model for a test of available modelling software. The state variables $u_i$ used are the following combinations of the species concentrations in mols/kg of the reaction mixture:

$$u_1 = [HA] + [A^-] \qquad\qquad u_6 = [M^-]$$
$$u_2 = [BM] \qquad\qquad\qquad u_7 = [H^+]$$
$$u_3 = [HABM] + [ABM^-] \qquad u_8 = [A^-]$$
$$u_4 = [AB] \qquad\qquad\qquad u_9 = [ABM^-]$$
$$u_5 = [MBMH] + [MBM^-] \quad u_{10} = [MBM^-]$$

Here the model will be solved for an isothermal batch reactor at 100°C.

The model includes six differential mass-balance equations

$$\frac{du_1}{dt} = f_1(t, \boldsymbol{u}) = -\mathcal{R}_2 \qquad\qquad \text{(B.10-1)}$$

$$\frac{du_2}{dt} = f_2(t, \boldsymbol{u}) = -\mathcal{R}_1 - \mathcal{R}_2 \qquad\qquad \text{(B.10-2)}$$

$$\frac{du_3}{dt} = f_3(t, \boldsymbol{u}) = \mathcal{R}_2 + \mathcal{R}_3 \qquad\qquad \text{(B.10-3)}$$

$$\frac{du_4}{dt} = f_4(t, \boldsymbol{u}) = -\mathcal{R}_3 \qquad\qquad \text{(B.10-4)}$$

$$\frac{du_5}{dt} = f_5(t, \boldsymbol{u}) = \mathcal{R}_1 \qquad\qquad \text{(B.10-5)}$$

$$\frac{du_6}{dt} = f_6(t, \boldsymbol{u}) = -\mathcal{R}_1 - \mathcal{R}_3 \qquad\qquad \text{(B.10-6)}$$

an electroneutrality condition,

$$f_7(t, \boldsymbol{u}) = 0.0131 - u_6 + u_7 - u_8 - u_9 - u_{10} = 0 \qquad\qquad \text{(B.10-7)}$$

and three equilibrium conditions for rapid acid-base reactions:

$$f_8(t, \boldsymbol{u}) = u_8 - \theta_7 u_1/(\theta_7 + u_7) = 0 \tag{B.10-8}$$

$$f_9(t, \boldsymbol{u}) = u_9 - \theta_8 u_3/(\theta_8 + u_7) = 0 \tag{B.10-9}$$

$$f_{10}(t, \boldsymbol{u}) = u_{10} - \theta_6 u_5/(\theta_6 + u_7) = 0 \tag{B.10-10}$$

The given rate expressions for the rate-controlling reactions are

$$\mathcal{R}_1 = \theta_1 u_2 u_6 - \theta_2 u_{10} \tag{B.10-11}$$

$$\mathcal{R}_2 = \theta_3 u_2 u_8 \tag{B.10-12}$$

$$\mathcal{R}_3 = \theta_4 u_4 u_6 - \theta_5 u_9 \tag{B.10-13}$$

in mols kg$^{-1}$ hr$^{-1}$, with the following parameter values at $100°$C:

$\theta_1 = 20.086$ $\qquad$ $\theta_5 = 4.05 \times 10^3$
$\theta_2 = 8.10 \times 10^3$ $\qquad$ $\theta_6 = 1.0 \times 10^{-17}$
$\theta_3 = 20.086$ $\qquad$ $\theta_7 = 1.0 \times 10^{-11}$
$\theta_4 = 20.086$ $\qquad$ $\theta_8 = 1.0 \times 10^{-17}$.

This model is to be integrated numerically, from the initial conditions

$u_1(t = 0-) = 1.5776$ $\qquad$ $u_6(t = 0-) = 0.0131$
$u_2(t = 0-) = 0.32$ $\qquad$ $u_7(t = 0-) = 4.0 \times 10^{-6}$
$u_3(t = 0-) = 0$ $\qquad$ $u_8(t = 0-) = 4.0 \times 10^{-6}$
$u_4(t = 0-) = 0$ $\qquad$ $u_9(t = 0-) = 0$
$u_5(t = 0-) = 0$ $\qquad$ $u_{10}(t = 0-) = 0$

until $t = 10$ hr. Since these initial conditions do not fully satisfy the algebraic constraints, we begin by computing the state at $t = 0$ with the constraints enforced. DDAPLUS does this by setting up and solving the system

$$
\begin{bmatrix}
1 & & & & & & & & & \\
& 1 & & & & & & & & \\
& & 1 & & & & & & & \\
& & & 1 & & & & & & \\
& & & & 1 & & & & & \\
& & & & & 1 & & & & \\
& & & & & & 0 & & & \\
& & & & & & & 0 & & \\
& & & & & & & & 0 & \\
& & & & & & & & & 0
\end{bmatrix}
\begin{bmatrix}
u_1 \\ u_2 \\ u_3 \\ u_4 \\ u_5 \\ u_6 \\ 0 \\ 0 \\ 0 \\ 0
\end{bmatrix}
=
\begin{bmatrix}
1.5776 \\ 0.32 \\ 0 \\ 0 \\ 0 \\ 0.0131 \\ f_7(0, \boldsymbol{u}) \\ f_8(0, \boldsymbol{u}) \\ f_9(0, \boldsymbol{u}) \\ f_{10}(0, \boldsymbol{u})
\end{bmatrix}
\tag{B.10-14}
$$

obtained from Eq. (B.1-2) at $t = 0$ for the present problem.

The following Athena Visual Studio code sets up this problem and solves it by calls to DDAPLUS. The chosen absolute tolerance, ATOL=1.D-10, is satisfactory until $t = 1.00$ hr, after which $u_1$ and $u_8$ attain much smaller values. Careful nesting of the multiplications and divisions proved necessary to avoid underflows in the calculations of $f_8$, $f_9$, and $f_{10}$; note the placement of parentheses in those lines of the *@Model Equations* section.

Small negative values of $u_1$ and $u_8$ appear near the end of the integration. These artifacts, caused by rounding and truncation errors, are negligible relative to the absolute tolerance as noted above. A fully nonnegative solution to this problem is obtained, though not quite as quickly, if one activates the NonNegativity conditions clause before calling DDAPLUS.

## Athena Visual Studio Code for Example B.3

```
! Appendix B: Example B.3
! Computer-Aided Modeling of Reactive Systems
!------------------------------------------------
 Global Theta1,Theta2,Theta3,Theta4,Theta5 As Real
 Global Theta6,Theta7,Theta8 As Real

! Enter the reaction rate constants
!---------------------------------------
 Theta1=20.086
 Theta2=8.10D3
 Theta3=20.086
 Theta4=20.086
 Theta5=4.05D3

! Enter the Equilibrium Constants
!-------------------------------------
 Theta6=1.0D-17
 Theta7=1.0D-11
 Theta8=1.0D-17

@Initial Conditions
 U(1)=1.5776
 U(2)=8.3200
 U(3)=0.0
 U(4)=0.0
 U(5)=0.0
 U(6)=0.0131
 U(7)=4.0D-6
 U(8)=4.0D-6
 U(9)=0.0
 U(10)=0.0

@Model Equations
 Dim r1,r2,r3 As Real

 r1=Theta1*U(2)*U(6)-Theta2*U(10)
 r2=Theta3*U(2)*U(8)
 r3=Theta4*U(4)*U(6)-Theta5*U(9)

! Formulate the Material Balance
!-----------------------------------
 F(01)=-r2
 F(02)=-r1-r2
 F(03)=r2+r3
 F(04)=-r3
 F(05)=r1
 F(06)=-r1-r3

! Algebraic Equations
!---------------------------
 F(07)=0.0131-U(6)+U(7)-U(8)-U(9)-U(10)
 F(08)=U(08)*(Theta7+U(7))-Theta7*U(1)
 F(09)=U(09)*(Theta8+U(7))-Theta8*U(3)
 F(10)=U(10)*(Theta6+U(7))-Theta6*U(5)

@Coefficient Matrix
 E(1:6)=1.0
```

```
@Solver Options
  Neq=10             ! Number of Equations
  Tend=8.0           ! End of Integration
  Npts=15            ! Number of Points for Intermediate Output
```

### Partial Numerical Results for Example B.3

```
Number of State Equations......................    10
Number of Sensitivity Parameters................     0
Number of Integration Output Points.............    17
```

| Time | U(1) | U(2) | U(3) | U(4) |
|---|---|---|---|---|
| 0.00000E+00 | 1.57760E+00 | 8.32000E+00 | 0.00000E+00 | 0.00000E+00 |
| 5.00000E-01 | 5.65059E-01 | 7.29434E+00 | 1.01252E+00 | 1.64540E-05 |
| 1.00000E+00 | 7.35524E-12 | 6.46279E+00 | 1.29835E+00 | 2.79245E-01 |
| 1.50000E+00 | 5.50329E-12 | 6.07353E+00 | 9.09030E-01 | 6.68570E-01 |
| 2.00000E+00 | 7.53678E-12 | 5.87770E+00 | 7.13181E-01 | 8.64419E-01 |
| 2.50000E+00 | 5.97515E-12 | 5.76947E+00 | 6.04936E-01 | 9.72664E-01 |
| 3.00000E+00 | 6.21453E-12 | 5.70717E+00 | 5.42634E-01 | 1.03497E+00 |
| 3.50000E+00 | 2.07159E-12 | 5.67057E+00 | 5.06027E-01 | 1.07157E+00 |
| 4.00000E+00 | 1.42078E-12 | 5.64882E+00 | 4.84273E-01 | 1.09333E+00 |
| 4.50000E+00 | 2.85205E-12 | 5.63581E+00 | 4.71261E-01 | 1.10634E+00 |
| 5.00000E+00 | 7.78057E-12 | 5.62800E+00 | 4.63449E-01 | 1.11415E+00 |
| 5.50000E+00 | 8.51554E-12 | 5.62330E+00 | 4.58748E-01 | 1.11885E+00 |
| 6.00000E+00 | 1.16238E-12 | 5.62046E+00 | 4.55915E-01 | 1.12169E+00 |
| 6.50000E+00 | 2.37696E-12 | 5.61876E+00 | 4.54206E-01 | 1.12339E+00 |
| 7.00000E+00 | 1.34937E-12 | 5.61772E+00 | 4.53176E-01 | 1.12442E+00 |
| 7.50000E+00 | 2.16271E-13 | 5.61710E+00 | 4.52554E-01 | 1.12505E+00 |
| 8.00000E+00 | 1.50032E-12 | 5.61673E+00 | 4.52178E-01 | 1.12542E+00 |

```
Exit DDAPLUS: SOLUTION FOUND

Number of Steps Taken Thus Far...................  1060
Number of Function Evaluations...................  3448
Number of Jacobian Evaluations...................   119
Number of Jacobian Factorizations...............   119
```

## Example B.4. Sensitivity Analysis of a Differential Equation

The evolution of the methylcyclohexane (M) profile in a catalytic fixed bed reactor, subject to dehydrogenation in excess of hydrogen to reduce coking, is given by the equation:

$$\frac{dM}{dw} = -\frac{k_r M}{1 + K_a M}$$

$$M(w = 0) = 1 \qquad 0 \le w \le 1$$

where $w$ is the normalized catalyst mass, $k_r = 6.67$ and $K_A = 5.0$ are the reaction rate and adsorption constants respectively, with the appropriate units. We wish to calculate the methylcyclohexane profile, and its sensitivity with respect to the reaction rate constant and the adsorption constant.

The following Athena Visual Studio code sets up this problem and solves it by call to DDAPLUS. The Athena nomenclature for the state and sensitivity functions is given by the following:

$$\mathbf{U}(1,1) = M$$

$$\mathbf{U}(1,2) = \frac{\partial M}{\partial k_r}$$

$$\mathbf{U}(1,3) = \frac{\partial M}{\partial K_a}$$

## Athena Visual Studio Code for Example B.4

```
! Appendix B: Example B.4
! Computer-Aided Modeling of Reactive Systems
!-------------------------------------------
Global kr,Ka As Real

! Enter the reaction rate constants
!-------------------------------------------
kr=6.67
Ka=5.00

@Initial Conditions
U(1)=1.0

@Model Equations
Dim rM As Real

rM=kr*U(1)/(1.0+Ka*U(1))

! Formulate the Material Balance
!-------------------------------------------
F(1)=-rM

@Solver Options
Neq=1                 ! Number of Equations
Tend=1.0              ! Normalized Catalyst Mass
Npts=15              ! Number of Points for Intermediate Output
Sens=1               ! Activate Sensitivity Analysis Option
SensParam=kr;Ka ! Specify the Sensitivity Function
                     ! Parameters
```

## Partial Numerical Results for Example B.4

```
Number of State Equations......................  1
Number of Sensitivity Parameters...............  2
Number of Integration Output Points............ 17
```

| Time | U(1,1) | U(1,2) | U(1,3) |
|------|--------|--------|--------|
| 0.00000E+00 | 1.00000E+00 | 0.00000E+00 | 0.00000E+00 |
| 6.25000E-02 | 9.30940E-01 | -1.02887E-02 | 1.13686E-02 |
| 1.25000E-01 | 8.62774E-01 | -2.02945E-02 | 2.22795E-02 |
| 1.87500E-01 | 7.95608E-01 | -2.99657E-02 | 3.26653E-02 |
| 2.50000E-01 | 7.29566E-01 | -3.92405E-02 | 4.24482E-02 |

```
3.12500E-01    6.64788E-01    -4.80434E-02    5.15355E-02
3.75000E-01    6.01442E-01    -5.62811E-02    5.98173E-02
4.37500E-01    5.39722E-01    -6.38395E-02    6.71638E-02
5.00000E-01    4.79860E-01    -7.05788E-02    7.34224E-02
5.62500E-01    4.22125E-01    -7.63299E-02    7.84172E-02
6.25000E-01    3.66831E-01    -8.08913E-02    8.19496E-02
6.87500E-01    3.14341E-01    -8.40297E-02    8.38058E-02
7.50000E-01    2.65066E-01    -8.54899E-02    8.37739E-02
8.12500E-01    2.19456E-01    -8.50161E-02    8.16735E-02
8.75000E-01    1.77978E-01    -8.24002E-02    7.74120E-02
9.37500E-01    1.41076E-01    -7.75524E-02    7.10531E-02
1.00000E+00    1.09102E-01    -7.05920E-02    6.28907E-02
```

EXIT DDAPLUS: SOLUTION FOUND

```
Number of Steps Taken Thus Far...................    48
Number of Function Evaluations...................   307
Number of Jacobian Evaluations...................    51
Number of Jacobian Factorizations................    51
```

## REFERENCES and FURTHER READING

Anderson, E., Z. Bai, C. Bischof, J. Demmel, J. J. Dongarra, J. Du Croz, A. Greenbaum, S. Hammarling, A. McKenney, S. Ostrouchov, and D. Sorensen, *LAPACK Users' Guide*, SIAM, Philadelphia (1992).

Biegler, L. T., J. J. Damiano, and G. E. Blau, Nonlinear parameter estimation: a case study comparison, *AIChE J.* **32**, 29–45 (1986).

Bird, R. B., W. E. Stewart, and E. N. Lightfoot, *Transport Phenomena*, Wiley, New York (2002).

Brenan, K. E., S. L. Campbell, and L. R. Petzold, *Numerical Solution of Initial-Value Problems in Differential-Algebraic Equations*, North-Holland, New York (1989).

Caracotsios, M., and W. E. Stewart, Sensitivity analysis of initial value problems with mixed ODEs and algebraic equations, *Comput. Chem. Eng.*, **9**, 359–365 (1985).

Gear, C. W., Simultaneous numerical solution of differential-algebraic equations, *IEEE Trans. Circuit Theory* **CT-18**, 89–95 (1971).

Petzold, L. R., A description of DASSL: a differential/algebraic system solver, *Sandia Tech. Rep. 82-8637* (1982).

Shampine, L. F., Starting variable-order Adams and BDF codes, *Appl. Num. Math.*, **3**, 331–337, North-Holland, Amsterdam (1987).

Suen, S. Y., and M. R. Etzel, A mathematical analysis of affinity membrane bioseparations, *Chem. Eng. Sci.*, **47**, 1355–1364 (1992).

# Appendix C
# GREGPLUS Documentation

## C.1 DESCRIPTION OF GREGPLUS

The package GREGPLUS, included in Athena Visual Studio, tests user-provided models against data on one or more observed response types and assists sequential planning of experiments. Nonlinear and overparameterized models are handled directly. The following types of information can be obtained:

1. Locally optimal values $\widehat{\theta}_r$, 95% HPD intervals, and covariances for model parameters that are estimable from the data; also the fitted function value and residual for each observation. Large-sample asymptotes for these quantities are obtainable by the least-squares solver LSGREG, though for multiresponse data GREGPLUS gives better information by considering coupled multivariate error distributions. GREGPLUS is used for the statistical examples in this book.

2. Corresponding estimates for auxiliary parametric functions that the user may define. Such estimates may include alternative parameters, predicted process states, and performance measures for processes designed with the given models and data.

3. Relative posterior probabilities and goodness of fit for rival models. Data containing replicates are strongly recommended for these calculations, but assumptions or specifications of experimental precision are accepted.

4. Optimal choice of an additional event INEXT from a list of candidates, to obtain better estimates of model parameters and auxiliary functions and/or better discrimination among a set of candidate models.

For calculations of Type 4, a menu of simulations of candidate events must be provided in the user's subroutine MODEL. GREGPLUS then selects the preferred conditions for the next event according to the chosen criterion of optimality, as described below and in Chapters 6 and 7.

Let $\theta$ denote the vector of parameters for the current model. A point estimate, $\widehat{\theta}$, with locally maximum posterior probability density in the parameter space, is obtained by minimizing a statistical objective function

$S(\boldsymbol{\theta})$ based on Bayes' theorem. For single-response data, $S(\boldsymbol{\theta})$ is a weighted sum of squares of deviations of the observations $y_u$ from the fitted functions $f_u(\boldsymbol{\theta})$ of the current model in events $u = 1, \ldots, \text{NEVT}$:

$$S(\boldsymbol{\theta}) = \sum_{u=1}^{\text{NEVT}} w_u [y_u - f_u(\boldsymbol{\theta})]^2 \qquad \text{(Single Response)} \qquad \text{(C.1-1)}$$

GREGPLUS uses this objective function when called via Athena or by a user-provided code at Level 10, with the observations $y_u$ (preferably adjusted to remove systematic errors such as mass-balance departures) provided in the array OBS. Each weight $w_u = \sigma^2/\sigma_u^2$ is the user-assigned ratio of a reference variance $\sigma^2$ to the variance expected for the observation $y_u$; this weighting is designed to induce a normal distribution of weighted residuals $[y_u - f_u(\hat{\boldsymbol{\theta}})]\sqrt{w_u}$ whenever the model form is true. GREGPLUS uses $\sqrt{w_u} = 1$ when called with JWT=0. The other permitted choice, JWT=1, calls for input values $\sqrt{w_u} = \text{SQRW}(u)$ to be provided by the user.

For multiresponse data, GREGPLUS uses an objective function

$$S(\boldsymbol{\theta}) = \sum_{b=1}^{\text{NBLK}} \text{LPOWR}_b \ln |\boldsymbol{v}_b(\boldsymbol{\theta})| \qquad \text{(Multiple Responses)} \qquad \text{(C.1-2)}$$

based on the user's arrangement of the responses into one or more blocks $b = 1, \ldots, \text{NBLK}$. Responses $i$ and $k$ should be placed in the same block if (1) they are observed in the same events, and (2) their errors of observation may be correlated. From each response block $b$, GREGPLUS selects $m_b$ working responses whose weighted deviations $[y_{iu} - f_{iu}(\boldsymbol{\theta})]\sqrt{w_u}$ from the current model are uncorrelated over the events. Each block submatrix function $\boldsymbol{v}_b(\boldsymbol{\theta})$, with elements

$$v_{bik}(\boldsymbol{\theta}) := \sum_{u=1}^{\text{NEVT}} w_u [y_{iu} - f_{iu}(\boldsymbol{\theta})][y_{ku} - f_{ku}(\boldsymbol{\theta})], \quad \begin{cases} i,k \text{ in working set} \\ \text{of response block } b \end{cases}$$

$$\text{(C.1-3)}$$

then has a positive determinant over the investigated range of $\boldsymbol{\theta}$, as Eq. (C.1-2) requires. GREGPLUS uses $\sqrt{w_u} = 1$ in Eq. (C.1-3) when called with JWT=0. The other permitted choice, JWT=1, calls for input values $\sqrt{w_u} = \text{SQRW}(u)$ to be specified by the user. The observations $y_{iu}$ and $y_{ku}$ are taken from the array OBS, preferably adjusted before the call of GREGPLUS (as in Example C.4) to remove systematic errors such as mass-balance departures.

Let $m_b$ denote the number of working responses in block $b$ and $n_b$ the number of events in that data block. Then the coefficients in Eq. (C.1-2) can take either of the following forms as directed by the user:

$$
\text{LPOWR}_b = 
\begin{cases}
n_b & \text{for the likelihood objective of Bard (1974)} \\
& \text{or the Bayesian objectives of Box and Draper} \\
& \text{(1965) and M. Box and N. Draper (1972);} \\
n_b + m_b + 1 & \text{for the Bayesian objective in Eqs. 15–16 of} \\
& \text{Stewart, Caracotsios, and Sørensen (1992)}
\end{cases}
$$

(C.1-4)

The right-hand value $n_b$ is used when GREGPLUS is called with LEVEL = 20, whereas $(n_b + m_b + 1)$ is used when LEVEL = 22. LEVEL 20 requires fuller data and gives a fuller covariance analysis; it gives expectation estimates of the covariance elements for each data block. LEVEL 22 gives maximum-density (most probable) covariance estimates; these are smaller than the expectation values, which are averages over the posterior probability distribution.

GREGPLUS uses successive quadratic programming to minimize $S(\boldsymbol{\theta})$ for the current model, starting from the user's guesses for the parameters. In each iteration $S(\boldsymbol{\theta})$ is approximated locally as a quadratic expansion $\widetilde{S}(\boldsymbol{\theta})$, constructed from the data and the user's subroutine MODEL. Then $\widetilde{S}(\boldsymbol{\theta})$ is minimized within a feasible region of $\boldsymbol{\theta}$ by GREGPLUS's quadratic programming subroutine GRQP. The resulting step $\Delta\boldsymbol{\theta}$ in parameter space is shortened when necessary to ensure descent of the true function $S(\boldsymbol{\theta})$. These steps recur until a convergence criterion for $\Delta\boldsymbol{\theta}$ is met or until ITMAX iterations are completed.

The estimates in Item 1 are then computed along with any of Items 2, 3, and 4 that the user requests. Thus, NPHIV(JMOD).GT.0 is used in the call of GREGPLUS to request auxiliary function computations for model JMOD; KVAR(0).NE.0 to request calculations of goodness of fit and posterior probability share for each model considered; and JNEXT, NMOD, and IDSIGN control selection of the next event condition. These calculations are described below, and their statistical foundations are presented in Chapters 6 and 7.

## C.2 LEVELS OF GREGPLUS

Each level of GREGPLUS has its own objective function form:
**LEVEL=10**: Eq. (C.1-1) is used.
**LEVEL=20**: Eq. (C.1-2) is used, with $\text{LPOWR}_b = n_b$.
**LEVEL=22**: Eq. (C.1-2) is used, with $\text{LPOWR}_b = n_b + m_b + 1$.

## C.3 CALLING GREGPLUS

The code GREGPLUS is normally called via Athena, by clicking on the desired options under PARAMETER ESTIMATION. For very detailed control of the available options, GREGPLUS may be called as follows from a user-provided MAIN program for the given problem:

```
     CALL GREGPLUS(NEVT,JWT,LEVEL,NRESP,NBLK,IBLK,IOBS,YTOL,
   A NMOD,NPARV,NPHIV,OBS,SQRW,PAR,BNDLW,BNDUP,CHMAX,
   B DEL, DW,LENDW,JGRDW, IW,LENIW,JGRIW, LUN,NREAD,NWRITE,
   C ADTOL,RPTOL,RSTOL,MODEL,IPROB,ITMAX,LISTS,IDIF,IRES2,
   D MODOUT,JNEXT,IDSIGN,KVAR,NUEB,VAR,IERR,LINMOD)
```

The arguments for this call are described below, along with their declarations in the MAIN program. Selected results (described under LISTS) are written by GREGPLUS into an output file LUNREP identified in the MAIN program. Athena gives extended reports in its output file LUNDBG.

**NEVT** . . . . . . . Number of events included in OBS.

**JWT** . . . . . . . . . Integer used to specify the desired kind of weighting.
Set JWT=0 for unit weighting; then SQRW(1) is a dummy 1.
Set JWT=1 if SQRW(u) is to be provided as input for each
reported event u=1,...NEVT, but taken as 1 for any candidate
events INEXT. GREGPLUS's default value is JWT = 0 .

**LEVEL** . . . . . . Objective function index, defined in Section C.2.

**NRESP** . . . . . . Number of responses per event; 1 at Level 10.

**NBLK** . . . . . . . Number of response blocks per event; 1 at Level 10.
At Levels 20 and 22 various block structures are available:
NBLK=1 gives a full covariance matrix,
1<NBLK<NRESP gives a block-diagonal covariance matrix,
and NBLK=NRESP gives a diagonal covariance matrix.
The latter form yields a least-squares computation with
block weightings optimized by GREGPLUS and multiplied
by the user's optional weights SQRW(u) for individual
events.

**IBLK** . . . . . . . . Dummy 0 at Level 10; array IBLK(NBLK,NMOD+1)
at Levels 20 and 22, in which IBLK(b,1) is the number
of responses in block b. IBLK(b, JMOD+1) is the number of
parameters of Model JMOD that apply only in block b.
GREGPLUS sets IBLK(1, JMOD+1)=NPARV(JMOD) for
each model if NBLK=1. The response values in a block must
occupy consecutive rows in OBS, and each response must be
in just one block. Error covariances between different blocks
are ignored. Partial observations of a response block are not
accepted by GREGPLUS; one must either omit such events
or regroup the responses so that, in each event, each resulting
block is either full or empty. GREGPLUS tests the errors
$y_{iu} - f_{iu}(\boldsymbol{\theta})$ for linear independence and selects working
responses in each block to give strictly positive determinants
$|v_b(\boldsymbol{\theta})|$ in Eq. (C.1-2).

**IOBS** . . . . . . . . Dummy 0 at Level 10; array at Level 20 or 22 indicating which
elements of OBS contain data. At Level 20 or 22, set IOBS(i,u)
to 1 if OBS(i,u) contains an observation, or to 0 otherwise.
Declare IOBS(NRESP,NEVT+|JNEXT|) at these levels, and
include a full set of elements for each candidate "next" event.

**YTOL(i)** . . . . . Fractional absolute error tolerance for residuals of response $i$.
The default fraction 0.01 is inserted by GREGPLUS at Level

10, and is used at Levels 20 and 22 for inputs between 0 and 0.01. The latter precaution avoids the catastrophic failures shown by Box et al. (1973) for multiresponse estimation with overfitting of the data.

NMOD ...... Signed number of rival models provided in MODEL. The sign convention is stated under JNEXT.

NPARV ...... Vector, declared NPARV(ABS(NMOD)). NPARV(j) is the number of parameters in Model $j$.

NPHIV ...... Vector, declared NPHIV(ABS(NMOD)). The $j$th element is the number of user-defined PHI functions of the parameters of Model JMOD. If these functions prove linearly dependent, GREGPLUS selects a good subset.

OBS ......... Double precision observation array, dimensioned OBS(NEVT) at Level 10 and OBS(NRESP,NEVT) at Levels 20 and 22.

SQRW ....... Double precision array used when JWT=1 to hold the values $\sqrt{w_u}$ of Eq. (C.1-1) or (C.1-3) for u=1,...NEVT. Declare SQRW(NEVT) if JWT=1, or SQRW(1) (a dummy 1.) if JWT=0. JWT is taken as 0 for candidate events INEXT.GE.1 ; such events can be given greater weight when desired by repeating them in your subroutine MODEL, described below.

PAR ......... Double precision vector of parameter values $\theta_i$ in models 1,...,NMOD. Declare it as PAR(0:NTPAR) in the MAIN program; here NTPAR is the total number of parameters.

Multiple initial guesses for parameters of a given model form can be entered in PAR in segment lengths NPARV(JMOD); then on return from GREGPLUS, each segment of PAR contains the parameter values found with that initial guess.

BNDLW ..... Double precision vector of lower bounds on the parameters of models 1,...,NMOD. Declared as BNDLW(NTPAR) in the MAIN program.
GREGPLUS's default values are BNDLW(i) = $-1.0$D30.

BNDUP ..... Double precision vector of upper bounds on the parameters of models 1,...,NMOD. Declared as BNDUP(NTPAR) in the MAIN program.
GREGPLUS's default values are: BNDUP(i) = $1.0$D30.

CHMAX .... Double precision vector of trust-region halfwidths for the local quadratic expansions $\widetilde{S}(\boldsymbol{\theta})$ for each model. Declared as CHMAX(NTPAR) in the user's MAIN program.
GREGPLUS adjusts these halfwidths according to the line search outcomes.
If CHMAX(i).GE.0D0, parameter PAR(i) is allowed to change by no more than $\pm$CHMAX(i) in an iteration.
If CHMAX(i).LT.0D0, parameter PAR(i) is allowed a maximum fractional change $\pm$CHMAX(i) in an iteration, but an additive change $\pm$CHMAX(i) is allowed when PAR(i)=0D0.
GREGPLUS provides default inputs CHMAX(i) = $-0.8$D0, but for nearly linear models larger negative values save computation.

DEL ......... Double precision vector of parameter perturbation sizes for models 1,...,NMOD. Declared as DEL(NTPAR) in the MAIN program.

Nonzero elements DEL(i) request finite-difference estimates of $\partial\mathbf{f}/\partial\theta_i$ and $\partial\mathbf{PHI}/\partial\theta_i$ for those parameters, and are updated often to balance truncation and rounding errors.

If DEL(i)>0D0 or PAR(i)=0D0, PAR(i) has initial *absolute* perturbation DEL(i).

If DEL(i)<0D0 and PAR(i).NE.0D0, PAR(i) has initial *relative* perturbation DEL(i).

If DEL(i)=0D0, MODEL is to compute (when IDER = 1) the vector $\partial\mathbf{f}/\partial\theta_i$.

GREGPLUS provides default values DEL(i) = $-1.0D-2$ , replaced by ABS(DEL(i)) whenever PAR(i)=0D0.

**DW** .......... Double precision work vector. Declare it as DW(LENDW) in the MAIN program.

**LENDW** ..... Length of DW. The minimum usable length is obtainable by calling GREGPLUS with LENDW=1.

**JGRDW** ..... First storage address to be used by GREGPLUS in DW. Locations 1,...JGDW-1 in DW are available for saving double precision constants to be used by MODEL.

**IW** .......... Integer work vector. Declare it as IW(LENIW) in the MAIN program.

**LENIW** ...... Length of IW. The minimum usable length is obtainable by calling GREGPLUS with LENIW=1.

**JGRIW** ...... First storage address to be used by GREGPLUS in Iw. Locations 1,...JGIW-1 in IW are available for saving integer constants to be used by MODEL.

**LUN** ......... Output unit number.

**NREAD,** .... Unit numbers of an optional pair of user-defined files

**NWRITE**.... For use by Subroutine MODEL. See Section C.5.

**ADTOL** ..... Threshold value for test divisors $A(i,i)/A(i,i)_{\mathrm{orig.}}$ used by GREGPLUS in selecting pivots in the normal-equation matrix $\mathbf{A}$ for each constrained minimization of $\widetilde{S}(\boldsymbol{\theta})$. GREGPLUS does not judge a parameter estimable unless its test divisor exceeds ADTOL at pivoting time. GREGPLUS provides the default value ADTOL = 1.0D$-4$ .

**RPTOL**...... Secondary tolerance for convergence of the estimation. This tolerance is satisfied when the relative change of each active basis parameter in a call of GRQP is smaller than RPTOL. GREGPLUS provides the default value RPTOL = 1.0D$-5$ if RSTOL is zero, or RPTOL=0D0 if RSTOL is nonzero. Thus, RSTOL is the usual criterion of convergence.

**RSTOL** ...... Primary tolerance for convergence of the estimation. This tolerance is satisfied when the current step of each basis parameter is no larger than RSTOL times its 95% HPD half-interval and no free parameter is at a CHMAX limit. At Level 10, if NPAR=NEVT (giving no residual degrees of freedom), convergence is declared when the sum of squares $S(\boldsymbol{\theta})$ reaches RSTOL or less.

GREGPLUS sets RSTOL$\geq 0.1$.

**MODEL** . . . . . User's subroutine; see Section C.5.

**IPROB** . . . . . . Problem identification number.

**ITMAX** . . . . . . Maximum number of iterations; this limit may be extended
by 1 or 2 when very detailed normal equations are in use;
see IRES2 below. An iteration consists of:
(i) the construction of a local quadratic expansion, $\widetilde{S}(\boldsymbol{\theta})$,
(ii) a solution by GRQP for a local minimum of $\widetilde{S}(\boldsymbol{\theta})$, and
(iii) a weak line search by GRS1 or GRS2 for a descent
of the true objective function $S(\boldsymbol{\theta})$.
GREGPLUS provides the default value ITMAX $= 100$ .

**LISTS** . . . . . . . User-selected integer that controls the amount of output:

$=1$ . . . . . . . . . . **PAR** and $S$ for the current model are listed at the start
of each iteration; at Levels 20 and 22 the current array
$v_b$ and working response set are also shown. The HPD
intervals for parameters $\theta_r$ and auxiliary functions $\phi_i(\boldsymbol{\theta})$
are listed at the end of the estimation, along with fitted
function values and residuals. Optimal conditions for an
added event are then chosen if requested (see JNEXT).

$=2$ . . . . . . . . . . In addition, the normal-equation matrix $\boldsymbol{A}$ is shown for the
first iteration, and the covariance matrix of the parameter
estimates is shown at the end of the estimation.

$=3$ . . . . . . . . . . In addition, for each iteration, the descent steps and
line search points are reported.

$=4$ or $5$ . . . . . . Very detailed information about the iterations is listed.
GREGPLUS provides the default value LISTS $= 2$ .

**IDIF** . . . . . . . . . This integer argument is used in finite-difference
calculations of $\boldsymbol{\theta}$-derivatives of $\widetilde{S}(\boldsymbol{\theta})$ and **PHI**$(\boldsymbol{\theta})$.

$=+1$ . . . . . . . . . One-sided first differences with positive perturbations $\Delta\theta_r$.

$=-1$ . . . . . . . . . . One-sided first differences with negative perturbations $\Delta\theta_r$.

$=0$ . . . . . . . . . . Two-sided first differences, centered when the bounds and
DEL(r) values permit.

$=2$ . . . . . . . . . . Difference approximations of second order for $\widetilde{S}(\boldsymbol{\theta})$ and
first order for **PHI**$(\boldsymbol{\theta})$, available at Level 10 only.
GREGPLUS provides the default value IDIF $= 1$ .

**IRES2** . . . . . . . This integer controls the timing of certain refinements of the
normal equations:

$=0$ . . . . . . . . . . At Level 10 when called with IDIF $= 2$, GREGPLUS starts with
IDIF $= 1$ and switches to IDIF $= 2$ near the end.
When called with IDIF $= 1$ at any level, GREGPLUS sets IDIF
to zero at the first indication of line search difficulty.
At Levels 20 and 22 with IRES2 $= 0$, GREGPLUS omits some
high-order terms from the normal equations until near the end.

$=1$ . . . . . . . . . . GREGPLUS uses its most detailed normal equations for the
given IDIF and LEVEL. IRES2 $= 0$ is recommended to
stabilize the minimizations.

**MODOUT** . . . This integer controls the option of a final MODEL call with
IDER $= -2$ for special user-specified output. Normally

       MODOUT = 0; set MODOUT=1 to request special output.

**JNEXT** . . . . . . This integer, when nonzero, requests that the next event be selected from a menu of simulated events INEXT defined in MODEL by the user.

With NMOD=1 and JNEXT.NE.0, GREGPLUS selects the event INEXT for optimal estimation of the parameters named in IDSIGN. The selection is for maximum determinant of the resulting estimation matrix if JNEXT>0, or minimum trace of the inverse of that matrix if JNEXT<0.

With NMOD< −1 and JNEXT.NE.0, GREGPLUS selects INEXT for best discrimination among the candidate models. The criterion, adapted from Box and Hill (1967), Hill and Hunter (1969), and Reilly (1970), is the expectation of entropy decrease (information increase) obtainable by adding event INEXT to the data set. The expectation calculations are described in Chapters 6 and 7, where the integral formula of Reilly (1970) is extended to multiresponse models and unknown covariance matrix.

With NMOD>1 and JNEXT.NE.0, GREGPLUS selects INEXT by a dual criterion of discrimination and precise estimation, adapted from that of Hill, Hunter, and Wichern (1968).

**IDSIGN** . . . . . This integer vector lists, when JNEXT.NE.0, the indices $i$ for the parameters or functions whose precision is to be enhanced by the choice of event INEXT.

Use GREGPLUS's default declaration, IDSIGN(0:0), to choose an optimal INEXT value for parameter and (optionally) $\phi$-function estimation for each model. All estimable parameters for each model are then considered in choosing INEXT, without further specification of IDSIGN.

IDSIGN(0).GT.0, used only with NMOD=1, marks IDSIGN as a list of parameters to be included (if IDSIGN(i)=i) or omitted (if IDSIGN(i)=0) in the estimation criterion when choosing INEXT. Here IDSIGN(0) is the number of parameters listed in IDSIGN, and the required declaration is IDSIGN(0:NPAR).

**KVAR** . . . . . . . This integer vector, along with NUEB and VAR, describes the information provided on the precision of observations.

If there is no such information, declare KVAR(0:0) and set KVAR(0)=0 in the MAIN program, and declare NUEB(1) and VAR(1,1) as zeros for the call to GREGPLUS.

If replicates are provided, declare KVAR(0:NEVT) and set KVAR(0)=1.
Then mark each group $k$ of replicate *observations* by setting KVAR(u)= +$k$ for those events, mark each group $r$ of replicate *residuals* by setting KVAR(u)= −$r$ for those events, and set KVAR(u)=0 for any unreplicated events. When called,

GREGPLUS will calculate the working elements of VAR and
NUEB according to the declarations VAR(NRESP,NRESP)
and NUEB(NRESP). If Level=10, the result obtained for
VAR(1,1) is a sample variance estimate with NUEB(1)
degrees of freedom, whereas if Level=20 or 22, each
working block b of VAR is a block covariance estimate
with NUEB(b) degrees of freedom. Tests of goodness of fit
will be performed, and posterior probability shares will
be calculated for any rival models provided.

If VAR(NRESP,NRESP) and NUEB(NBLK) have been inferred,
as in Example C.3, from a very detailed model parameterized
to the point of minimum residual variance, then this minimum
point provides estimates of VAR and NUEB. If replicates are
not available, we may proceed as in that example: declare and
insert VAR and NUEB in the MAIN program, declare KVAR(0:0)
to show the absence of replicates, and declare KVAR(0)=−1
before calling GREGPLUS. GREGPLUS then proceeds as for
KVAR(0)=1, except for the different source of VAR and NUEB.

If VAR and NUEB are provided from other information
or belief, specify them before the call of GREGPLUS. Also
declare KVAR(0:0) and set KVAR(0)=2, to indicate that
the variance information is from another source.

If VAR is exactly known (as in some computer simulations),
specify it before the call of GREGPLUS. Declare KVAR(0:0)
and set KVAR(0)=3, to indicate an exact specification
of VAR. NUEB then becomes a dummy zero array of order
NRESP, VAR is an asymptote for infinitely many degrees of
freedom, and the F-test is replaced by a $\chi^2$ test.

If KVAR(0) is not in the interval $[-1,3]$, an error message is
written to LUN and to LUNREP, and control is returned to
the user's MAIN program.

**NUEB** ....... An integer list of the experimental error degrees of freedom
assigned to the elements of VAR for each response block b.
Declare NUEB(NBLK) in the MAIN program unless KVAR(0)=0;
in that case a dummy declaration NUEB(1) suffices.
GREGPLUS computes NUEB if KVAR(0)=1, as in
Example C.3; the user provides it if KVAR(0)=2, and it is a dummy
0 if KVAR(0)=3.

**VAR** ......... A double precision error variance (for Level 10), or an error
covariance matrix of order NRESP for Level 20 or 22.
If KVAR(0)=1, GREGPLUS computes VAR from the replicate
observations and/or replicate residuals.

If KVAR(0)=2, the user provides VAR as a covariance matrix
estimate with given degrees of freedom NUEB(b) for each block.
Symmetry is required; VAR(i,j) must equal VAR(j,i).
Declare VAR(NRESP,NRESP) unless KVAR(0)=0;
in that case a dummy declaration VAR(1,1) suffices.

**IERR** ........ This integer is zero on successful return from GREGPLUS.

Otherwise, the following message has been written to LUN and LUNREP: "GREGPLUS error exit: from the VIEW menu choose Debug and error messages."

LINMOD .... An integer list, dimensioned LINMOD(ABS(NMOD)) in the MAIN program. GREGPLUS sets this list to zero if Level.NE.10. At Level 10, set LINMOD(j)=1 if Model j is linear with ranges +-1D30 for all its parameters; set LINMOD(j)=-1 if Model j is linear and its only parameters with range less than +-1D30 are those with CHMAX(i)=0. For nonlinear models, and whenever Level.NE.10, set LINMOD(j)=0.GREGPLUS uses magnitudes 1D30 in BNDLW and BNDUP at Level 10 if LINMOD(j)=+-1, and does the same for CHMAX when LINMOD(j)=-1. Levels 20 and 22 use nonlinear algorithms, so do not need LINMOD except as a placeholder in the call of GREGPLUS.

## C.4 WORK SPACE REQUIREMENTS FOR GREGPLUS

The work array lengths for a given problem are computed by GREGPLUS. Lengths LENDW=300 and LENIW=100 suffice for simple problems. If more space is needed for either or both of these arrays, GREGPLUS will report the additional lengths required.

## C.5 SPECIFICATIONS FOR THE USER-PROVIDED MODEL

The following subroutine is declared **external** in GREGPLUS:

SUBROUTINE MODEL(JMOD,PAR,NPAR,NEVT,NRESP,NEVT1,
A OBS,f,lenf,dfdp,PHI,NPHI,DPHIDP,IDER,IOBS,DW,IW,
B LUN,NREAD,NWRITE,INEXT,NNEXT,IERR)

This subroutine computes values and requested derivatives of the unweighted arrays **f** and/or **PHI** for the current model. When called with INEXT.EQ.0, MODEL must simulate events 1,...NEVT. When called with IN-EXT.GT.0, MODEL must simulate candidate event INEXT and write its calculations of **f** and/or **PHI** in the space for u=NEVT+1, leaving intact the values for events 1,...,NEVT.

The arguments in the call of MODEL are as follows:

JMOD ..... is the model number (counting from 1 to NMOD) specified in the current call of MODEL. This value is provided by GREGPLUS and must not be changed by MODEL. The current segments of PAR, f, PHI, and the output arrays are addressed directly by GREGPLUS; therefore, MODEL must address them with subscripts numbered as if the current model were the only one.

PAR........must be declared in MODEL as PAR(0:*); then PAR(1) is the first parameter of model JMOD and PAR(0) is the last parameter of the previous model. This argument must not be changed by subroutine MODEL.

NPAR......is the number of parameters $\theta_r = \mathrm{PAR}(r)$ in model JMOD. It is set by GREGPLUS as NPARV(JMOD) and must not be changed by MODEL.

**NEVT**......is the same as in the call of GREGPLUS.

**NRESP** .... is the same as in the call of GREGPLUS.

**NEVT1** .... = NEVT+1.

**OBS** ........ is the same as in the call of GREGPLUS. This argument has proved useful in starting iterative solutions of implicit models.

**f** ............ is the array of unweighted function values for model JMOD. MODEL should declare f(*) at Level 10 or f(**NRESP**,*) at Levels 20 and 22. GREGPLUS reserves space for all arrays, so an asterisk suffices as the trailing dimension in MODEL's array declarations.

**lenf** ......... = NRESP*NEVT1 is the length of work space reserved by GREGPLUS for function values for the current model.

**dfdp** ........ is the array of first derivatives of the current model. If MODEL provides any of these directly, it must declare **dfdp**(**NEVT1**,*) for Level 10 or **dfdp**(**NRESP**,**NEVT1**,*) for Level 20 or 22. **dfdp** is not to be altered by MODEL unless IDER = 1. When called with IDER=1, MODEL must return the following derivatives (see IDER description):

None if LEVEL = 10 with IDIF = 2.
$\mathbf{dfdp}(u, r) = \partial f_u(\boldsymbol{\theta})/\partial \theta_r$ at Level10.
$\mathbf{dfdp}(i, u, r) = \partial f_{iu}(\boldsymbol{\theta})/\partial \theta_r$ atLevel 20 or 22.
These values are to be reported for each event when INEXT=0, and for u=NEVT+1 when analyzing any candidate event INEXT.

**PHI** ........ is an optional vector of auxiliary functions $\phi_i(\boldsymbol{\theta})$ for the current model, to be analyzed when the final parameter vector $\widehat{\boldsymbol{\theta}}$ of the model has been computed; see IDER.

**NPHI** ...... =NPHIV(JMOD) is the number of auxiliary functions $\phi_i(\boldsymbol{\theta})$ for the current model.

**DPHIDP** .. This array, if used, must be declared **DPHIDP**(**NPHI**,*) in MODEL. It is to be accessed by MODEL only when IDER = −2; then each derivative $\partial \text{PHI}(i)/\partial \theta_r$ available directly from MODEL must be returned as DPHIDP(i,r). If any DEL(i).NE.0. GREGPLUS will complete DPHIDP with central-differencevalues if needed, using **f** values from calls of MODEL with IDER = −1.

**IDER** ...... is set by GREGPLUS to specify the tasks for MODEL:

when **0:** ..... MODEL must compute the array **f** for the current model and no derivatives.

when **1:** ..... MODEL must compute the array **f** and any requested derivatives of **f** for the current model (see **dfdp**), but must not alter PHI or DPHIDP.

when **−1:**.... MODEL must compute the optional vector PHI of auxiliary functions for the current model and no derivatives.

when **−2:**.... MODEL is being called to compute the functions PHI(i) of model JMOD and any directly available derivatives $\partial \text{PHI}(i)/\partial \theta_r$, and/or to write the user's special final output into LUN or NWRITE.

**IOBS** ....... is the same as in the call of GREGPLUS. IOBS is used in

MODEL at LEVEL=20 or 22 to indicate which responses are to be simulated in the current call.

**DW** ........ Double precision work array in the call of GREGPLUS. MODEL must not write in DW, but may read saved values that were stored before the call of GREGPLUS; see JGDW.

**IW** ......... Integer work array in the call of GREGPLUS. MODEL must not write in IW, but may read saved values that were stored before the call of GREGPLUS; see JGIW.

**LUN** ....... is the same as in the call of GREGPLUS.

**NREAD,** ... Optional temporary files which MODEL may use to save
**NWRITE** needed information, such as conditions and inputs for the next execution of an implicit algorithm. To use these files, the user must insert conditions and initial guesses into NREAD before calling GREGPLUS. In execution, MODEL must read from NREAD and write the updated values into NWRITE for each event. GREGPLUS does the needed manipulations of these files and their labels during each iteration to save the latest confirmed values in NREAD until an updated file NWRITE is ready to serve asNREAD for the next iteration.

**INEXT** ..... Ordinal index for the optional candidate events. This integer is set by GREGPLUS; MODEL must not change it. If INEXT = 0, MODEL must limit its calculations to the events u=1,...NEVT. If INEXT.GT.0, MODEL must limit its calculations to event INEXT from the user-defined candidate event list, and must compute **f** (along with **dfdp** if IDER=1) with event index u=NEVT1. For Levels 20 and 22, the input array IOBS(i,u) must be initialized through u=NEVT+NNEXT in the MAIN program to indicate which responses are to be simulated in each candidate event.

**NNEXT** ... =|JNEXT| is the number of candidate events considered in choosing conditions for the next event. These events are simulated after the estimation if JNEXT is nonzero in the GREGPLUS call.

**IERR** ...... is zero on entry to MODEL. If an imminent task cannot be performed (for instance, if a divisor is zero or an argument for SQRT, EXP or LOG is out of range), MODEL should set ERR=1 and give a RETURN statement. This procedure enables GREGPLUS to recover if a line search overshoots and to resume the search from an intermediate point.

## C.6 SINGLE-RESPONSE EXAMPLES

Some examples are given here of nonlinear modeling with single-response data. Examples C.1 and C.2 demonstrate GREGPLUS on the model

$$f_u \; = \; \theta_1 \; + \; \theta_2 \, x_u^2 \; + \; 3 \, x_u^{\theta_3} \qquad \text{(C.6-1)}$$

using "observations" $\text{OBS(u)} \equiv Y_u$ simulated by the expression

$$\text{OBS}(u) \; = \; 1 \; + \; 2 \, x_u^2 \; + \; 3 \, x_u^3 + 0.0001 * \text{OBSERR}(u) \qquad \text{(C.6-2)}$$

with OBSERR(u) taken from a table of random normal deviates.

Example C.3 applies GREGPLUS with differential reactor data to assess four rival models of hydrogenation of iso-octenes for production of aviation gasoline. The variance estimate needed for discrimination of these models was estimated from the residuals of high-order polynomial regressions of the same data set.

## Example C.1. Estimation with Numerical Derivatives

This example uses a subroutine MODEL that computes only the values $f_u(\boldsymbol{\theta})$ of the expectation function in Eq. (C.6-1). Input values DEL(i)=-1D-2 are used, with IDIF=1 so that GREGPLUS uses forward-difference approximations of the derivatives $\partial f_u(\boldsymbol{\theta})/\partial\theta_j$. To make these approximations accurate, the elements DEL(I) are refined by GREGPLUS in each iteration, using Eq. (6.B-7). Abbreviated output is requested by using LISTS = 1. Finally, an additional event condition is selected from five candidates, to minimize the volume of the three-parameter HPD region. Details of the implementation of this example in Athena Visual Studio can be found by running the software and selecting *Book Examples* under the *Help* menu item.

The user-written Athena Visual Studio code given below was compiled and linked, giving the executable code to run this problem. The numerical results are shown below. Noteworthy features are (1) the rapid convergence of $S(\boldsymbol{\theta})$ and the parameter vector $\boldsymbol{\theta}$, (2) the closeness of the final parameter estimates to the true values 1, 2, and 3 of Eq. (C.6-2), (3) the narrowness of the 95% HPD intervals for the individual parameters, and (4) the probabilistic assessment of goodness of fit, which is more informative than the customary pass/fail decision based on a particular significance level.

### Athena Visual Studio Code for Example C.1

```
! Appendix C: Example C.1
! Computer-Aided Modeling of Reactive Systems
!--------------------------------------------

@Response Model
 Dim Theta1, Theta2, Theta3 As Real
 Theta1=Par(1)
 Theta2=Par(2)
 Theta3=Par(3)

 Y(1)=Theta1+Theta2*Xu(1)^2+3.0*Xu(1)^Theta3
```

### Numerical Results for Example C.1

```
Number of Experiments.......................... 30
Number of Parameters...........................  3
Number of Responses............................  1
Number of Settings.............................  1

EXIT GREGPLUS: SOLUTION FOUND. MODEL No. 1

STATISTICAL ANALYSIS

OBJECTIVE FUNCTION....................... 2.02655E-03
SUM OF SQUARES OF RESIDUALS.............. 2.02655E-03
```

```
ESTIMATED PARAMETERS FROM DATA...........        3
TOTAL OBSERVATIONS DATA COUNT IS.........       30

                 OPTIMAL
PARAMETER       ESTIMATES     95% MARGINAL HPD INTERVALS    PARAMETER STATUS
  PAR( 1)      1.001948E+00   1.001948E+00 +- 6.410E-03     Estimated
  PAR( 2)      1.997633E+00   1.997633E+00 +- 4.139E-03     Estimated
  PAR( 3)      3.000263E+00   3.000263E+00 +- 4.546E-04     Estimated

NORMALIZED PARAMETER COVARIANCE MATRIX
   1.000
  -0.777  1.000
   0.636 -0.958  1.000

PARAMETER COVARIANCE MATRIX
   1.3000E-01
  -6.5210E-02   5.4193E-02
   5.8671E-03  -5.7008E-03   6.5384E-04

EVENT OBSERVED     PREDICTED     RESIDUAL    DERIVATIVES--->
  1  1.0176E+00   1.0249E+00   -7.3222E-03  1.0000E+00  1.0000E-02  -6.9036E-03
  2  1.1082E+00   1.1058E+00    2.3572E-03  1.0000E+00  4.0000E-02  -3.8610E-02
  3  1.2584E+00   1.2627E+00   -4.3090E-03  1.0000E+00  9.0000E-02  -9.7491E-02
  4  1.5324E+00   1.5135E+00    1.8877E-02  1.0000E+00  1.6000E-01  -1.7589E-01
  5  1.8833E+00   1.8763E+00    7.0124E-03  1.0000E+00  2.5000E-01  -2.5988E-01
  6  2.3703E+00   2.3690E+00    1.2914E-03  1.0000E+00  3.6000E-01  -3.3097E-01
  7  3.0042E+00   3.0097E+00   -5.4914E-03  1.0000E+00  4.9000E-01  -3.6698E-01
  8  3.8176E+00   3.8163E+00    1.2572E-03  1.0000E+00  6.4000E-01  -3.4273E-01
  9  4.8049E+00   4.8070E+00   -2.0701E-03  1.0000E+00  8.1000E-01  -2.3042E-01
 10  5.9941E+00   5.9996E+00   -5.4810E-03  1.0000E+00  1.0000E+00   0.0000E+00
 11  7.4028E+00   7.4122E+00   -9.3842E-03  1.0000E+00  1.2100E+00   3.8058E-01
 12  9.0532E+00   9.0628E+00   -9.5885E-03  1.0000E+00  1.4400E+00   9.4520E-01
 13  1.0977E+01   1.0969E+01    7.3968E-03  1.0000E+00  1.6900E+00   1.7294E+00
 14  1.3153E+01   1.3150E+01    2.8619E-03  1.0000E+00  1.9600E+00   2.7701E+00
 15  1.5614E+01   1.5623E+01   -9.0032E-03  1.0000E+00  2.2500E+00   4.1058E+00
 16  1.8402E+01   1.8405E+01   -3.5091E-03  1.0000E+00  2.5600E+00   5.7761E+00
 17  2.1513E+01   2.1516E+01   -3.1665E-03  1.0000E+00  2.8900E+00   7.8220E+00
 18  2.4983E+01   2.4973E+01    9.7135E-03  1.0000E+00  3.2400E+00   1.0286E+01
 19  2.8793E+01   2.8794E+01   -9.8028E-04  1.0000E+00  3.6100E+00   1.3210E+01
 20  3.3017E+01   3.2997E+01    1.9840E-02  1.0000E+00  4.0000E+00   1.6639E+01
 21  3.7599E+01   3.7600E+01   -6.3623E-04  1.0000E+00  4.4100E+00   2.0617E+01
 22  4.2612E+01   4.2621E+01   -9.4223E-03  1.0000E+00  4.8400E+00   2.5192E+01
 23  4.8086E+01   4.8078E+01    7.5699E-03  1.0000E+00  5.2900E+00   3.0409E+01
 24  5.3999E+01   5.3990E+01    9.5279E-03  1.0000E+00  5.7600E+00   3.6316E+01
 25  6.0359E+01   6.0373E+01   -1.4361E-02  1.0000E+00  6.2500E+00   4.2961E+01
 26  6.7249E+01   6.7247E+01    1.3898E-03  1.0000E+00  6.7600E+00   5.0395E+01
 27  7.4617E+01   7.4629E+01   -1.2332E-02  1.0000E+00  7.2900E+00   5.8666E+01
 28  8.2539E+01   8.2537E+01    1.5594E-03  1.0000E+00  7.8400E+00   6.7825E+01
 29  9.0990E+01   9.0990E+01    5.1197E-05  1.0000E+00  8.4100E+00   7.7924E+01
 30  9.0996E+01   9.0990E+01    6.3512E-03  1.0000E+00  8.4100E+00   7.7924E+01


    NUMBER OF ITERATIONS..............   4
    NUMBER OF FUNCTION CALLS..........  17

EXIT GREGPLUS: MODEL DISCRIMINATION AND CRITICISM

SOURCE OF VARIANCE      SUM OF SQUARES   DEG. OF FREEDOM   MEAN SQUARE
Residuals                 2.02655E-03          27
Lack of Fit               1.78996E-03          25          7.16E-05
Experimental              2.36596E-04           2          1.18E-04

Variance Ratio (Lack-of-Fit/Experimental)                    0.605
Sampling Probability of Greater Ratio                        0.788
Log10 of Posterior Unnormalized Probability Density          2.242

EXIT GREGPLUS: STATISTICAL OPTIMAL DESIGN

FOR THE DATA SET  0 THE INVERSE HESSIAN OF ORDER  3 HAS LOG(DET) = 7.010
FOR THE DATA SET  1 THE INVERSE HESSIAN OF ORDER  3 HAS LOG(DET) = 7.063
FOR THE DATA SET  2 THE INVERSE HESSIAN OF ORDER  3 HAS LOG(DET) = 7.062
FOR THE DATA SET  3 THE INVERSE HESSIAN OF ORDER  3 HAS LOG(DET) = 7.035
FOR THE DATA SET  4 THE INVERSE HESSIAN OF ORDER  3 HAS LOG(DET) = 7.069
FOR THE DATA SET  5 THE INVERSE HESSIAN OF ORDER  3 HAS LOG(DET) = 8.150

OPTIMAL: THE SET  5 MAY BE SELECTED AS THE BEST AMONG THE CANDIDATES
```

## Example C.2. Nonlinear Estimation with Analytical Derivatives

This example uses a subroutine MODEL that computes the expectation function values $f_u(\boldsymbol{\theta})$ and, when requested, their parametric sensitivities $\partial f_u(\boldsymbol{\theta})/\partial\theta_j$. To notify GREGPLUS which sensitivities MODEL will provide, the corresponding elements DEL(i) are set to zero. The same task can be accomplished by selecting the appropriate option in the *Gradient Calculation* method in Athena Visual Studio. After convergence of the parameter estimation, a goodness-of-fit analysis is performed, giving a very satisfactory result. A posterior probability is also calculated, which could be used in choosing a preferred model from a set of alternative models of these data.

Estimation is then performed for two auxiliary functions of the parameters, namely, the parameter $\theta_2$ and the expectation function $f(x,\theta)$ at $x = 10$. Finally, conditions for an additional event are chosen for optimal estimation of the parameters (and thus, their auxiliary functions) in the sense of minimum volume of the three-parameter HPD region.

The user-written Athena Visual Studio code and results are shown below. The results agree closely with those of the previous example, except that the number of calls of MODEL is reduced from 17 to 8. The number of iterations needed is unchanged.

### Athena Visual Studio Code for Example C.2

```
! Appendix C: Example C.2
! Computer-Aided Modeling of Reactive Systems
!--------------------------------------------

@Response Model
 Dim Theta1, Theta2, Theta3 As Real
 Theta1=Par(1)
 Theta2=Par(2)
 Theta3=Par(3)

 Y(1)=Theta1+Theta2*Xu(1)^2+3.0*Xu(1)^Theta3

@Gradient Vector
 dY(1,1)=1.0
 dY(1,2)=Xu(1)^2
 dY(1,3)=3.0*Xu(1)^Par(3)*log(Xu(1))

@Phi Functions
 Phi(1)=Par(2)
 Phi(2)=Par(1)+Par(2)*10.0^2+3.0*10.0^Par(3)
```

### Numerical Results for Example C.2

```
Number of Experiments.......................... 30
Number of Parameters...........................  3
Number of Responses............................  1
Number of Settings.............................  1

EXIT GREGPLUS: SOLUTION FOUND. MODEL No. 1

STATISTICAL ANALYSIS

OBJECTIVE FUNCTION....................... 2.02655E-03
SUM OF SQUARES OF RESIDUALS.............. 2.02655E-03
ESTIMATED PARAMETERS FROM DATA...........          3
TOTAL OBSERVATIONS DATA COUNT IS.........         30
```

```
                  OPTIMAL
PARAMETER       ESTIMATES      95% MARGINAL HPD INTERVALS   PARAMETER STATUS
  PAR( 1)      1.001948E+00    1.001948E+00 +- 6.410E-03    Estimated
  PAR( 2)      1.997633E+00    1.997633E+00 +- 4.139E-03    Estimated
  PAR( 3)      3.000263E+00    3.000263E+00 +- 4.546E-04    Estimated

NORMALIZED PARAMETER COVARIANCE MATRIX
   1.000
  -0.777   1.000
   0.636  -0.958   1.000

PARAMETER COVARIANCE MATRIX
   1.3000E-01
  -6.5210E-02    5.4193E-02
   5.8671E-03   -5.7008E-03    6.5384E-04

EVENT OBSERVED     PREDICTED        RESIDUAL   DERIVATIVES--->
  1  1.0176E+00   1.0249E+00     -7.3222E-03   1.0000E+00   1.0000E-02   -6.9036E-03
  2  1.1082E+00   1.1058E+00      2.3572E-03   1.0000E+00   4.0000E-02   -3.8610E-02
  3  1.2584E+00   1.2627E+00     -4.3090E-03   1.0000E+00   9.0000E-02   -9.7491E-02
  4  1.5324E+00   1.5135E+00      1.8877E-02   1.0000E+00   1.6000E-01   -1.7589E-01
  5  1.8833E+00   1.8763E+00      7.0124E-03   1.0000E+00   2.5000E-01   -2.5988E-01
  6  2.3703E+00   2.3690E+00      1.2914E-03   1.0000E+00   3.6000E-01   -3.3097E-01
  7  3.0042E+00   3.0097E+00     -5.4914E-03   1.0000E+00   4.9000E-01   -3.6698E-01
  8  3.8176E+00   3.8163E+00      1.2572E-03   1.0000E+00   6.4000E-01   -3.4273E-01
  9  4.8049E+00   4.8070E+00     -2.0701E-03   1.0000E+00   8.1000E-01   -2.3042E-01
 10  5.9941E+00   5.9996E+00     -5.4810E-03   1.0000E+00   1.0000E+00    0.0000E+00
 11  7.4028E+00   7.4122E+00     -9.3842E-03   1.0000E+00   1.2100E+00    3.8058E-01
 12  9.0532E+00   9.0628E+00     -9.5885E-03   1.0000E+00   1.4400E+00    9.4520E-01
 13  1.0977E+01   1.0969E+01      7.3968E-03   1.0000E+00   1.6900E+00    1.7294E+00
 14  1.3153E+01   1.3150E+01      2.8619E-03   1.0000E+00   1.9600E+00    2.7701E+00
 15  1.5614E+01   1.5623E+01     -9.0032E-03   1.0000E+00   2.2500E+00    4.1058E+00
 16  1.8402E+01   1.8405E+01     -3.5091E-03   1.0000E+00   2.5600E+00    5.7761E+00
 17  2.1513E+01   2.1516E+01     -3.1665E-03   1.0000E+00   2.8900E+00    7.8220E+00
 18  2.4983E+01   2.4973E+01      9.7135E-03   1.0000E+00   3.2400E+00    1.0286E+01
 19  2.8793E+01   2.8794E+01     -9.8028E-04   1.0000E+00   3.6100E+00    1.3210E+01
 20  3.3017E+01   3.2997E+01      1.9840E-02   1.0000E+00   4.0000E+00    1.6639E+01
 21  3.7599E+01   3.7600E+01     -6.3623E-04   1.0000E+00   4.4100E+00    2.0617E+01
 22  4.2612E+01   4.2621E+01     -9.4223E-03   1.0000E+00   4.8400E+00    2.5192E+01
 23  4.8086E+01   4.8078E+01      7.5699E-03   1.0000E+00   5.2900E+00    3.0409E+01
 24  5.3999E+01   5.3990E+01      9.5279E-03   1.0000E+00   5.7600E+00    3.6316E+01
 25  6.0359E+01   6.0373E+01     -1.4361E-02   1.0000E+00   6.2500E+00    4.2961E+01
 26  6.7249E+01   6.7247E+01      1.3898E-03   1.0000E+00   6.7600E+00    5.0395E+01
 27  7.4617E+01   7.4629E+01     -1.2332E-02   1.0000E+00   7.2900E+00    5.8666E+01
 28  8.2539E+01   8.2537E+01      1.5594E-03   1.0000E+00   7.8400E+00    6.7825E+01
 29  9.0990E+01   9.0990E+01      5.1197E-05   1.0000E+00   8.4100E+00    7.7924E+01
 30  9.0996E+01   9.0990E+01      6.3512E-03   1.0000E+00   8.4100E+00    7.7924E+01


   NUMBER OF ITERATIONS..............   4
   NUMBER OF FUNCTION CALLS..........   8

EXIT GREGPLUS: MODEL DISCRIMINATION AND CRITICISM

SOURCE OF VARIANCE     SUM OF SQUARES    DEG. OF FREEDOM    MEAN SQUARE
Residuals               2.02655E-03            27
Lack of Fit             1.78996E-03            25           7.16E-05
Experimental            2.36596E-04             2           1.18E-04

Variance Ratio (Lack-of-Fit/Experimental)                      0.605
Sampling Probability of Greater Ratio                          0.788
Log10 of Posterior Unnormalized Probability Density            2.242

EXIT GREGPLUS: USER PREDICTED FUNCTIONS

FUNCTIONS         PREDICTED         95% MARGINAL HPD INTERVALS
  PHI( 1)       1.997633E+00        1.997633E+00   +- 4.139E-03
  PHI( 2)       3.202584E+03        3.202584E+03   +- 2.752E+00

NORMALIZED COVARIANCE MATRIX
   1.000
  -0.945   1.000

EXIT GREGPLUS: STATISTICAL OPTIMAL DESIGN

FOR THE DATA SET   0 THE INVERSE HESSIAN OF ORDER   3 HAS LOG(DET) = 7.010
FOR THE DATA SET   1 THE INVERSE HESSIAN OF ORDER   3 HAS LOG(DET) = 7.063
```

```
FOR THE DATA SET  2 THE INVERSE HESSIAN OF ORDER   3 HAS LOG(DET) = 7.062
FOR THE DATA SET  3 THE INVERSE HESSIAN OF ORDER   3 HAS LOG(DET) = 7.035
FOR THE DATA SET  4 THE INVERSE HESSIAN OF ORDER   3 HAS LOG(DET) = 7.069
FOR THE DATA SET  5 THE INVERSE HESSIAN OF ORDER   3 HAS LOG(DET) = 8.150

OPTIMAL: THE SET  5 MAY BE SELECTED AS THE BEST AMONG THE CANDIDATES
```

### Example C.3. Analysis of Four Rival Models of a Catalytic Process

Tschernitz et al. (1946) reported a detailed study of the kinetics of hydrogenation of mixed iso-octenes over a supported nickel catalyst. Their article gives 40 unreplicated observations of hydrogenation rate as a function of interfacial temperature $T$ and interfacial partial pressures $p_H$, $p_U$, and $p_S$ of hydrogen, unsaturated and saturated hydrocarbons. Eighteen rival mechanistic models of the reaction were formulated and fitted in linearized forms to isothermal subsets of the experimental data. Four of their models are compared here on the same data but with improved weighting in the manner of Stewart, Henson, and Box (1996), using the statistical procedures and criteria provided in GREGPLUS and in Chapter 6.

The user-written Athena Visual Studio code is shown below. It expresses the observations as $y_u \equiv \ln \mathcal{R}_u$, where $\mathcal{R}$ is the hydrogen uptake rate in lb-mols per hour per lb of catalyst, adjusted to a standard level of catalyst activity. Weights are assigned to these adjusted observations according to the formula

$$\sigma_u^2 = (0.00001/\Delta n_{RI})^2 + (0.1)^2$$

based on assigned standard deviations of 0.00001 for the observations of refractive-index difference $\Delta n_{RI}$ [used in calculating the reactant conversions] and 0.1 for the catalyst activity corrections. Models c, d, g, and h of the original paper are included in the code as models IMODEL=1, 2, 3, and 4, parameterized in the form $\theta_{i,A} \exp[\theta_{iB}(1/538.9 - 1/T)]$ for each rate coefficient $k_i(T)$ and equilibrium coefficient $K_i(T)$. With this parameterization, the data at all temperatures can be analyzed at once, rather than as separate, nearly isothermal data sets. A lower bound of zero is used for each preexponential parameter $\theta_{iA}$ to keep the rate and equilibrium coefficients nonnegative, in accord with physicochemical theory. The alternative form $\exp[\theta_{iC} + \theta_{iB}[(1/538.9 - 1/T)]]$ would give nonnegative $k_i(T)$ and $K_i(T)$ automatically, but would need huge parameter values and some of its functions would become negligible over portions of the experimental temperature range. A generic initial guess was used for each type of parameter; an iteration allowance ITMAX=100 sufficed for convergence for every model from these rough initial guesses.

In the absence of replicate observations, *replicate residuals* were used. The observational variance was estimated as the smallest residual mean square, $S(\widehat{\boldsymbol{\theta}})/(n - p)$, obtained when the weighted observations of $\log(\mathcal{R}/p_H p_U)$ were fitted with detailed polynomial functions of the interfacial temperature and the three partial pressures.

The numerical results, summarized below, show the fourth model to be the most probable one according to the data, and also show that this model does better than the others on the goodness-of-fit test. These results are consistent with those of Stewart, Henson, and Box (1996), who found this model to be the most probable *a posteriori* of the 18 models considered by Tschernitz et al. (1946). The linearized model forms used by Tschernitz et al. yield the same conclusion if one uses the appropriate variable weighting for each linearized model form.

### Athena Visual Studio Code for Example C.3

```
! Appendix C: Example C.3
! Computer-Aided Modeling of Reactive Systems
!-------------------------------------------
Parameter TBM = 538.9D0 As Real

Dim i,k As Integer
Do i=1,NRESP
 Do k=1,NEXP
  SQRW(i,k)=1D0/sqrt( (1.0D-5/SQRW(i,k))**2 + (1.0D-1)**2 )
 EndDo
EndDo

If(iModel.EQ.3)Then
 CHMAX(7)=Zero
 CHMAX(8)=Zero
EndIf

@Response Model
 Dim FT,RK,EQKHRK,EQKURK,EQKSRK,DEN,RMOD As Real
 Dim TU,PU,PS,PH As Real

 TU=Xu(1)+273.15
 PU=Xu(2)
 PS=Xu(3)
 PH=Xu(4)

 If(iModel.EQ.1) Then
  FT = 1D0/TBM - 1D0/TU
  RK = Par(1)*exp(FT*Par(4))
  EQKHRK = Par(2)*exp(FT*Par(5))
  EQKURK = Par(3)*exp(FT*Par(6))
  DEN = RK + EQKHRK*PH + EQKURK*PU
  RMOD = PH*PU/DEN
  Y(1) = log(RMOD)

 Else If(iModel.EQ.2) Then
  FT = 1D0/TBM - 1D0/TU
  RK     = Par(1)*exp(FT*Par(5))
  EQKHRK = Par(2)*exp(FT*Par(6))
  EQKURK = Par(3)*exp(FT*Par(7))
  EQKSRK = Par(4)*exp(FT*Par(8))
  DEN = (RK + EQKHRK*PH + EQKURK*PU + EQKSRK*PS)**2
  RMOD = PH*PU/DEN
  Y(1) = log(RMOD)

 Else If(iModel.EQ.3) Then
  FT = 1D0/TBM - 1D0/TU
  RK     = Par(1)*exp(FT*Par(4))
  EQKHRK = Par(2)*exp(FT*Par(5))
  EQKURK = Par(3)*exp(FT*Par(6))
  DEN = (RK + sqrt(EQKHRK*PH) + EQKURK*PU)
  RMOD = PH*PU/DEN
  Y(1) = log(RMOD)

 Else If (iModel.EQ.4) Then
  FT = 1D0/TBM - 1D0/TU
  RK     = Par(1)*exp(FT*Par(5))
```

```
EQKHRK = Par(2)*exp(FT*Par(6))
EQKURK = Par(3)*exp(FT*Par(7))
EQKSRK = Par(4)*exp(FT*Par(8))
DEN = (RK + sqrt(EQKHRK*PH) + EQKURK*PU + EQKSRK*PS)**3
RMOD = PH*PU/DEN
Y(1) = log(RMOD)
End If
```

## Numerical Results for Example C.3

Number of Experiments........................... 40
Number of Parameters........................... 8
Number of Responses........................... 1
Number of Settings........................... 4

EXIT GREGPLUS: SOLUTION FOUND. MODEL No. 1

STATISTICAL ANALYSIS

OBJECTIVE FUNCTION...................... 2.79506E+02
SUM OF SQUARES OF RESIDUALS.............. 3.13221E+00
ESTIMATED PARAMETERS FROM DATA........... 6
TOTAL OBSERVATIONS DATA COUNT IS......... 40

|  | OPTIMAL |  |  |
| --- | --- | --- | --- |
| PARAMETER | ESTIMATES | 95% MARGINAL HPD INTERVALS | PARAMETER STATUS |
| PAR( 1) | 6.957375E+00 | 6.957375E+00 +- 5.299E+00 | Estimated |
| PAR( 2) | 8.787257E+00 | 8.787257E+00 +- 4.638E+00 | Estimated |
| PAR( 3) | 1.807569E+01 | 1.807569E+01 +- 5.631E+00 | Estimated |
| PAR( 4) | 2.925434E+03 | 2.925434E+03 +- 4.081E+03 | Estimated |
| PAR( 5) | -2.714606E+03 | -2.714606E+03 +- 2.301E+03 | Estimated |
| PAR( 6) | 1.859572E+01 | 1.859572E+01 +- 1.517E+03 | Estimated |
| PAR( 7) | 0.000000E+00 |  | Indeterminate |
| PAR( 8) | 0.000000E+00 |  | Indeterminate |

NORMALIZED PARAMETER COVARIANCE MATRIX
```
  1.000
 -0.519  1.000
 -0.568  0.014  1.000
 -0.668  0.170  0.146  1.000
 -0.094  0.550  0.037 -0.256  1.000
  0.061 -0.025  0.238 -0.468 -0.161  1.000
  0.000 -0.000  0.000 -0.000  0.000  0.000  0.000
  0.000 -0.000  0.000 -0.000  0.000  0.000  0.000  0.000
```

PARAMETER COVARIANCE MATRIX
```
 8.268E-01
-3.755E-01  6.334E-01
-4.991E-01  1.086E-02 9.339E-01
-4.251E+02  9.471E+01 9.847E+01  4.904E+05
-3.371E+01  1.728E+02 1.396E+01 -7.091E+04  1.559E+05
 1.433E+01 -5.234E+00 5.979E+01 -8.538E+04 -1.654E+04 6.774E+04
 0.000E+00 -0.000E+00 0.000E+00 -0.000E+00  0.000E+00 0.000E+00 0.000E+00
 0.000E+00 -0.000E+00 0.000E+00 -0.000E+00  0.000E+00 0.000E+00 0.000E+00 0.000E+00
```

| EVENT | OBSERVED | PREDICTED | RESIDUAL | DERIVATIVES (COLUMNS 1 THROUGH 3)---> | | |
| --- | --- | --- | --- | --- | --- | --- |
| 1 | -5.6465E+00 | -5.6425E+00 | -3.9778E-03 | -2.8435E-01 | -5.5646E-01 | -5.7140E-02 |
| 2 | -3.6889E+00 | -3.7628E+00 | 7.3962E-02 | -8.5226E-02 | -8.8202E-01 | -9.0254E-02 |
| 3 | -3.4420E+00 | -3.7670E+00 | 3.2497E-01 | -8.7900E-02 | -8.7436E-01 | -9.3616E-02 |
| 4 | -5.2344E+00 | -4.4707E+00 | -7.6368E-01 | -2.1067E-01 | -4.0799E-01 | -2.0913E-01 |

```
 5  -4.7444E+00  -4.3765E+00  -3.6797E-01  -2.1085E-01  -4.4002E-01  -2.2880E-01
 6  -4.2744E+00  -4.4198E+00   1.4538E-01  -2.1609E-01  -4.3573E-01  -2.4559E-01
 7  -4.6460E+00  -4.4178E+00  -2.2817E-01  -2.1073E-01  -4.2222E-01  -2.4210E-01
 8  -5.2707E+00  -5.5624E+00   2.9168E-01  -2.8022E-01  -1.1832E-01  -3.1647E-01
 9  -3.9528E+00  -3.8879E+00  -6.4978E-02  -7.7486E-02  -1.4506E-01  -4.4760E-01
10  -3.8825E+00  -3.9734E+00   9.0918E-02  -7.9266E-02  -1.3487E-01  -4.5304E-01
11  -4.3351E+00  -4.0969E+00  -2.3823E-01  -1.5077E-01  -3.1109E-01  -3.3371E-01
12  -3.9846E+00  -4.2089E+00   2.2432E-01  -1.2800E-01  -1.8553E-01  -4.0898E-01
13  -5.8158E+00  -5.5994E+00  -2.1644E-01  -6.8181E-01  -2.9569E-01  -6.6255E-02
14  -3.2468E+00  -3.3561E+00   1.0935E-01  -2.9121E-01  -6.1534E-01  -1.3859E-01
15  -3.1011E+00  -3.3614E+00   2.6031E-01  -2.8847E-01  -6.1946E-01  -1.3826E-01
16  -3.8825E+00  -3.9518E+00   6.9290E-02  -1.8058E-01  -6.2445E-02  -4.5138E-01
17  -3.9900E+00  -3.9314E+00  -5.8566E-02  -1.7605E-01  -6.3153E-02  -4.4884E-01
18  -4.0174E+00  -4.3724E+00   3.5506E-01  -4.9973E-01  -2.0244E-01  -2.5572E-01
19  -4.5756E+00  -4.3379E+00  -2.3767E-01  -4.9449E-01  -2.3191E-01  -2.3196E-01
20  -4.4104E+00  -4.4087E+00  -1.6855E-03  -4.9029E-01  -1.9751E-01  -2.5272E-01
21  -4.9547E+00  -5.8201E+00   8.6538E-01  -5.5857E-01  -4.7117E-02  -2.7712E-01
22  -4.7988E+00  -4.5995E+00  -1.9923E-01  -5.5331E-01  -1.9696E-01  -2.3486E-01
23  -4.9144E+00  -4.4870E+00  -4.2739E-01  -5.0661E-01  -2.0322E-01  -2.2779E-01
24  -3.4451E+00  -3.6355E+00   1.9032E-01  -3.7057E-01  -4.7586E-01  -1.7555E-01
25  -3.1350E+00  -3.5038E+00   3.6883E-01  -3.4430E-01  -3.5490E-01  -2.4546E-01
26  -3.9070E+00  -3.8513E+00  -5.5737E-02  -2.5943E-01  -4.6352E-02  -4.2500E-01
27  -4.4741E+00  -4.4948E+00   2.0675E-02  -6.6431E-01  -1.2335E-01  -2.2967E-01
28  -4.3125E+00  -4.5139E+00   2.0141E-01  -6.7625E-01  -1.2865E-01  -2.1835E-01
29  -4.6607E+00  -4.5345E+00  -1.2615E-01  -6.6205E-01  -1.1583E-01  -2.2262E-01
30  -4.7433E+00  -4.5231E+00  -2.2020E-01  -6.5137E-01  -1.1773E-01  -2.2359E-01
31  -5.9915E+00  -5.9944E+00   2.8958E-03  -5.4676E-01  -2.0969E-02  -1.8498E-01
32  -3.3873E+00  -3.2868E+00  -1.0052E-01  -4.9475E-01  -4.0664E-01  -1.6177E-01
33  -3.5543E+00  -3.3783E+00  -1.7609E-01  -4.9346E-01  -4.6165E-01  -1.3563E-01
34  -3.5166E+00  -3.3023E+00  -2.1429E-01  -4.8489E-01  -4.2952E-01  -1.5450E-01
35  -5.4262E+00  -5.8150E+00   3.8887E-01  -9.7737E-01  -1.8428E-01  -5.9586E-02
36  -4.5044E+00  -4.6146E+00   1.1014E-01  -7.0172E-01  -1.2219E-01  -2.0327E-01
37  -5.2344E+00  -4.8065E+00  -4.2788E-01  -7.6749E-01  -1.5689E-01  -1.3954E-01
38  -3.9712E+00  -4.0116E+00   4.0361E-02  -4.8885E-01  -1.1290E-01  -3.0290E-01
39  -4.2587E+00  -4.1375E+00  -1.2120E-01  -3.8978E-01  -6.3843E-02  -3.6371E-01
40  -4.1414E+00  -4.0671E+00  -7.4335E-02  -3.9858E-01  -7.0172E-02  -3.5870E-01

EVENT  DERIVATIVES (COLUMNS 4 THROUGH 8)--->
 1   5.1933E-04   1.2583E-03   2.6628E-04   0.0000E+00   0.0000E+00
 2   1.5565E-04   1.9944E-03   4.2060E-04   0.0000E+00   0.0000E+00
 3   1.6054E-04   1.9771E-03   4.3626E-04   0.0000E+00   0.0000E+00
 4   3.8476E-04   9.2255E-04   9.7456E-04   0.0000E+00   0.0000E+00
 5   3.8509E-04   9.9498E-04   1.0662E-03   0.0000E+00   0.0000E+00
 6   4.0840E-04   1.0195E-03   1.1843E-03   0.0000E+00   0.0000E+00
 7   3.9827E-04   9.8793E-04   1.1675E-03   0.0000E+00   0.0000E+00
 8   5.1180E-04   2.6756E-04   1.4748E-03   0.0000E+00   0.0000E+00
 9   1.4398E-04   3.3370E-04   2.1221E-03   0.0000E+00   0.0000E+00
10   1.4728E-04   3.1026E-04   2.1479E-03   0.0000E+00   0.0000E+00
11   2.8495E-04   7.2790E-04   1.6092E-03   0.0000E+00   0.0000E+00
12   2.3378E-04   4.1953E-04   1.9059E-03   0.0000E+00   0.0000E+00
13  -8.6456E-05  -4.6422E-05  -2.1437E-05   0.0000E+00   0.0000E+00
14  -6.4586E-05  -1.6897E-04  -7.8428E-05   0.0000E+00   0.0000E+00
15  -5.7166E-05  -1.5198E-04  -6.9912E-05   0.0000E+00   0.0000E+00
16  -4.8532E-05  -2.0778E-05  -3.0954E-04   0.0000E+00   0.0000E+00
17  -4.3189E-05  -1.9181E-05  -2.8095E-04   0.0000E+00   0.0000E+00
18  -1.2259E-04  -6.1486E-05  -1.6006E-04   0.0000E+00   0.0000E+00
19  -9.7994E-05  -5.6899E-05  -1.1729E-04   0.0000E+00   0.0000E+00
20  -9.7161E-05  -4.8460E-05  -1.2779E-04   0.0000E+00   0.0000E+00
21  -1.2388E-04  -1.2938E-05  -1.5682E-04   0.0000E+00   0.0000E+00
22  -1.2272E-04  -5.4083E-05  -1.3291E-04   0.0000E+00   0.0000E+00
```

```
23  -1.0040E-04  -4.9860E-05  -1.1518E-04   0.0000E+00   0.0000E+00
24  -7.3437E-05  -1.1675E-04  -8.8768E-05   0.0000E+00   0.0000E+00
25  -7.6361E-05  -9.7452E-05  -1.3891E-04   0.0000E+00   0.0000E+00
26  -3.4799E-04  -7.6976E-05  -1.4546E-03   0.0000E+00   0.0000E+00
27  -8.3847E-04  -1.9275E-04  -7.3965E-04   0.0000E+00   0.0000E+00
28  -8.5354E-04  -2.0103E-04  -7.0320E-04   0.0000E+00   0.0000E+00
29  -8.6191E-04  -1.8669E-04  -7.3952E-04   0.0000E+00   0.0000E+00
30  -8.2214E-04  -1.8397E-04  -7.2007E-04   0.0000E+00   0.0000E+00
31  -6.9010E-04  -3.2767E-05  -5.9572E-04   0.0000E+00   0.0000E+00
32  -6.7335E-04  -6.8518E-04  -5.6177E-04   0.0000E+00   0.0000E+00
33  -6.4243E-04  -7.4409E-04  -4.5053E-04   0.0000E+00   0.0000E+00
34  -6.3127E-04  -6.9231E-04  -5.1323E-04   0.0000E+00   0.0000E+00
35  -1.2141E-03  -2.8341E-04  -1.8886E-04   0.0000E+00   0.0000E+00
36  -9.2743E-04  -1.9994E-04  -6.8549E-04   0.0000E+00   0.0000E+00
37  -9.9918E-04  -2.5287E-04  -4.6353E-04   0.0000E+00   0.0000E+00
38  -6.0725E-04  -1.7364E-04  -9.6005E-04   0.0000E+00   0.0000E+00
39  -4.9971E-04  -1.0134E-04  -1.1898E-03   0.0000E+00   0.0000E+00
40  -5.1890E-04  -1.1311E-04  -1.1915E-03   0.0000E+00   0.0000E+00
NUMBER OF ITERATIONS..............   16
NUMBER OF FUNCTION CALLS..........  145


EXIT GREGPLUS: MODEL DISCRIMINATION AND CRITICISM

SOURCE OF VARIANCE      SUM OF SQUARES   DEG. OF FREEDOM   MEAN SQUARE
Residuals                 2.79506E+02          34
Lack of Fit               2.15485E+02          13           1.66E+01
Experimental              6.40210E+01          21           3.05E+00


Variance Ratio (Lack-of-Fit/Experimental)                    5.437
Sampling Probability of Greater Ratio                        0.000
Log10 of Posterior Unnormalized Probability Density        -26.590


EXIT GREGPLUS: SOLUTION FOUND. MODEL No. 2

STATISTICAL ANALYSIS

OBJECTIVE FUNCTION.......................  1.92175E+02
SUM OF SQUARES OF RESIDUALS..............  2.12450E+00
ESTIMATED PARAMETERS FROM DATA...........        8
TOTAL OBSERVATIONS DATA COUNT IS.........       40

                  OPTIMAL
PARAMETER        ESTIMATES     95% MARGINAL HPD INTERVALS   PARAMETER STATUS
PAR( 1)       2.827598E+00      2.827598E+00 +- 5.391E-01   Estimated
PAR( 2)       8.561766E-01      8.561766E-01 +- 3.209E-01   Estimated
PAR( 3)       1.677330E+00      1.677330E+00 +- 3.819E-01   Estimated
PAR( 4)       4.374090E-01      4.374090E-01 +- 3.962E-01   Estimated
PAR( 5)       1.157582E+03      1.157582E+03 +- 9.991E+02   Estimated
PAR( 6)      -1.587460E+03     -1.587460E+03 +- 1.805E+03   Estimated
PAR( 7)      -3.780040E+01     -3.780040E+01 +- 1.129E+03   Estimated
PAR( 8)      -4.886321E+03     -4.886321E+03 +- 3.564E+03   Estimated

NORMALIZED PARAMETER COVARIANCE MATRIX
  1.000
 -0.527  1.000
 -0.617  0.133  1.000
 -0.522  0.115  0.219  1.000
 -0.427  0.029  0.052  0.076  1.000
 -0.092  0.405  0.050  0.019 -0.389  1.000
```

```
   0.026   0.014   0.255   0.025  -0.578   0.062   1.000
  -0.237   0.076   0.151   0.806  -0.224  -0.050   0.062   1.000
```

PARAMETER COVARIANCE MATRIX
```
 1.166E-02
-3.660E-03  4.133E-03
-5.098E-03  6.523E-04  5.854E-03
-4.474E-03  5.843E-04  1.329E-03  6.300E-03
-9.221E+00  3.770E-01  7.966E-01  1.212E+00   4.006E+04
-3.582E+00  9.422E+00  1.377E+00  5.506E-01  -2.818E+04   1.308E+05
 6.307E-01  2.046E-01  4.409E+00  4.480E-01  -2.617E+04   5.064E+03  5.112E+04
-1.826E+01  3.490E+00  8.233E+00  4.569E+01  -3.203E+04  -1.301E+04  1.001E+04  5.098E+05
```

```
EVENT   OBSERVED     PREDICTED     RESIDUAL    DERIVATIVES (COLUMNS 1 THROUGH 3)--->
   1   -5.6465E+00  -5.6339E+00  -1.2562E-02  -3.2126E+00  -3.1262E+00  -4.3585E-01
   2   -3.6889E+00  -3.6169E+00  -7.2028E-02  -2.1009E+00  -1.0812E+01  -1.5021E+00
   3   -3.4420E+00  -3.6158E+00   1.7380E-01  -2.1405E+00  -1.0588E+01  -1.5391E+00
   4   -5.2344E+00  -5.4663E+00   2.3188E-01  -1.7704E+00  -1.7049E+00  -1.1865E+00
   5   -4.7444E+00  -4.2344E+00  -5.1006E-01  -3.2257E+00  -3.3474E+00  -2.3632E+00
   6   -4.2744E+00  -4.3025E+00   2.8066E-02  -3.2836E+00  -3.2133E+00  -2.4853E+00
   7   -4.6460E+00  -4.3009E+00  -3.4510E-01  -3.2054E+00  -3.1169E+00  -2.4525E+00
   8   -5.2707E+00  -5.5422E+00   2.7148E-01  -3.3666E+00  -7.0687E-01  -2.5669E+00
   9   -3.9528E+00  -3.8744E+00  -7.8400E-02  -1.8699E+00  -1.7197E+00  -7.2429E+00
  10   -3.8825E+00  -3.9778E+00   9.5310E-02  -1.8768E+00  -1.5687E+00  -7.1928E+00
  11   -4.3351E+00  -4.2529E+00  -8.2244E-02  -2.3963E+00  -2.3995E+00  -3.5321E+00
  12   -3.9846E+00  -3.9711E+00  -1.3485E-02  -2.6897E+00  -1.9386E+00  -5.8021E+00
  13   -5.8158E+00  -5.6191E+00  -1.9672E-01  -4.7955E+00  -2.1831E+00  -4.7889E-01
  14   -3.2468E+00  -3.3320E+00   8.5252E-02  -3.4257E+00  -7.8797E+00  -1.7100E+00
  15   -3.1011E+00  -3.3334E+00   2.3231E-01  -3.4207E+00  -7.9241E+00  -1.7109E+00
  16   -3.8825E+00  -4.0080E+00   1.2554E-01  -2.5902E+00  -9.9271E-01  -6.8599E+00
  17   -3.9900E+00  -3.9872E+00  -2.7497E-03  -2.5508E+00  -1.0050E+00  -6.8556E+00
  18   -4.0174E+00  -4.2088E+00   1.9140E-01  -4.8065E+00  -2.1387E+00  -2.5928E+00
  19   -4.5756E+00  -4.2431E+00  -3.3250E-01  -4.5777E+00  -2.3160E+00  -2.2409E+00
  20   -4.4104E+00  -4.3153E+00  -9.5088E-02  -4.5651E+00  -1.9839E+00  -2.4556E+00
  21   -4.9547E+00  -5.7370E+00   7.8225E-01  -4.7469E+00  -4.3587E-01  -2.4701E+00
  22   -4.7988E+00  -4.8663E+00   6.7546E-02  -4.0698E+00  -1.5770E+00  -1.8119E+00
  23   -4.9144E+00  -4.7398E+00  -1.7461E-01  -3.8437E+00  -1.6633E+00  -1.8035E+00
  24   -3.4451E+00  -3.5628E+00   1.1762E-01  -3.9652E+00  -5.4929E+00  -1.9603E+00
  25   -3.1350E+00  -3.3757E+00   2.4074E-01  -3.9299E+00  -4.4096E+00  -2.9387E+00
  26   -3.9070E+00  -3.9528E+00   4.5755E-02  -2.9230E+00  -8.7239E-01  -6.3898E+00
  27   -4.4741E+00  -4.4053E+00  -6.8854E-02  -5.1511E+00  -1.5501E+00  -2.3363E+00
  28   -4.3125E+00  -4.4579E+00   1.4536E-01  -5.0905E+00  -1.5694E+00  -2.1563E+00
  29   -4.6607E+00  -4.4765E+00  -1.8376E-01  -4.9988E+00  -1.4390E+00  -2.2241E+00
  30   -4.7433E+00  -4.4651E+00  -2.7823E-01  -4.9562E+00  -1.4518E+00  -2.2319E+00
  31   -5.9915E+00  -5.9183E+00  -7.3187E-02  -4.0034E+00  -2.4882E-01  -1.7769E+00
  32   -3.3873E+00  -3.3419E+00  -4.5380E-02  -4.1348E+00  -5.7197E+00  -1.8117E+00
  33   -3.5543E+00  -3.4341E+00  -1.2024E-01  -4.1366E+00  -6.3677E+00  -1.5043E+00
  34   -3.5166E+00  -3.3558E+00  -1.6078E-01  -4.1053E+00  -5.9836E+00  -1.7308E+00
  35   -5.4262E+00  -5.8304E+00   4.0429E-01  -5.8228E+00  -1.7657E+00  -4.6371E-01
  36   -4.5044E+00  -4.7117E+00   2.0727E-01  -4.7512E+00  -1.3717E+00  -1.8288E+00
  37   -5.2344E+00  -4.8673E+00  -3.6711E-01  -4.9616E+00  -1.6689E+00  -1.1936E+00
  38   -3.9712E+00  -3.9905E+00   1.9258E-02  -4.2737E+00  -1.5875E+00  -3.4590E+00
  39   -4.2587E+00  -4.0954E+00  -1.6336E-01  -3.8483E+00  -1.0293E+00  -4.7311E+00
  40   -4.1414E+00  -4.0255E+00  -1.1594E-01  -3.8929E+00  -1.1277E+00  -4.6356E+00
```

```
EVENT DERIVATIVES (COLUMNS 4 THROUGH 8)--->
   1  -7.6402E+00   2.3451E-03   6.8963E-04   1.8850E-04   8.6453E-04
   2  -5.0652E+00   1.5336E-03   2.3850E-03   6.4962E-04   5.7315E-04
   3  -5.1608E+00   1.5625E-03   2.3356E-03   6.6566E-04   5.8397E-04
   4  -2.1052E+01   1.2924E-03   3.7610E-04   5.1316E-04   2.3821E-03
```

| | | | | |
|---|---|---|---|---|
| 5 -6.8710E+00 | 2.3547E-03 | 7.3842E-04 | 1.0220E-03 | 7.7748E-04 |
| 6 -7.6723E+00 | 2.4804E-03 | 7.3350E-04 | 1.1123E-03 | 8.9837E-04 |
| 7 -7.4896E+00 | 2.4213E-03 | 7.1149E-04 | 1.0976E-03 | 8.7697E-04 |
| 8 -6.9347E+00 | 2.4575E-03 | 1.5593E-04 | 1.1101E-03 | 7.8469E-04 |
| 9 -2.0681E+00 | 1.3887E-03 | 3.8594E-04 | 3.1868E-03 | 2.3808E-04 |
| 10 -2.6081E+00 | 1.3938E-03 | 3.5206E-04 | 3.1648E-03 | 3.0025E-04 |
| 11 -1.1151E+01 | 1.8101E-03 | 5.4773E-04 | 1.5807E-03 | 1.3056E-03 |
| 12 -1.9266E+00 | 1.9634E-03 | 4.2765E-04 | 2.5093E-03 | 2.1800E-04 |
| 13 -2.2618E+00 | -2.4304E-04 | -3.3436E-05 | -1.4380E-05 | -1.7769E-05 |
| 14 -1.4116E+00 | -3.0367E-04 | -2.1108E-04 | -8.9807E-05 | -1.9396E-05 |
| 15 -1.4032E+00 | -2.7094E-04 | -1.8967E-04 | -8.0288E-05 | -1.7228E-05 |
| 16 -6.3980E-01 | -2.7824E-04 | -3.2225E-05 | -4.3658E-04 | -1.0653E-05 |
| 17 -5.5976E-01 | -2.5011E-04 | -2.9780E-05 | -3.9825E-04 | -8.5076E-06 |
| 18 -1.9923E-01 | -4.7127E-04 | -6.3370E-05 | -1.5062E-04 | -3.0280E-06 |
| 19 -1.7268E+00 | -3.6258E-04 | -5.5434E-05 | -1.0516E-04 | -2.1201E-05 |
| 20 -1.8340E+00 | -3.6158E-04 | -4.7486E-05 | -1.1523E-04 | -2.2518E-05 |
| 21 -1.8261E+00 | -4.2079E-04 | -1.1676E-05 | -1.2973E-04 | -2.5092E-05 |
| 22 -8.8234E+00 | -3.6077E-04 | -4.2245E-05 | -9.5158E-05 | -1.2124E-04 |
| 23 -8.3062E+00 | -3.0445E-04 | -3.9812E-05 | -8.4633E-05 | -1.0198E-04 |
| 24 -1.6601E+00 | -3.1407E-04 | -1.3148E-04 | -9.1989E-05 | -2.0382E-05 |
| 25 -3.3559E-01 | -3.4837E-04 | -1.1813E-04 | -1.5434E-04 | -4.6113E-06 |
| 26 -2.5400E-01 | -1.5671E-03 | -1.4134E-04 | -2.0296E-03 | -2.1109E-05 |
| 27 -1.0990E-01 | -2.5986E-03 | -2.3630E-04 | -6.9828E-04 | -8.5942E-06 |
| 28 -8.1632E-01 | -2.5680E-03 | -2.3925E-04 | -6.4447E-04 | -6.3835E-05 |
| 29 -7.8347E-01 | -2.6011E-03 | -2.2628E-04 | -6.8566E-04 | -6.3194E-05 |
| 30 -8.1525E-01 | -2.5002E-03 | -2.2132E-04 | -6.6706E-04 | -6.3751E-05 |
| 31 -6.2133E-01 | -2.0196E-03 | -3.7933E-05 | -5.3107E-04 | -4.8587E-05 |
| 32 -7.9841E-01 | -2.2492E-03 | -9.4022E-04 | -5.8388E-04 | -6.7322E-05 |
| 33 -7.0201E-01 | -2.1524E-03 | -1.0013E-03 | -4.6376E-04 | -5.6624E-05 |
| 34 -7.8139E-01 | -2.1361E-03 | -9.4089E-04 | -5.3357E-04 | -6.3026E-05 |
| 35 -1.0454E+00 | -2.8909E-03 | -2.6492E-04 | -1.3640E-04 | -8.0457E-05 |
| 36 -3.9648E+00 | -2.5098E-03 | -2.1896E-04 | -5.7236E-04 | -3.2465E-04 |
| 37 -2.7159E+00 | -2.5817E-03 | -2.6242E-04 | -3.6797E-04 | -2.1907E-04 |
| 38 -1.3566E+00 | -2.1218E-03 | -2.3818E-04 | -1.0175E-03 | -1.0440E-04 |
| 39 -1.7710E-01 | -1.9719E-03 | -1.5939E-04 | -1.4363E-03 | -1.4067E-05 |
| 40 -1.8654E-01 | -2.0257E-03 | -1.7733E-04 | -1.4291E-03 | -1.5046E-05 |

```
NUMBER OF ITERATIONS.............    4
NUMBER OF FUNCTION CALLS..........   37
```

EXIT GREGPLUS: MODEL DISCRIMINATION AND CRITICISM

| SOURCE OF VARIANCE | SUM OF SQUARES | DEG. OF FREEDOM | MEAN SQUARE |
|---|---|---|---|
| Residuals | 1.92175E+02 | 32 | |
| Lack of Fit | 1.28154E+02 | 11 | 1.17E+01 |
| Experimental | 6.40210E+01 | 21 | 3.05E+00 |

```
Variance Ratio (Lack-of-Fit/Experimental)                  3.822
Sampling Probability of Greater Ratio                      0.004
Log10 of Posterior Unnormalized Probability Density      -25.183
```

EXIT GREGPLUS: SOLUTION FOUND. MODEL No. 3

STATISTICAL ANALYSIS

```
OBJECTIVE FUNCTION......................   2.45993E+02
SUM OF SQUARES OF RESIDUALS.............   2.77839E+00
ESTIMATED PARAMETERS FROM DATA..........         4
TOTAL OBSERVATIONS DATA COUNT IS........        40
```

```
                OPTIMAL
PARAMETER       ESTIMATES      95% MARGINAL HPD INTERVALS    PARAMETER STATUS
PAR( 1)       0.000000E+00                                   Lower Bound
PAR( 2)       3.624160E+02    3.624160E+02 +- 1.565E+02      Estimated
PAR( 3)       1.756237E+01    1.756237E+01 +- 4.579E+00      Estimated
PAR( 4)       6.069134E+02                                   Indeterminate
PAR( 5)      -2.098353E+03   -2.098353E+03 +- 2.289E+03      Estimated
PAR( 6)       1.214369E+03    1.214369E+03 +- 1.455E+03      Estimated
PAR( 7)       0.000000E+00                                   Fixed
PAR( 8)       0.000000E+00                                   Fixed
```

```
NORMALIZED PARAMETER COVARIANCE MATRIX
   0.000
   0.000  1.000
   0.000 -0.711  1.000
   0.000  0.000 -0.000  0.000
   0.000  0.217 -0.037  0.000  1.000
   0.000 -0.028 -0.174  0.000 -0.695  1.000
   0.000  0.000 -0.000  0.000  0.000  0.000  0.000
   0.000  0.000 -0.000  0.000  0.000  0.000  0.000  0.000
```

```
PARAMETER COVARIANCE MATRIX
0.000E+00
3.761E+01  8.715E+02
3.495E-01 -1.812E+01  7.458E-01
0.000E+00  0.000E+00 -0.000E+00 0.000E+00
1.362E+02  2.760E+03 -1.398E+01 0.000E+00  1.864E+05
1.975E+00 -2.279E+02 -4.117E+01 0.000E+00 -8.230E+04 7.530E+04
0.000E+00  0.000E+00 -0.000E+00 0.000E+00  0.000E+00 0.000E+00 0.000E+00
0.000E+00  0.000E+00 -0.000E+00 0.000E+00  0.000E+00 0.000E+00 0.000E+00 0.000E+00
```

```
EVENT  OBSERVED    PREDICTED    RESIDUAL    DERIVATIVES (COLUMNS 1 THROUGH 3)--->
   1  -5.6465E+00  -5.9559E+00   3.0948E-01  -3.6607E-01  -1.0304E-02  -3.1351E-02
   2  -3.6889E+00  -3.5465E+00  -1.4234E-01  -1.8369E-01  -1.1891E-02  -8.2908E-02
   3  -3.4420E+00  -3.5644E+00   1.2239E-01  -1.8689E-01  -1.1861E-02  -8.4834E-02
   4  -5.2344E+00  -4.6057E+00  -6.2871E-01  -3.2193E-01  -9.0147E-03  -1.3620E-01
   5  -4.7444E+00  -4.4935E+00  -2.5095E-01  -3.2776E-01  -9.5274E-03  -1.5158E-01
   6  -4.2744E+00  -4.5347E+00   2.6031E-01  -3.4311E-01  -9.7109E-03  -1.6133E-01
   7  -4.6460E+00  -4.5312E+00  -1.1483E-01  -3.3511E-01  -9.4544E-03  -1.5929E-01
   8  -5.2707E+00  -5.5651E+00   2.9436E-01  -4.9026E-01  -6.4102E-03  -2.3597E-01
   9  -3.9528E+00  -3.7266E+00  -2.2622E-01  -1.5904E-01  -4.3577E-03  -3.8578E-01
  10  -3.8825E+00  -3.8105E+00  -7.1921E-02  -1.6295E-01  -4.2568E-03  -3.9111E-01
  11  -4.3351E+00  -4.0930E+00  -2.4211E-01  -2.6842E-01  -7.6847E-03  -2.4580E-01
  12  -3.9846E+00  -4.1309E+00   1.4629E-01  -2.4025E-01  -5.8202E-03  -3.2716E-01
  13  -5.8158E+00  -5.7120E+00  -1.0382E-01  -5.9400E-01  -1.0537E-02  -6.1244E-02
  14  -3.2468E+00  -3.3929E+00   1.4616E-01  -2.6416E-01  -1.0492E-02  -1.3944E-01
  15  -3.1011E+00  -3.3973E+00   2.9618E-01  -2.6380E-01  -1.0525E-02  -1.3870E-01
  16  -3.8825E+00  -3.9779E+00   9.5396E-02  -1.6243E-01  -2.6281E-03  -4.6030E-01
  17  -3.9900E+00  -3.9563E+00  -3.3643E-02  -1.5967E-01  -2.6220E-03  -4.5648E-01
  18  -4.0174E+00  -4.4328E+00   4.1545E-01  -4.3983E-01  -7.6756E-03  -2.5238E-01
  19  -4.5756E+00  -4.4155E+00  -1.6014E-01  -4.3424E-01  -8.0966E-03  -2.2345E-01
  20  -4.4104E+00  -4.4692E+00   5.8822E-02  -4.3779E-01  -7.5652E-03  -2.4754E-01
  21  -4.9547E+00  -5.6710E+00   7.1622E-01  -6.1029E-01  -4.8429E-03  -3.3581E-01
  22  -4.7988E+00  -4.6503E+00  -1.4850E-01  -4.9567E-01  -8.0801E-03  -2.3335E-01
  23  -4.9144E+00  -4.5505E+00  -3.6388E-01  -4.5125E-01  -7.7813E-03  -2.2257E-01
  24  -3.4451E+00  -3.7315E+00   2.8634E-01  -3.1927E-01  -9.8502E-03  -1.6591E-01
  25  -3.1350E+00  -3.5947E+00   4.5969E-01  -2.9571E-01  -8.2028E-03  -2.3383E-01
  26  -3.9070E+00  -3.9566E+00   4.9521E-02  -1.5575E-01  -2.1262E-03  -4.7734E-01
  27  -4.4741E+00  -4.4417E+00  -3.2480E-02  -4.7742E-01  -6.5658E-03  -2.9758E-01
```

```
28  -4.3125E+00  -4.4564E+00   1.4394E-01  -4.8808E-01  -6.7942E-03  -2.8412E-01
29  -4.6607E+00  -4.4706E+00  -1.9009E-01  -4.7495E-01  -6.3779E-03  -2.9332E-01
30  -4.7433E+00  -4.4656E+00  -2.7767E-01  -4.7012E-01  -6.3788E-03  -2.9093E-01
31  -5.9915E+00  -5.7254E+00  -2.6603E-01  -4.8647E-01  -3.0405E-03  -2.9672E-01
32  -3.3873E+00  -3.3327E+00  -5.4594E-02  -3.1345E-01  -9.2043E-03  -1.9350E-01
33  -3.5543E+00  -3.4156E+00  -1.3874E-01  -3.2126E-01  -9.9761E-03  -1.6217E-01
34  -3.5166E+00  -3.3480E+00  -1.6858E-01  -3.1300E-01  -9.4577E-03  -1.8317E-01
35  -5.4262E+00  -5.6187E+00   1.9260E-01  -8.1449E-01  -1.1254E-02  -8.8692E-02
36  -4.5044E+00  -4.5322E+00   2.7758E-02  -5.0948E-01  -6.8458E-03  -2.7357E-01
37  -5.2344E+00  -4.7008E+00  -5.3363E-01  -5.7378E-01  -8.3288E-03  -1.9160E-01
38  -3.9712E+00  -4.0400E+00   6.8721E-02  -3.2624E-01  -4.9890E-03  -3.6106E-01
39  -4.2587E+00  -4.1832E+00  -7.5560E-02  -2.5262E-01  -3.2728E-03  -4.2896E-01
40  -4.1414E+00  -4.1153E+00  -2.6119E-02  -2.5610E-01  -3.4500E-03  -4.2330E-01


EVENT DERIVATIVES (COLUMNS 4 THROUGH 8)--->
 1   0.0000E+00   9.5078E-04   1.4232E-04   0.0000E+00   0.0000E+00
 2   0.0000E+00   1.0972E-03   3.7635E-04   0.0000E+00   0.0000E+00
 3   0.0000E+00   1.0944E-03   3.8509E-04   0.0000E+00   0.0000E+00
 4   0.0000E+00   8.3180E-04   6.1828E-04   0.0000E+00   0.0000E+00
 5   0.0000E+00   8.7910E-04   6.8809E-04   0.0000E+00   0.0000E+00
 6   0.0000E+00   9.2721E-04   7.5784E-04   0.0000E+00   0.0000E+00
 7   0.0000E+00   9.0272E-04   7.4822E-04   0.0000E+00   0.0000E+00
 8   0.0000E+00   5.9148E-04   1.0712E-03   0.0000E+00   0.0000E+00
 9   0.0000E+00   4.0907E-04   1.7816E-03   0.0000E+00   0.0000E+00
10   0.0000E+00   3.9960E-04   1.8062E-03   0.0000E+00   0.0000E+00
11   0.0000E+00   7.3375E-04   1.1546E-03   0.0000E+00   0.0000E+00
12   0.0000E+00   5.3704E-04   1.4851E-03   0.0000E+00   0.0000E+00
13   0.0000E+00  -6.7506E-05  -1.9302E-05   0.0000E+00   0.0000E+00
14   0.0000E+00  -1.1756E-04  -7.6863E-05   0.0000E+00   0.0000E+00
15   0.0000E+00  -1.0537E-04  -6.8315E-05   0.0000E+00   0.0000E+00
16   0.0000E+00  -3.5685E-05  -3.0748E-04   0.0000E+00   0.0000E+00
17   0.0000E+00  -3.2497E-05  -2.7833E-04   0.0000E+00   0.0000E+00
18   0.0000E+00  -9.5129E-05  -1.5388E-04   0.0000E+00   0.0000E+00
19   0.0000E+00  -8.1062E-05  -1.1006E-04   0.0000E+00   0.0000E+00
20   0.0000E+00  -7.5742E-05  -1.2192E-04   0.0000E+00   0.0000E+00
21   0.0000E+00  -5.4264E-05  -1.8512E-04   0.0000E+00   0.0000E+00
22   0.0000E+00  -9.0537E-05  -1.2863E-04   0.0000E+00   0.0000E+00
23   0.0000E+00  -7.7905E-05  -1.0963E-04   0.0000E+00   0.0000E+00
24   0.0000E+00  -9.8619E-05  -8.1719E-05   0.0000E+00   0.0000E+00
25   0.0000E+00  -9.1913E-05  -1.2890E-04   0.0000E+00   0.0000E+00
26   0.0000E+00  -1.4408E-04  -1.5914E-03   0.0000E+00   0.0000E+00
27   0.0000E+00  -4.1868E-04  -9.3352E-04   0.0000E+00   0.0000E+00
28   0.0000E+00  -4.3324E-04  -8.9129E-04   0.0000E+00   0.0000E+00
29   0.0000E+00  -4.1950E-04  -9.4913E-04   0.0000E+00   0.0000E+00
30   0.0000E+00  -4.0675E-04  -9.1267E-04   0.0000E+00   0.0000E+00
31   0.0000E+00  -1.9388E-04  -9.3082E-04   0.0000E+00   0.0000E+00
32   0.0000E+00  -6.3287E-04  -6.5455E-04   0.0000E+00   0.0000E+00
33   0.0000E+00  -6.5616E-04  -5.2474E-04   0.0000E+00   0.0000E+00
34   0.0000E+00  -6.2206E-04  -5.9268E-04   0.0000E+00   0.0000E+00
35   0.0000E+00  -7.0626E-04  -2.7383E-04   0.0000E+00   0.0000E+00
36   0.0000E+00  -4.5710E-04  -8.9865E-04   0.0000E+00   0.0000E+00
37   0.0000E+00  -5.4781E-04  -6.1997E-04   0.0000E+00   0.0000E+00
38   0.0000E+00  -3.1310E-04  -1.1147E-03   0.0000E+00   0.0000E+00
39   0.0000E+00  -2.1198E-04  -1.3669E-03   0.0000E+00   0.0000E+00
40   0.0000E+00  -2.2692E-04  -1.3697E-03   0.0000E+00   0.0000E+00


NUMBER OF ITERATIONS..............    9
NUMBER OF FUNCTION CALLS..........   64

EXIT GREGPLUS: MODEL DISCRIMINATION AND CRITICISM
```

```
SOURCE OF VARIANCE       SUM OF SQUARES   DEG. OF FREEDOM   MEAN SQUARE
Residuals                2.45993E+02            36
Lack of Fit              1.81972E+02            15            1.21E+01
Experimental             6.40210E+01            21            3.05E+00


Variance Ratio (Lack-of-Fit/Experimental)                   3.979
Sampling Probability of Greater Ratio                       0.002
Log10 of Posterior Unnormalized Probability Density        -25.707


EXIT GREGPLUS: SOLUTION FOUND. MODEL No. 4

STATISTICAL ANALYSIS

OBJECTIVE FUNCTION...................... 1.65214E+02
SUM OF SQUARES OF RESIDUALS............. 1.80525E+00
ESTIMATED PARAMETERS FROM DATA..........        8
TOTAL OBSERVATIONS DATA COUNT IS........       40

                OPTIMAL
PARAMETER      ESTIMATES      95% MARGINAL HPD INTERVALS   PARAMETER STATUS
  PAR( 1)    1.756774E+00     1.756774E+00 +- 2.631E-01    Estimated
  PAR( 2)    4.982570E-01     4.982570E-01 +- 3.300E-01    Estimated
  PAR( 3)    6.133069E-01     6.133069E-01 +- 1.253E-01    Estimated
  PAR( 4)    1.696235E-01     1.696235E-01 +- 1.389E-01    Estimated
  PAR( 5)    1.109990E+03     1.109990E+03 +- 6.776E+02    Estimated
  PAR( 6)   -3.869252E+03    -3.869252E+03 +- 2.723E+03    Estimated
  PAR( 7)   -4.749702E+01    -4.749702E+01 +- 1.015E+03    Estimated
  PAR( 8)   -4.314068E+03    -4.314068E+03 +- 3.344E+03    Estimated

NORMALIZED PARAMETER COVARIANCE MATRIX
  1.000
 -0.754  1.000
 -0.528  0.160  1.000
 -0.437  0.096  0.241  1.000
 -0.422  0.081  0.064  0.111  1.000
 -0.205  0.530  0.081 -0.040 -0.500  1.000
  0.020 -0.003  0.242  0.039 -0.457 -0.039  1.000
 -0.208  0.034  0.148  0.779 -0.089 -0.188  0.104  1.000

PARAMETER COVARIANCE MATRIX
 3.232E-03
-3.056E-03  5.083E-03
-8.125E-04  3.082E-04 7.330E-04
-7.450E-04  2.056E-04 1.955E-04  9.001E-04
-3.509E+00  8.417E-01 2.518E-01  4.862E-01  2.144E+04
-6.846E+00  2.223E+01 1.282E+00 -7.051E-01 -4.310E+04  3.462E+05
 2.497E-01 -4.014E-02 1.434E+00  2.579E-01 -1.468E+04 -4.991E+03 4.808E+04
-8.547E+00  1.747E+00 2.897E+00  1.688E+01 -9.364E+03 -8.006E+04 1.649E+04 5.220E+05

EVENT  OBSERVED     PREDICTED     RESIDUAL     DERIVATIVES (COLUMNS 1 THROUGH 3)--->
  1   -5.6465E+00  -5.7216E+00   7.5125E-02   -7.2763E+00  -7.9066E+00  -9.8271E-01
  2   -3.6889E+00  -3.6848E+00  -4.0996E-03   -5.9427E+00  -1.4850E+01  -4.2297E+00
  3   -3.4420E+00  -3.6911E+00   2.4908E-01   -6.0055E+00  -1.4712E+01  -4.2987E+00
  4   -5.2344E+00  -5.4024E+00   1.6799E-01   -5.3318E+00  -5.7635E+00  -3.5573E+00
  5   -4.7444E+00  -4.2869E+00  -4.5757E-01   -7.8366E+00  -8.7936E+00  -5.7153E+00
  6   -4.2744E+00  -4.3409E+00   6.6421E-02   -8.0260E+00  -8.8809E+00  -6.0464E+00
  7   -4.6460E+00  -4.3377E+00  -3.0833E-01   -7.8427E+00  -8.6506E+00  -5.9725E+00
  8   -5.2707E+00  -5.2994E+00   2.8696E-02   -8.6690E+00  -4.3757E+00  -6.5800E+00
```

```
 9  -3.9528E+00  -3.9017E+00  -5.1156E-02  -5.5643E+00  -5.9231E+00  -2.1454E+01
10  -3.8825E+00  -3.9895E+00   1.0704E-01  -5.6111E+00  -5.6943E+00  -2.1406E+01
11  -4.3351E+00  -4.2372E+00  -9.7988E-02  -6.6294E+00  -7.4204E+00  -9.7262E+00
12  -3.9846E+00  -3.9413E+00  -4.3336E-02  -7.2455E+00  -6.7757E+00  -1.5559E+01
13  -5.8158E+00  -5.6578E+00  -1.5804E-01  -1.0956E+01  -5.0808E+00  -1.0944E+00
14  -3.2468E+00  -3.2859E+00   3.9115E-02  -9.4940E+00  -9.6722E+00  -4.7417E+00
15  -3.1011E+00  -3.2877E+00   1.8662E-01  -9.4756E+00  -9.7431E+00  -4.7417E+00
16  -3.8825E+00  -3.9960E+00   1.1353E-01  -7.8107E+00  -3.2115E+00  -2.0700E+01
17  -3.9900E+00  -3.9767E+00  -1.3306E-02  -7.7002E+00  -3.2284E+00  -2.0708E+01
18  -4.0174E+00  -4.2026E+00   1.8524E-01  -1.1746E+01  -5.2332E+00  -6.3399E+00
19  -4.5756E+00  -4.2475E+00  -3.2816E-01  -1.1224E+01  -5.3933E+00  -5.4970E+00
20  -4.4104E+00  -4.3039E+00  -1.0651E-01  -1.1279E+01  -5.0230E+00  -6.0697E+00
21  -4.9547E+00  -5.5805E+00   6.2581E-01  -1.1997E+01  -2.4421E+00  -6.2464E+00
22  -4.7988E+00  -4.8507E+00   5.1905E-02  -1.0512E+01  -4.3954E+00  -4.6823E+00
23  -4.9144E+00  -4.7308E+00  -1.8365E-01  -9.9457E+00  -4.4199E+00  -4.6689E+00
24  -3.4451E+00  -3.5822E+00   1.3708E-01  -1.0231E+01  -8.1347E+00  -5.0601E+00
25  -3.1350E+00  -3.3925E+00   2.5746E-01  -1.0197E+01  -7.2556E+00  -7.6292E+00
26  -3.9070E+00  -3.9897E+00   8.2641E-02  -8.7641E+00  -2.4543E+00  -1.9222E+01
27  -4.4741E+00  -4.4225E+00  -5.1622E-02  -1.2826E+01  -3.6762E+00  -5.8355E+00
28  -4.3125E+00  -4.4809E+00   1.6841E-01  -1.2668E+01  -3.6754E+00  -5.3831E+00
29  -4.6607E+00  -4.4973E+00  -1.6336E-01  -1.2493E+01  -3.4689E+00  -5.5765E+00
30  -4.7433E+00  -4.4836E+00  -2.5969E-01  -1.2388E+01  -3.5032E+00  -5.5960E+00
31  -5.9915E+00  -5.8388E+00  -1.5263E-01  -1.0144E+01  -1.3214E+00  -4.5166E+00
32  -3.3873E+00  -3.3294E+00  -5.7890E-02  -1.1271E+01  -6.7631E+00  -4.9553E+00
33  -3.5543E+00  -3.4040E+00  -1.5033E-01  -1.1309E+01  -7.2612E+00  -4.1259E+00
34  -3.5166E+00  -3.3359E+00  -1.8074E-01  -1.1213E+01  -7.0055E+00  -4.7425E+00
35  -5.4262E+00  -5.8889E+00   4.6277E-01  -1.3564E+01  -3.9215E+00  -1.0835E+01
36  -4.5044E+00  -4.7508E+00   2.4642E-01  -1.2005E+01  -3.3223E+00  -4.6362E+00
37  -5.2344E+00  -4.9179E+00  -3.1649E-01  -1.2136E+01  -3.6426E+00  -2.9291E+00
38  -3.9712E+00  -4.0073E+00   3.6052E-02  -1.1311E+01  -3.6193E+00  -9.1829E+00
39  -4.2587E+00  -4.0880E+00  -1.7070E-01  -1.0641E+01  -2.8618E+00  -1.3124E+01
40  -4.1414E+00  -4.0238E+00  -1.1764E-01  -1.0724E+01  -2.9870E+00  -1.2811E+01


EVENT  DERIVATIVES (COLUMNS 4 THROUGH 8)--->
 1  -1.4809E+01   3.2943E-03   1.0149E-03   1.5552E-04   6.5209E-04
 2  -1.2261E+01   2.6905E-03   1.9061E-03   6.6937E-04   5.3992E-04
 3  -1.2391E+01   2.7189E-03   1.8884E-03   6.8029E-04   5.4562E-04
 4  -5.4256E+01   2.4139E-03   7.3978E-04   5.6296E-04   2.3892E-03
 5  -1.4285E+01   3.5480E-03   1.1287E-03   9.0447E-04   6.2904E-04
 6  -1.5961E+01   3.7602E-03   1.1796E-03   9.9016E-04   7.2731E-04
 7  -1.5597E+01   3.6743E-03   1.1490E-03   9.7806E-04   7.1070E-04
 8  -1.5281E+01   3.9249E-03   5.6164E-04   1.0413E-03   6.7291E-04
 9  -5.2522E+00   2.5629E-03   7.7346E-04   3.4541E-03   2.3529E-04
10  -6.6551E+00   2.5845E-03   7.4359E-04   3.4464E-03   2.9814E-04
11  -2.6257E+01   3.1059E-03   9.8559E-04   1.5928E-03   1.1964E-03
12  -4.4412E+00   3.2803E-03   8.6971E-04   2.4623E-03   1.9557E-04
13  -5.2235E+00  -3.4439E-04  -4.5279E-05  -1.2025E-05  -1.5970E-05
14  -3.9867E+00  -5.2198E-04  -1.5076E-04  -9.1124E-05  -2.1318E-05
15  -3.9532E+00  -4.6549E-04  -1.3569E-04  -8.1422E-05  -1.8888E-05
16  -1.9740E+00  -5.2038E-04  -6.0659E-05  -4.8205E-04  -1.2792E-05
17  -1.7255E+00  -4.6826E-04  -5.5659E-05  -4.4017E-04  -1.0205E-05
18  -4.9715E-01  -7.1428E-04  -9.0223E-05  -1.3476E-04  -2.9404E-06
19  -4.3060E+00  -5.5137E-04  -7.5113E-05  -9.4391E-05  -2.0574E-05
20  -4.6084E+00  -5.5407E-04  -6.9957E-05  -1.0423E-04  -2.2019E-05
21  -4.7034E+00  -6.5961E-04  -3.8064E-05  -1.2004E-04  -2.5151E-05
22  -2.3225E+01  -5.7792E-04  -6.8511E-05  -8.9984E-05  -1.2419E-04
23  -2.1859E+01  -4.8858E-04  -6.1557E-05  -8.0172E-05  -1.0444E-04
24  -4.3562E+00  -5.0258E-04  -1.1329E-04  -8.6890E-05  -2.0814E-05
25  -8.8738E-01  -5.6062E-04  -1.1309E-04  -1.4662E-04  -4.7451E-06
26  -8.5389E-01  -2.9141E-03  -2.3136E-04  -2.2342E-03  -2.7615E-05
```

```
27  -3.0475E-01  -4.0130E-03  -3.2609E-04  -6.3820E-04  -9.2739E-06
28  -2.2624E+00  -3.9637E-03  -3.2602E-04  -5.8872E-04  -6.8848E-05
29  -2.1880E+00  -4.0318E-03  -3.1738E-04  -6.2907E-04  -6.8678E-05
30  -2.2693E+00  -3.8760E-03  -3.1074E-04  -6.1201E-04  -6.9057E-05
31  -1.7534E+00  -3.1740E-03  -1.1722E-04  -4.9396E-04  -5.3356E-05
32  -2.4442E+00  -3.8026E-03  -6.4688E-04  -5.8436E-04  -8.0204E-05
33  -2.1446E+00  -3.6497E-03  -6.6437E-04  -4.6544E-04  -6.7316E-05
34  -2.3848E+00  -3.6186E-03  -6.4097E-04  -5.3499E-04  -7.4857E-05
35  -2.7073E+00  -4.1767E-03  -3.4234E-04  -1.1662E-04  -8.1084E-05
36  -1.1213E+01  -3.9332E-03  -3.0858E-04  -5.3093E-04  -3.5732E-04
37  -7.4235E+00  -3.9168E-03  -3.3328E-04  -3.3042E-04  -2.3302E-04
38  -3.9916E+00  -3.4829E-03  -3.1596E-04  -9.8841E-04  -1.1955E-04
39  -5.4630E-01  -3.3820E-03  -2.5786E-04  -1.4580E-03  -1.6887E-05
40  -5.7420E-01  -3.4609E-03  -2.7330E-04  -1.4452E-03  -1.8024E-05


NUMBER OF ITERATIONS..............   5
NUMBER OF FUNCTION CALLS..........  46

EXIT GREGPLUS: MODEL DISCRIMINATION AND CRITICISM
```

| SOURCE OF VARIANCE | SUM OF SQUARES | DEG. OF FREEDOM | MEAN SQUARE |
|---|---|---|---|
| Residuals | 1.65214E+02 | 32 | |
| Lack of Fit | 1.01193E+02 | 11 | 9.20E+00 |
| Experimental | 6.40210E+01 | 21 | 3.05E+00 |

```
Variance Ratio (Lack-of-Fit/Experimental)                     3.018
Sampling Probability of Greater Ratio                         0.014
Log10 of Posterior Unnormalized Probability Density         -24.494
```

## C.7 MULTIRESPONSE EXAMPLES

Simulated data on a perfectly stirred three-component reaction system are to be used to test the following models $f_i(t, \boldsymbol{\theta})$ for the species concentrations:

$$\text{Model 1}: \frac{df_1(t,\boldsymbol{\theta})}{dt} = -\exp(\theta_1)\, f_1(t,\boldsymbol{\theta})$$
$$\frac{df_2(t,\boldsymbol{\theta})}{dt} = +\exp(\theta_1)\, f_1(t,\boldsymbol{\theta}) - \exp(\theta_2)\, f_2(t,\boldsymbol{\theta})$$
$$\frac{df_3(t,\boldsymbol{\theta})}{dt} = +\exp(\theta_2)\, f_2(t,\boldsymbol{\theta})$$

$$\text{Model 2}: \frac{df_1(t,\boldsymbol{\theta})}{dt} = -\exp(\theta_1)\, f_1(t,\boldsymbol{\theta}) + \exp(\theta_3)\, f_2(t,\boldsymbol{\theta})$$
$$\frac{df_2(t,\boldsymbol{\theta})}{dt} = +\exp(\theta_1)\, f_1(t,\boldsymbol{\theta})$$
$$\qquad\qquad - \exp(\theta_2)\, f_2(t,\boldsymbol{\theta}) - \exp(\theta_3)\, f_2(t,\boldsymbol{\theta})$$
$$\frac{df_3(t,\boldsymbol{\theta})}{dt} = +\exp(\theta_2)\, f_2(t,\boldsymbol{\theta})$$

$$\text{Model 3}: \frac{df_1(t,\boldsymbol{\theta})}{dt} = -\exp(\theta_1)\ f_1(t,\boldsymbol{\theta}) - \exp(\theta_3)\ f_1(t,\boldsymbol{\theta})$$

$$\frac{df_2(t,\boldsymbol{\theta})}{dt} = -\exp(\theta_2)\ f_2(t,\boldsymbol{\theta}) + \exp(\theta_1)\ f_1(t,\boldsymbol{\theta})$$

$$\frac{df_3(t,\boldsymbol{\theta})}{dt} = +\exp(\theta_2)\ f_2(t,\boldsymbol{\theta}) + \exp(\theta_3)\ f_1(t,\boldsymbol{\theta})$$

$$\text{Model 4}: \frac{df_1(t,\boldsymbol{\theta})}{dt} = -\exp(\theta_1)\ f_1(t,\boldsymbol{\theta}) + \exp(\theta_3)\ f_2(t,\boldsymbol{\theta})$$

$$\frac{df_2(t,\boldsymbol{\theta})}{dt} = -\exp(\theta_2)\ f_2(t,\boldsymbol{\theta}) + \exp(\theta_4)\ f_3(t,\boldsymbol{\theta})$$

$$\frac{df_3(t,\boldsymbol{\theta})}{dt} = -\exp(\theta_4)\ f_3(t,\boldsymbol{\theta}) + \exp(\theta_2)\ f_2(t,\boldsymbol{\theta})$$

$$\text{Model 5}: \frac{df_1(t,\boldsymbol{\theta})}{dt} = -\exp(\theta_1)\ f_1(t,\boldsymbol{\theta})$$

$$\frac{df_2(t,\boldsymbol{\theta})}{dt} = +\exp(\theta_1)\ f_1(t,\boldsymbol{\theta}) - \exp(\theta_2)\ f_2(t,\boldsymbol{\theta})$$

$$+ \exp(\theta_3)\ f_3(t,\boldsymbol{\theta})$$

$$\frac{df_3(t,\boldsymbol{\theta})}{dt} = -\exp(\theta_3)\ f_3(t,\boldsymbol{\theta}) + \exp(\theta_2)\ f_2(t,\boldsymbol{\theta})$$

with the initial conditions

$$f_1(0,\boldsymbol{\theta}) = 1 \qquad f_2(0,\boldsymbol{\theta}) = f_3(0,\boldsymbol{\theta}) = 0$$

and uniform weighting $w(i,u) = 1$. Each weighted observation $Y_{iu}$ is represented as the corresponding expectation function plus a random variable $\mathcal{E}_{iu}$ from a multiresponse normal error distribution as in Eq. (4.4-3). Replicates included in the data permit a sample estimate of the covariance matrix of the three responses. Box and Draper (1965) applied Model 1 to a simulated data set in their pioneer paper on parameter estimation from multiresponse observations.

The following examples demonstrate estimation of the parameters of Model 1 from the data; selection of a preferred condition for an additional event $t_u$ from among a set of user-defined candidates $t_u = \{0.25, 3.0, 6.0, 12.0, 32.0\}$; and discrimination and criticism of the five models via their posterior probability shares and goodness of fit.

## Example C.4. Estimation with Numerical Derivatives

The Athena Visual Studio code for this example is shown below. The array IOBS is set to show which locations in OBS contain data. NBLK is set to 1 to put the three responses in one block, and the elements of YTOL are set to +0.01, indicating that deviations less than 0.01 are insignificant for these data. LPAR(1) and LPAR(2) are set to $-1$ to request calculation of the posterior probabilities of positive reaction rate coefficients rather than of positive PAR(1) and PAR(2). Detailed output is requested by calling GREGPLUS with LISTS = 1. Further information on the implementation of this example in Athena Visual Studio can be found by running the software and selecting *Book Examples* under the *Help* menu item and then clicking on the corresponding example.

The output from GREGPLUS is shown below. More detailed output is available by running this example in Athena Visual Studio and selecting *View Debug Output* from the *View* Menu. The following comments can be made by looking at the debug output and the report file. Responses 1 and 3 suffice as the working set in view of the mass balancing. Descent of S is obtained at the first point of each line search, and the logarithmic objective function S of Eq. (C.1-3) converges rapidly to the final value -131.058. The parameters converge well to their final values; their estimation intervals (0.95 HPD intervals or 95% confidence intervals) are somewhat broad, and could be improved by strategic addition of events. The candidate event at $t = 6.0$ gives the largest predicted normal-equation determinant $|\hat{A}|$ at the modal parameter values; experiments starting with pure component 2 would do still better, as noted in Example 7.1.

The goodness of fit obtained is very satisfactory, as indicated by the sampling probability 0.71 exceeding the value 1.81 found here.

### Athena Visual Studio Code for Example C.4

```
! Appendix C: Example C.4
! Computer-Aided Modeling of Reactive Systems
!-------------------------------------------
 Global k1,k2 As Real

@Connect Parameters
 k1=exp(Par(1))
 k2=exp(Par(2))

@Initial Conditions
 U(1)=1.0
 U(2)=0.0
 U(3)=0.0

@Model Equations
 F(1)=-k1*U(1)
 F(2)= k1*U(1)-k2*U(2)
 F(3)= k2*U(2)

@Response Model
 Y(1)=U(1)
 Y(2)=U(3)
```

### Numerical Results for Example C.4

```
Number of Experiments.......................... 12
Number of Parameters...........................  2
Number of Responses............................  2
Number of Settings.............................  1

EXIT GREGPLUS: SOLUTION FOUND. MODEL No. 1
```

STATISTICAL ANALYSIS

```
OBJECTIVE FUNCTION..................... -1.31058E+02
SUM OF SQUARES OF RESIDUALS.............  3.01545E-02
SUM OF SQUARES OF WEIGHTED RESIDUALS.....  3.01545E-02
ESTIMATED PARAMETERS FROM DATA...........         2
TOTAL OBSERVATIONS DATA COUNT IS.........        24
```

```
                  OPTIMAL
PARAMETER     ESTIMATES     95% MARGINAL HPD INTERVALS   PARAMETER STATUS
  PAR( 1)  -1.568617E+00    -1.568617E+00 +- 8.318E-02   Estimated
  PAR( 2)  -6.869812E-01    -6.869812E-01 +- 1.636E-01   Estimated
```

NORMALIZED PARAMETER COVARIANCE MATRIX
```
   1.000
  -0.691   1.000
```

PARAMETER COVARIANCE MATRIX
```
   1.8011E-03
  -2.4482E-03    6.9714E-03
```

| EVENT | OBSERVED | PREDICTED | RESIDUAL | DERIVATIVES---> | |
|---|---|---|---|---|---|
| 1 | 9.4763E-01 | 9.0108E-01 | 4.6550E-02 | -9.3801E-02 | 3.7437E-07 |
| 2 | 9.3744E-01 | 9.0108E-01 | 3.6358E-02 | -9.3801E-02 | 3.7437E-07 |
| 3 | 7.9535E-01 | 8.1194E-01 | -1.6590E-02 | -1.6906E-01 | -1.5336E-07 |
| 4 | 7.2820E-01 | 8.1194E-01 | -8.3739E-02 | -1.6906E-01 | -1.5336E-07 |
| 5 | 7.4057E-01 | 6.5924E-01 | 8.1323E-02 | -2.7457E-01 | 4.9616E-08 |
| 6 | 6.2703E-01 | 6.5924E-01 | -3.2210E-02 | -2.7457E-01 | 4.9616E-08 |
| 7 | 4.6243E-01 | 4.3460E-01 | 2.7828E-02 | -3.6212E-01 | 7.0364E-07 |
| 8 | 3.9203E-01 | 4.3460E-01 | -4.2570E-02 | -3.6212E-01 | 7.0364E-07 |
| 9 | 1.7220E-01 | 1.8888E-01 | -1.6678E-02 | -3.1495E-01 | -2.8642E-07 |
| 10 | 2.1832E-01 | 1.8888E-01 | 2.9440E-02 | -3.1495E-01 | -2.8642E-07 |
| 11 | 3.6442E-02 | 3.5676E-02 | 7.6647E-04 | -1.1911E-01 | -2.2217E-06 |
| 12 | 4.9451E-02 | 3.5676E-02 | 1.3775E-02 | -1.1911E-01 | -2.2217E-06 |
| | | | | | |
| 1 | 2.7668E-02 | 1.1643E-02 | 1.6025E-02 | 1.1262E-02 | 1.0717E-02 |
| 2 | 0.0000E+00 | 1.1643E-02 | -1.1643E-02 | 1.1262E-02 | 1.0717E-02 |
| 3 | 6.3256E-02 | 4.1557E-02 | 2.1699E-02 | 3.8694E-02 | 3.5066E-02 |
| 4 | 8.9054E-02 | 4.1557E-02 | 4.7497E-02 | 3.8694E-02 | 3.5066E-02 |
| 5 | 1.0613E-01 | 1.3322E-01 | -2.7084E-02 | 1.1469E-01 | 9.4208E-02 |
| 6 | 1.1992E-01 | 1.3322E-01 | -1.3297E-02 | 1.1469E-01 | 9.4208E-02 |
| 7 | 3.6031E-01 | 3.5271E-01 | 7.6017E-03 | 2.5526E-01 | 1.7277E-01 |
| 8 | 3.3179E-01 | 3.5271E-01 | -2.0917E-02 | 2.5526E-01 | 1.7277E-01 |
| 9 | 6.7531E-01 | 6.9026E-01 | -1.4948E-02 | 3.3112E-01 | 1.5514E-01 |
| 10 | 7.2844E-01 | 6.9026E-01 | 3.8176E-02 | 3.3112E-01 | 1.5514E-01 |
| 11 | 9.6356E-01 | 9.3934E-01 | 2.4220E-02 | 1.6044E-01 | 4.0710E-02 |
| 12 | 9.0751E-01 | 9.3934E-01 | -3.1829E-02 | 1.6044E-01 | 4.0710E-02 |

```
NUMBER OF ITERATIONS..............    4
NUMBER OF FUNCTION CALLS..........   15
```

EXIT GREGPLUS: MODEL DISCRIMINATION AND CRITICISM

```
Two-Sided Criterion of Lack-of-Fit in Response Block  1 is   1.807E+00
The Sampling Probability of a Larger Value is                    0.707
Log10 of the Posterior Unnormalized Probability Density is       5.391
```

EXIT GREGPLUS: STATISTICAL OPTIMAL DESIGN

```
FOR THE DATA SET  0 THE INVERSE HESSIAN OF ORDER   2 HAS LOG(DET) = 5.183
FOR THE DATA SET  1 THE INVERSE HESSIAN OF ORDER   2 HAS LOG(DET) = 5.184
FOR THE DATA SET  2 THE INVERSE HESSIAN OF ORDER   2 HAS LOG(DET) = 5.305
FOR THE DATA SET  3 THE INVERSE HESSIAN OF ORDER   2 HAS LOG(DET) = 5.343
FOR THE DATA SET  4 THE INVERSE HESSIAN OF ORDER   2 HAS LOG(DET) = 5.243
FOR THE DATA SET  5 THE INVERSE HESSIAN OF ORDER   2 HAS LOG(DET) = 5.183
```

OPTIMAL: THE SET  3 MAY BE SELECTED AS THE BEST AMONG THE CANDIDATES

## Example C.5. Multiresponse Estimation with Analytical Derivatives

This example uses parametric sensitivity analysis to estimate the analytic formulas for all the derivatives $\partial Y_i(t_u, \boldsymbol{\theta})/\partial \theta_j$. Therefore, the array DEL is set to zero before GREGPLUS is called, and the IDIF value is read by GREGPLUS but not used. The settings of IOBS, NBLK, YTOL, and LISTS are the same as in Example C.4.

The Athena Visual Studio code and the output with LISTS = 1 are shown below. The results are similar to those of Example C.4, except that the number of function calls for the estimation phase is reduced from 15 to 8.

### Athena Visual Studio Code for Example C.5

```
! Appendix C: Example C.5
! Computer-Aided Modeling of Reactive Systems
!-----------------------------------------------
  Global k1,k2 As Real

@Connect Parameters
  k1=exp(Par(1))
  k2=exp(Par(2))

@Initial Conditions
  U(1)=1.0
  U(2)=0.0
  U(3)=0.0

@Model Equations
  F(1)=-k1*U(1)
  F(2)= k1*U(1)-k2*U(2)
  F(3)= k2*U(2)

@Response Model
  Y(1)=U(1)
  Y(2)=U(3)

@Gradient Vector
  dY(1,1)=U(1,2)
  dY(1,2)=U(1,3)
  dY(2,1)=U(3,2)
  dY(2,2)=U(3,3)
```

### Numerical Results for Example C.5

```
Number of Experiments..........................   12
Number of Parameters...........................    2
Number of Responses............................    2
Number of Settings.............................    1

EXIT GREGPLUS: SOLUTION FOUND. MODEL No. 1

STATISTICAL ANALYSIS

OBJECTIVE FUNCTION...................... -1.31058E+02
SUM OF SQUARES OF RESIDUALS.............  3.01545E-02
SUM OF SQUARES OF WEIGHTED RESIDUALS.....  3.01545E-02
ESTIMATED PARAMETERS FROM DATA..........          2
TOTAL OBSERVATIONS DATA COUNT IS........         24

                OPTIMAL
PARAMETER     ESTIMATES     95% MARGINAL HPD INTERVALS   PARAMETER STATUS
PAR( 1)   -1.568616E+00    -1.568616E+00 +- 8.318E-02   Estimated
PAR( 2)   -6.869966E-01    -6.869966E-01 +- 1.637E-01   Estimated

NORMALIZED PARAMETER COVARIANCE MATRIX
  1.000
 -0.691   1.000
```

```
PARAMETER COVARIANCE MATRIX
   1.8012E-03
  -2.4484E-03   6.9720E-03

EVENT  OBSERVED    PREDICTED     RESIDUAL  DERIVATIVES--->
   1   9.4763E-01  9.0108E-01  4.6550E-02  -9.3800E-02   0.0000E+00
   2   9.3744E-01  9.0108E-01  3.6358E-02  -9.3800E-02   0.0000E+00
   3   7.9535E-01  8.1194E-01 -1.6590E-02  -1.6905E-01   0.0000E+00
   4   7.2820E-01  8.1194E-01 -8.3739E-02  -1.6905E-01   0.0000E+00
   5   7.4057E-01  6.5924E-01  8.1324E-02  -2.7457E-01   0.0000E+00
   6   6.2703E-01  6.5924E-01 -3.2209E-02  -2.7457E-01   0.0000E+00
   7   4.6243E-01  4.3460E-01  2.7829E-02  -3.6212E-01   0.0000E+00
   8   3.9203E-01  4.3460E-01 -4.2569E-02  -3.6212E-01   0.0000E+00
   9   1.7220E-01  1.8888E-01 -1.6677E-02  -3.1494E-01   0.0000E+00
  10   2.1832E-01  1.8888E-01  2.9441E-02  -3.1494E-01   0.0000E+00
  11   3.6442E-02  3.5675E-02  7.6662E-04  -1.1911E-01   0.0000E+00
  12   4.9451E-02  3.5675E-02  1.3776E-02  -1.1911E-01   0.0000E+00


   1   2.7668E-02  1.1642E-02  1.6026E-02   1.1238E-02   1.0690E-02
   2   0.0000E+00  1.1642E-02 -1.1642E-02   1.1238E-02   1.0690E-02
   3   6.3256E-02  4.1556E-02  2.1700E-02   3.8685E-02   3.5054E-02
   4   8.9054E-02  4.1556E-02  4.7498E-02   3.8685E-02   3.5054E-02
   5   1.0613E-01  1.3321E-01 -2.7083E-02   1.1468E-01   9.4197E-02
   6   1.1992E-01  1.3321E-01 -1.3296E-02   1.1468E-01   9.4197E-02
   7   3.6031E-01  3.5270E-01  7.6041E-03   2.5525E-01   1.7277E-01
   8   3.3179E-01  3.5270E-01 -2.0915E-02   2.5525E-01   1.7277E-01
   9   6.7531E-01  6.9026E-01 -1.4946E-02   3.3112E-01   1.5515E-01
  10   7.2844E-01  6.9026E-01  3.8178E-02   3.3112E-01   1.5515E-01
  11   9.6356E-01  9.3934E-01  2.4221E-02   1.6044E-01   4.0714E-02
  12   9.0751E-01  9.3934E-01 -3.1828E-02   1.6044E-01   4.0714E-02


NUMBER OF ITERATIONS..............   4
NUMBER OF FUNCTION CALLS..........   8

EXIT GREGPLUS: MODEL DISCRIMINATION AND CRITICISM

Two-Sided Criterion of Lack-of-Fit in Response Block  1 is   1.807E+00
The Sampling Probability of a Larger Value is                0.707
Log10 of the Posterior Unnormalized Probability Density is   5.391

EXIT GREGPLUS: STATISTICAL OPTIMAL DESIGN

FOR THE DATA SET  0 THE INVERSE HESSIAN OF ORDER   2 HAS LOG(DET) =  5.183
FOR THE DATA SET  1 THE INVERSE HESSIAN OF ORDER   2 HAS LOG(DET) =  5.184
FOR THE DATA SET  2 THE INVERSE HESSIAN OF ORDER   2 HAS LOG(DET) =  5.305
FOR THE DATA SET  3 THE INVERSE HESSIAN OF ORDER   2 HAS LOG(DET) =  5.343
FOR THE DATA SET  4 THE INVERSE HESSIAN OF ORDER   2 HAS LOG(DET) =  5.243
FOR THE DATA SET  5 THE INVERSE HESSIAN OF ORDER   2 HAS LOG(DET) =  5.183

OPTIMAL: THE SET  3 MAY BE SELECTED AS THE BEST AMONG THE CANDIDATES
```

## Example C.6. Multiresponse Discrimination and Goodness of Fit

This example compares the five reaction models given at the start of Section C.7 as to their posterior probabilities and goodness of fit to a data set generated with Model 1 and simulated random errors. The following Athena Visual Studio code sets up the problem, using vectors PAR, BNDLW, BNDUP, CHMAX, and DEL with capacity for all five models. Special values are inserted, including the vector YTOL of neglectable error ranges for the various responses and a vector KVAR indicating the available replicate event groups. GREGPLUS then analyzes and compares the models in a single call, using the same set of working responses for every model, and writes a summary table of comparisons at the end. Further information on the implementation of this example in Athena Visual Studio can be found by running the software and selecting *Book Examples* under the *Help* menu item and then clicking on the corresponding example.

Here, as in Example C.3, the subroutine MODEL contains IF blocks to access the equations and parameters for the current reaction model. DDAPLUS is called to solve the current model for its expectation values $F_i(t_u, \boldsymbol{\theta})$ and (when IDER.EQ.1) its parametric sensitivities $\partial F_i(t_u, \boldsymbol{\theta})/\partial \theta_j$.

Abbreviated numerical output for this problem is shown explixitly for each model. The summary comparison can be accessed from Athena Visual Studio by selecting *Discrimination and lack-of-Fit* from the *View* menu. Remarkably, Models 2, 4, and 5 reduce to Model 1 as the estimation proceeds. Model 3 yields an additional estimated reaction parameter, but the penalty for this addition outweighs the very modest improvement obtained in $\hat{S}$. Model 1 (which generated the data used here) is the most probable in this candidate set, and its goodness of fit probability $\Pr(\mathcal{M} > \mathcal{M}_{bj})$ of 0.71 is very satisfactory, as noted in the first example of this section.

Since Models 2, 4, and 5 have reduced to Model 1, the posterior probability should be normalized over Models 1 and 3. The resulting probability share for Model 1 is $0.211/(0.211 + 0.154) = 0.58$.

### Athena Visual Studio Code for Example C.6

```
! Appendix C: Example C.6
! Computer-Aided Modeling of Reactive Systems
!----------------------------------------------
Global k1,k2,k3,k4 As Real

If(iModel.EQ.1)Then
  CHMAX(3)=Zero
  CHMAX(4)=Zero
ElseIf(iModel.EQ.2)Then
  CHMAX(4)=Zero
ElseIf(iModel.EQ.3)Then
  CHMAX(4)=Zero
ElseIf(iModel.EQ.5)Then
  CHMAX(4)=Zero
EndIf

@Connect Parameters
k1=exp(Par(1))
k2=exp(Par(2))
k3=exp(Par(3))
k4=exp(Par(4))

@Initial Conditions
U(1)=1.0
U(2)=0.0
U(3)=0.0

@Model Equations
Dim RR(3) As Real

If(iModel.EQ.1) Then
  RR(1)=k1*U(1)
  RR(2)=k2*U(2)
  F(1)=-RR(1)
  F(2)= RR(1) - RR(2)
  F(3)= RR(2)

Else If(iModel.EQ.2) Then
  RR(1)=k1*U(1) - k3*U(2)
  RR(2)=k2*U(2)
  F(1)=-RR(1)
```

```
 F(2)= RR(1) - RR(2)
 F(3)= RR(2)

Else If(iModel.EQ.3) Then
 RR(1)=k1*U(1)
 RR(2)=k2*U(2)
 RR(3)=k3*U(1)
 F(1)=-RR(1) - RR(3)
 F(2)= RR(1) - RR(2)
 F(3)= RR(2) + RR(3)

Else If(iModel.EQ.4) Then
 RR(1)=k1*U(1) - k3*U(2)
 RR(2)=k2*U(2) - k4*U(3)
 F(1)=-RR(1)
 F(2)= RR(1) - RR(2)
 F(3)= RR(2)

Else If(iModel.EQ.5) Then
 RR(1)=k1*U(1)
 RR(2)=k2*U(2) - k3*U(3)
 F(1)=-RR(1)
 F(2)= RR(1) - RR(2)
 F(3)= RR(2)
End If

@Response Model
 Y(1)=U(1)
 Y(2)=U(3)
```

## Numerical Results for Example C.6

```
Number of Experiments..........................  12
Number of Parameters...........................   4
Number of Responses............................   2
Number of Settings.............................   1


EXIT GREGPLUS: SOLUTION FOUND. MODEL No. 1

STATISTICAL ANALYSIS


OBJECTIVE FUNCTION...................... -1.31058E+02
SUM OF SQUARES OF RESIDUALS.............  3.01545E-02
SUM OF SQUARES OF WEIGHTED RESIDUALS.....  3.01545E-02
ESTIMATED PARAMETERS FROM DATA...........       2
TOTAL OBSERVATIONS DATA COUNT IS.........      24


                OPTIMAL
PARAMETER     ESTIMATES    95% MARGINAL HPD INTERVALS   PARAMETER STATUS
 PAR( 1)   -1.568622E+00   -1.568622E+00 +- 8.320E-02   Estimated
 PAR( 2)   -6.869604E-01   -6.869604E-01 +- 1.636E-01   Estimated
 PAR( 3)   -1.500000E+00                                 Fixed
 PAR( 4)   -7.500000E-01                                 Fixed


NORMALIZED PARAMETER COVARIANCE MATRIX
   1.000
  -0.691  1.000
   0.000 -0.000  0.000
   0.000 -0.000  0.000  0.000


PARAMETER COVARIANCE MATRIX
   1.8020E-03
  -2.4481E-03   6.9658E-03
   0.0000E+00  -0.0000E+00   0.0000E+00
   0.0000E+00  -0.0000E+00   0.0000E+00   0.0000E+00
EVENT OBSERVED    PREDICTED     RESIDUAL DERIVATIVES--->
  1 9.4763E-01  9.0108E-01   4.6550E-02 -9.3826E-02  3.6904E-07  0.0000E+00  0.0000E+00
```

```
 2  9.3744E-01   9.0108E-01   3.6358E-02  -9.3826E-02   3.6904E-07   0.0000E+00   0.0000E+00
 3  7.9535E-01   8.1194E-01  -1.6591E-02  -1.6910E-01  -1.4124E-07   0.0000E+00   0.0000E+00
 4  7.2820E-01   8.1194E-01  -8.3740E-02  -1.6910E-01  -1.4124E-07   0.0000E+00   0.0000E+00
 5  7.4057E-01   6.5924E-01   8.1322E-02  -2.7462E-01   5.0117E-08   0.0000E+00   0.0000E+00
 6  6.2703E-01   6.5924E-01  -3.2211E-02  -2.7462E-01   5.0117E-08   0.0000E+00   0.0000E+00
 7  4.6243E-01   4.3460E-01   2.7826E-02  -3.6214E-01   7.0619E-07   0.0000E+00   0.0000E+00
 8  3.9203E-01   4.3460E-01  -4.2572E-02  -3.6214E-01   7.0619E-07   0.0000E+00   0.0000E+00
 9  1.7220E-01   1.8888E-01  -1.6679E-02  -3.1488E-01  -2.7450E-07   0.0000E+00   0.0000E+00
10  2.1832E-01   1.8888E-01   2.9439E-02  -3.1488E-01  -2.7450E-07   0.0000E+00   0.0000E+00
11  3.6442E-02   3.5676E-02   7.6583E-04  -1.1903E-01  -2.1949E-06   0.0000E+00   0.0000E+00
12  4.9451E-02   3.5676E-02   1.3775E-02  -1.1903E-01  -2.1949E-06   0.0000E+00   0.0000E+00

 1  2.7668E-02   1.1643E-02   1.6025E-02   1.1258E-02   1.0714E-02   0.0000E+00   0.0000E+00
 2  0.0000E+00   1.1643E-02  -1.1643E-02   1.1258E-02   1.0714E-02   0.0000E+00   0.0000E+00
 3  6.3256E-02   4.1557E-02   2.1699E-02   3.8682E-02   3.5060E-02   0.0000E+00   0.0000E+00
 4  8.9054E-02   4.1557E-02   4.7497E-02   3.8682E-02   3.5060E-02   0.0000E+00   0.0000E+00
 5  1.0613E-01   1.3322E-01  -2.7086E-02   1.1466E-01   9.4209E-02   0.0000E+00   0.0000E+00
 6  1.1992E-01   1.3322E-01  -1.3299E-02   1.1466E-01   9.4209E-02   0.0000E+00   0.0000E+00
 7  3.6031E-01   3.5271E-01   7.5995E-03   2.5521E-01   1.7283E-01   0.0000E+00   0.0000E+00
 8  3.3179E-01   3.5271E-01  -2.0920E-02   2.5521E-01   1.7283E-01   0.0000E+00   0.0000E+00
 9  6.7531E-01   6.9026E-01  -1.4950E-02   3.3107E-01   1.5527E-01   0.0000E+00   0.0000E+00
10  7.2844E-01   6.9026E-01   3.8174E-02   3.3107E-01   1.5527E-01   0.0000E+00   0.0000E+00
11  9.6356E-01   9.3934E-01   2.4220E-02   1.6038E-01   4.0754E-02   0.0000E+00   0.0000E+00
12  9.0751E-01   9.3934E-01  -3.1829E-02   1.6038E-01   4.0754E-02   0.0000E+00   0.0000E+00
NUMBER OF ITERATIONS..............   4
NUMBER OF FUNCTION CALLS..........  15
```

EXIT GREGPLUS: MODEL DISCRIMINATION AND CRITICISM

```
Two-Sided Criterion of Lack-of-Fit in Response Block  1 is   1.807E+00
The Sampling Probability of a Larger Value is                  0.707
Log10 of the Posterior Unnormalized Probability Density is     5.391
```

EXIT GREGPLUS: SOLUTION FOUND. MODEL No. 2

STATISTICAL ANALYSIS

```
OBJECTIVE FUNCTION....................... -1.31058E+02
SUM OF SQUARES OF RESIDUALS..............  3.01544E-02
SUM OF SQUARES OF WEIGHTED RESIDUALS.....  3.01544E-02
ESTIMATED PARAMETERS FROM DATA...........          2
TOTAL OBSERVATIONS DATA COUNT IS.........         24
```

```
                OPTIMAL
PARAMETER     ESTIMATES     95% MARGINAL HPD INTERVALS    PARAMETER STATUS
 PAR( 1)    -1.568631E+00    -1.568631E+00 +- 8.323E-02   Estimated
 PAR( 2)    -6.869293E-01    -6.869293E-01 +- 1.635E-01   Estimated
 PAR( 3)    -2.000000E+01                                 Lower Bound
 PAR( 4)    -7.500000E-01                                 Fixed
```

NORMALIZED PARAMETER COVARIANCE MATRIX
```
  1.000
 -0.691  1.000
  0.000 -0.000  0.000
  0.000 -0.000  0.000  0.000
```

PARAMETER COVARIANCE MATRIX
```
  1.8034E-03
 -2.4480E-03    6.9579E-03
```

```
    0.0000E+00  -0.0000E+00   0.0000E+00
    0.0000E+00  -0.0000E+00   0.0000E+00   0.0000E+00
```

```
EVENT OBSERVED   PREDICTED     RESIDUAL  DERIVATIVES--->
   1  9.4763E-01  9.0108E-01  4.6549E-02 -9.3861E-02  3.6773E-07  5.5190E-11  0.0000E+00
   2  9.3744E-01  9.0108E-01  3.6357E-02 -9.3861E-02  3.6773E-07  5.5190E-11  0.0000E+00
   3  7.9535E-01  8.1194E-01 -1.6592E-02 -1.6915E-01 -1.5436E-07  1.8993E-10  0.0000E+00
   4  7.2820E-01  8.1194E-01 -8.3741E-02 -1.6915E-01 -1.5436E-07  1.8993E-10  0.0000E+00
   5  7.4057E-01  6.5925E-01  8.1319E-02 -2.7468E-01  4.9939E-08  5.6307E-10  0.0000E+00
   6  6.2703E-01  6.5925E-01 -3.2214E-02 -2.7468E-01  4.9939E-08  5.6307E-10  0.0000E+00
   7  4.6243E-01  4.3460E-01  2.7823E-02 -3.6217E-01  7.1050E-07  1.2535E-09  0.0000E+00
   8  3.9203E-01  4.3460E-01 -4.2575E-02 -3.6217E-01  7.1050E-07  1.2535E-09  0.0000E+00
   9  1.7220E-01  1.8888E-01 -1.6682E-02 -3.1480E-01 -2.7921E-07  1.6263E-09  0.0000E+00
  10  2.1832E-01  1.8888E-01  2.9436E-02 -3.1480E-01 -2.7921E-07  1.6263E-09  0.0000E+00
  11  3.6442E-02  3.5677E-02  7.6481E-04 -1.1892E-01 -2.1602E-06  7.8764E-10  0.0000E+00
  12  4.9451E-02  3.5677E-02  1.3774E-02 -1.1892E-01 -2.1602E-06  7.8764E-10  0.0000E+00
```

```
   1  2.7668E-02  1.1643E-02  1.6025E-02  1.1252E-02  1.0709E-02 -4.4407E-12  0.0000E+00
   2  0.0000E+00  1.1643E-02 -1.1643E-02  1.1252E-02  1.0709E-02 -4.4407E-12  0.0000E+00
   3  6.3256E-02  4.1558E-02  2.1698E-02  3.8662E-02  3.5049E-02 -3.0264E-11  0.0000E+00
   4  8.9054E-02  4.1558E-02  4.7496E-02  3.8662E-02  3.5049E-02 -3.0264E-11  0.0000E+00
   5  1.0613E-01  1.3322E-01 -2.7088E-02  1.1461E-01  9.4207E-02 -1.7105E-10  0.0000E+00
   6  1.1992E-01  1.3322E-01 -1.3301E-02  1.1461E-01  9.4207E-02 -1.7105E-10  0.0000E+00
   7  3.6031E-01  3.5271E-01  7.5963E-03  2.5514E-01  1.7292E-01 -6.8941E-10  0.0000E+00
   8  3.3179E-01  3.5271E-01 -2.0923E-02  2.5514E-01  1.7292E-01 -6.8941E-10  0.0000E+00
   9  6.7531E-01  6.9026E-01 -1.4952E-02  3.3100E-01  1.5545E-01 -1.4720E-09  0.0000E+00
  10  7.2844E-01  6.9026E-01  3.8172E-02  3.3100E-01  1.5545E-01 -1.4720E-09  0.0000E+00
  11  9.6356E-01  9.3934E-01  2.4220E-02  1.6031E-01  4.0821E-02 -1.0020E-09  0.0000E+00
  12  9.0751E-01  9.3934E-01 -3.1829E-02  1.6031E-01  4.0821E-02 -1.0020E-09  0.0000E+00
```

```
NUMBER OF ITERATIONS..............    8
NUMBER OF FUNCTION CALLS..........   36
```

EXIT GREGPLUS: MODEL DISCRIMINATION AND CRITICISM

```
Two-Sided Criterion of Lack-of-Fit in Response Block  1 is   1.807E+00
The Sampling Probability of a Larger Value is                0.707
Log10 of the Posterior Unnormalized Probability Density is   5.391
```

EXIT GREGPLUS: SOLUTION FOUND. MODEL No. 3

STATISTICAL ANALYSIS

```
OBJECTIVE FUNCTION.......................  -1.31213E+02
SUM OF SQUARES OF RESIDUALS.............   3.01035E-02
SUM OF SQUARES OF WEIGHTED RESIDUALS.....  3.01035E-02
ESTIMATED PARAMETERS FROM DATA..........          3
TOTAL OBSERVATIONS DATA COUNT IS........         24
```

```
              OPTIMAL
PARAMETER    ESTIMATES     95% MARGINAL HPD INTERVALS   PARAMETER STATUS
 PAR( 1)  -1.598672E+00   -1.598672E+00 +- 1.771E-01    Estimated
 PAR( 2)  -7.334353E-01   -7.334353E-01 +- 3.036E-01    Estimated
 PAR( 3)  -4.962249E+00   -4.962249E+00 +- 5.166E+00    Estimated
 PAR( 4)  -7.500000E-01                                 Fixed
```

```
NORMALIZED PARAMETER COVARIANCE MATRIX
  1.000
  0.551  1.000
```

```
    -0.871 -0.842  1.000
     0.000  0.000 -0.000  0.000

PARAMETER COVARIANCE MATRIX
    8.1629E-03
    7.7063E-03   2.3986E-02
   -2.0734E-01  -3.4377E-01   6.9471E+00
    0.0000E+00   0.0000E+00  -0.0000E+00   0.0000E+00

EVENT OBSERVED    PREDICTED    RESIDUAL  DERIVATIVES--->
  1  9.4763E-01   9.0071E-01   4.6914E-02 -9.1046E-02  1.4402E-07 -3.1511E-03  0.0000E+00
  2  9.3744E-01   9.0071E-01   3.6722E-02 -9.1046E-02  1.4402E-07 -3.1511E-03  0.0000E+00
  3  7.9535E-01   8.1128E-01  -1.5934E-02 -1.6401E-01  2.2504E-08 -5.6765E-03  0.0000E+00
  4  7.2820E-01   8.1128E-01  -8.3083E-02 -1.6401E-01  2.2504E-08 -5.6765E-03  0.0000E+00
  5  7.4057E-01   6.5818E-01   8.2388E-02 -2.6611E-01  8.1013E-08 -9.2103E-03  0.0000E+00
  6  6.2703E-01   6.5818E-01  -3.1145E-02 -2.6611E-01  8.1013E-08 -9.2103E-03  0.0000E+00
  7  4.6243E-01   4.3320E-01   2.9232E-02 -3.5028E-01  7.4037E-07 -1.2124E-02  0.0000E+00
  8  3.9203E-01   4.3320E-01  -4.1166E-02 -3.5028E-01  7.4037E-07 -1.2124E-02  0.0000E+00
  9  1.7220E-01   1.8766E-01  -1.5460E-02 -3.0345E-01 -1.7665E-07 -1.0503E-02  0.0000E+00
 10  2.1832E-01   1.8766E-01   3.0658E-02 -3.0345E-01 -1.7665E-07 -1.0503E-02  0.0000E+00
 11  3.6442E-02   3.5217E-02   1.2249E-03 -1.1387E-01 -9.4262E-07 -3.9411E-03  0.0000E+00
 12  4.9451E-02   3.5217E-02   1.4234E-02 -1.1387E-01 -9.4262E-07 -3.9411E-03  0.0000E+00


  1  2.7668E-02   1.4130E-02   1.3538E-02  1.0303E-02  9.9905E-03  3.3034E-03  0.0000E+00
  2  0.0000E+00   1.4130E-02  -1.4130E-02  1.0303E-02  9.9905E-03  3.3034E-03  0.0000E+00
  3  6.3256E-02   4.5050E-02   1.8206E-02  3.5530E-02  3.2944E-02  6.2018E-03  0.0000E+00
  4  8.9054E-02   4.5050E-02   4.4004E-02  3.5530E-02  3.2944E-02  6.2018E-03  0.0000E+00
  5  1.0613E-01   1.3640E-01  -3.0263E-02  1.0595E-01  8.9775E-02  1.0777E-02  0.0000E+00
  6  1.1992E-01   1.3640E-01  -1.6476E-02  1.0595E-01  8.9775E-02  1.0777E-02  0.0000E+00
  7  3.6031E-01   3.5300E-01   7.3125E-03  2.3825E-01  1.6896E-01  1.5646E-02  0.0000E+00
  8  3.3179E-01   3.5300E-01  -2.1207E-02  2.3825E-01  1.6896E-01  1.5646E-02  0.0000E+00
  9  6.7531E-01   6.8842E-01  -1.3109E-02  3.1341E-01  1.5808E-01  1.5137E-02  0.0000E+00
 10  7.2844E-01   6.8842E-01   4.0015E-02  3.1341E-01  1.5808E-01  1.5137E-02  0.0000E+00
 11  9.6356E-01   9.3887E-01   2.4685E-02  1.5356E-01  4.3255E-02  6.2115E-03  0.0000E+00
 12  9.0751E-01   9.3887E-01  -3.1364E-02  1.5356E-01  4.3255E-02  6.2115E-03  0.0000E+00


NUMBER OF ITERATIONS.............     7
NUMBER OF FUNCTION CALLS..........   35


EXIT GREGPLUS: MODEL DISCRIMINATION AND CRITICISM

Two-Sided Criterion of Lack-of-Fit in Response Block  1 is   1.693E+00
The Sampling Probability of a Larger Value is                    0.751
Log10 of the Posterior Unnormalized Probability Density is       5.247


EXIT GREGPLUS: SOLUTION FOUND. MODEL No. 4

STATISTICAL ANALYSIS

OBJECTIVE FUNCTION...................... -1.31058E+02
SUM OF SQUARES OF RESIDUALS.............  3.01544E-02
SUM OF SQUARES OF WEIGHTED RESIDUALS.....  3.01544E-02
ESTIMATED PARAMETERS FROM DATA...........          2
TOTAL OBSERVATIONS DATA COUNT IS.........         24

                OPTIMAL
PARAMETER   ESTIMATES     95% MARGINAL HPD INTERVALS   PARAMETER STATUS
 PAR( 1)  -1.568631E+00  -1.568631E+00 +- 8.323E-02   Estimated
 PAR( 2)  -6.869292E-01  -6.869292E-01 +- 1.635E-01   Estimated
```

```
     PAR( 3)  -2.000000E+01                            Lower Bound
     PAR( 4)  -2.000000E+01                            Lower Bound


     NORMALIZED PARAMETER COVARIANCE MATRIX
       1.000
      -0.691  1.000
       0.000 -0.000  0.000
       0.000  0.000  0.000  0.000


     PARAMETER COVARIANCE MATRIX
       1.8034E-03
      -2.4480E-03    6.9579E-03
       0.0000E+00  -0.0000E+00   0.0000E+00
       0.0000E+00   0.0000E+00   0.0000E+00   0.0000E+00


     EVENT OBSERVED    PREDICTED    RESIDUAL  DERIVATIVES--->
      1  9.4763E-01  9.0108E-01  4.6549E-02 -9.3861E-02  3.8146E-07  1.2043E-10  8.7605E-15
      2  9.3744E-01  9.0108E-01  3.6357E-02 -9.3861E-02  3.8146E-07  1.2043E-10  8.7605E-15
      3  7.9535E-01  8.1194E-01 -1.6592E-02 -1.6915E-01 -1.3170E-07  4.1444E-10  7.0698E-15
      4  7.2820E-01  8.1194E-01 -8.3741E-02 -1.6915E-01 -1.3170E-07  4.1444E-10  7.0698E-15
      5  7.4057E-01  6.5925E-01  8.1319E-02 -2.7468E-01  5.4495E-08  1.2287E-09  5.9940E-15
      6  6.2703E-01  6.5925E-01 -3.2214E-02 -2.7468E-01  5.4495E-08  1.2287E-09  5.9940E-15
      7  4.6243E-01  4.3460E-01  2.7823E-02 -3.6217E-01  7.0389E-07  2.7353E-09  6.0708E-15
      8  3.9203E-01  4.3460E-01 -4.2575E-02 -3.6217E-01  7.0389E-07  2.7353E-09  6.0708E-15
      9  1.7220E-01  1.8888E-01 -1.6682E-02 -3.1480E-01 -2.7475E-07  3.5488E-09  1.9212E-16
     10  2.1832E-01  1.8888E-01  2.9436E-02 -3.1480E-01 -2.7475E-07  3.5488E-09  1.9212E-16
     11  3.6442E-02  3.5677E-02  7.6481E-04 -1.1892E-01 -2.1577E-06  1.7188E-09 -8.3090E-15
     12  4.9451E-02  3.5677E-02  1.3774E-02 -1.1892E-01 -2.1577E-06  1.7188E-09 -8.3090E-15


      1  2.7668E-02  1.1643E-02  1.6025E-02  1.1252E-02  1.0709E-02 -9.6886E-12 -5.6882E-12
      2  0.0000E+00  1.1643E-02 -1.1643E-02  1.1252E-02  1.0709E-02 -9.6886E-12 -5.6882E-12
      3  6.3256E-02  4.1558E-02  2.1698E-02  3.8662E-02  3.5049E-02 -6.6036E-11 -3.9138E-11
      4  8.9054E-02  4.1558E-02  4.7496E-02  3.8662E-02  3.5049E-02 -6.6036E-11 -3.9138E-11
      5  1.0613E-01  1.3322E-01 -2.7088E-02  1.1461E-01  9.4207E-02 -3.7326E-10 -2.3441E-10
      6  1.1992E-01  1.3322E-01 -1.3301E-02  1.1461E-01  9.4207E-02 -3.7326E-10 -2.3441E-10
      7  3.6031E-01  3.5271E-01  7.5963E-03  2.5514E-01  1.7292E-01 -1.5044E-09 -1.0802E-09
      8  3.3179E-01  3.5271E-01 -2.0923E-02  2.5514E-01  1.7292E-01 -1.5044E-09 -1.0802E-09
      9  6.7531E-01  6.9026E-01 -1.4952E-02  3.3100E-01  1.5545E-01 -3.2122E-09 -3.2129E-09
     10  7.2844E-01  6.9026E-01  3.8172E-02  3.3100E-01  1.5545E-01 -3.2122E-09 -3.2129E-09
     11  9.6356E-01  9.3934E-01  2.4220E-02  1.6031E-01  4.0822E-02 -2.1864E-09 -5.3979E-09
     12  9.0751E-01  9.3934E-01 -3.1829E-02  1.6031E-01  4.0822E-02 -2.1864E-09 -5.3979E-09


     NUMBER OF ITERATIONS..............    11
     NUMBER OF FUNCTION CALLS..........    60


     EXIT GREGPLUS: MODEL DISCRIMINATION AND CRITICISM


     Two-Sided Criterion of Lack-of-Fit in Response Block  1 is    1.807E+00
     The Sampling Probability of a Larger Value is                    0.707
     Log10 of the Posterior Unnormalized Probability Density is       5.391



     EXIT GREGPLUS: SOLUTION FOUND. MODEL No. 5


     STATISTICAL ANALYSIS


     OBJECTIVE FUNCTION...................... -1.31058E+02
     SUM OF SQUARES OF RESIDUALS.............  3.01544E-02
     SUM OF SQUARES OF WEIGHTED RESIDUALS.....  3.01544E-02
     ESTIMATED PARAMETERS FROM DATA..........          2
```

```
TOTAL OBSERVATIONS DATA COUNT IS........        24

                OPTIMAL
PARAMETER     ESTIMATES    95% MARGINAL HPD INTERVALS   PARAMETER STATUS
 PAR( 1)  -1.568631E+00    -1.568631E+00 +- 8.323E-02   Estimated
 PAR( 2)  -6.869301E-01    -6.869301E-01 +- 1.635E-01   Estimated
 PAR( 3)  -2.000000E+01                                 Lower Bound
 PAR( 4)  -7.500000E-01                                 Fixed

NORMALIZED PARAMETER COVARIANCE MATRIX
   1.000
  -0.691  1.000
   0.000  0.000  0.000
   0.000 -0.000  0.000  0.000

PARAMETER COVARIANCE MATRIX
   1.8033E-03
  -2.4480E-03   6.9581E-03
   0.0000E+00   0.0000E+00   0.0000E+00
   0.0000E+00  -0.0000E+00   0.0000E+00   0.0000E+00

EVENT OBSERVED   PREDICTED      RESIDUAL   DERIVATIVES--->
 1  9.4763E-01  9.0108E-01   4.6549E-02 -9.3859E-02  3.5856E-07  0.0000E+00  0.0000E+00
 2  9.3744E-01  9.0108E-01   3.6357E-02 -9.3859E-02  3.5856E-07  0.0000E+00  0.0000E+00
 3  7.9535E-01  8.1194E-01  -1.6592E-02 -1.6915E-01 -1.5432E-07  0.0000E+00  0.0000E+00
 4  7.2820E-01  8.1194E-01  -8.3741E-02 -1.6915E-01 -1.5432E-07  0.0000E+00  0.0000E+00
 5  7.4057E-01  6.5925E-01   8.1319E-02 -2.7468E-01  5.4465E-08 -1.7914E-13  0.0000E+00
 6  6.2703E-01  6.5925E-01  -3.2214E-02 -2.7468E-01  5.4465E-08 -1.7914E-13  0.0000E+00
 7  4.6243E-01  4.3460E-01   2.7823E-02 -3.6216E-01  7.1485E-07 -8.9569E-14  0.0000E+00
 8  3.9203E-01  4.3460E-01  -4.2575E-02 -3.6216E-01  7.1485E-07 -8.9569E-14  0.0000E+00
 9  1.7220E-01  1.8888E-01  -1.6682E-02 -3.1480E-01 -2.8027E-07 -2.4631E-13  0.0000E+00
10  2.1832E-01  1.8888E-01   2.9436E-02 -3.1480E-01 -2.8027E-07 -2.4631E-13  0.0000E+00
11  3.6442E-02  3.5677E-02   7.6481E-04 -1.1892E-01 -2.1585E-06 -2.0153E-13  0.0000E+00
12  4.9451E-02  3.5677E-02   1.3774E-02 -1.1892E-01 -2.1585E-06 -2.0153E-13  0.0000E+00


 1  2.7668E-02  1.1643E-02   1.6025E-02  1.1252E-02  1.0709E-02 -8.3138E-11  0.0000E+00
 2  0.0000E+00  1.1643E-02  -1.1643E-02  1.1252E-02  1.0709E-02 -8.3138E-11  0.0000E+00
 3  6.3256E-02  4.1558E-02   2.1698E-02  3.8663E-02  3.5049E-02 -5.7180E-10  0.0000E+00
 4  8.9054E-02  4.1558E-02   4.7496E-02  3.8663E-02  3.5049E-02 -5.7180E-10  0.0000E+00
 5  1.0613E-01  1.3322E-01  -2.7088E-02  1.1461E-01  9.4206E-02 -3.4256E-09  0.0000E+00
 6  1.1992E-01  1.3322E-01  -1.3301E-02  1.1461E-01  9.4206E-02 -3.4256E-09  0.0000E+00
 7  3.6031E-01  3.5271E-01   7.5964E-03  2.5514E-01  1.7291E-01 -1.5789E-08  0.0000E+00
 8  3.3179E-01  3.5271E-01  -2.0923E-02  2.5514E-01  1.7291E-01 -1.5789E-08  0.0000E+00
 9  6.7531E-01  6.9026E-01  -1.4951E-02  3.3100E-01  1.5545E-01 -4.6964E-08  0.0000E+00
10  7.2844E-01  6.9026E-01   3.8173E-02  3.3100E-01  1.5545E-01 -4.6964E-08  0.0000E+00
11  9.6356E-01  9.3934E-01   2.4220E-02  1.6031E-01  4.0821E-02 -7.8903E-08  0.0000E+00
12  9.0751E-01  9.3934E-01  -3.1829E-02  1.6031E-01  4.0821E-02 -7.8903E-08  0.0000E+00

NUMBER OF ITERATIONS..............    9
NUMBER OF FUNCTION CALLS..........   49

EXIT GREGPLUS: MODEL DISCRIMINATION AND CRITICISM

Two-Sided Criterion of Lack-of-Fit in Response Block  1 is   1.807E+00
The Sampling Probability of a Larger Value is                0.707
Log10 of the Posterior Unnormalized Probability Density is   5.391
```

## REFERENCES and FURTHER READING

Bard, Y., *Nonlinear Parameter Estimation*, Academic Press, New York (1974).

Blakemore, J. W., and A. W. Hoerl, Fitting non-linear reaction rate equations to data, *Chem. Eng. Prog. Symp. Ser.*, **59**, 14–27 (1963).

Box, G. E. P., A general distribution theory for a class of likelihood criteria, *Biometrika*, **36**, 317–346 (1949).

Box, G. E. P., and N. R. Draper, The Bayesian estimation of common parameters from several responses, *Biometrika*, **52**, 355–365 (1965).

Box, M. J., and N. R. Draper, Estimation and design criteria for multiresponse non-linear models with non-homogeneous variance, *J. Roy. Statist. Soc., Ser. C, (Appl. Statist.)*, **21**, 13–24 (1972).

Box, G. E. P., and W. J. Hill, Discrimination among mechanistic models, *Technometrics*, **9**, 57–71 (1967).

Box, G. E. P., W. G. Hunter, J. F. MacGregor, and J. Erjavec, Some problems associated with the analysis of multiresponse data, *Technometrics*, **15**, 33–51 (1973).

Bradshaw, R. W., and B. Davidson, A new approach to the analysis of heterogeneous reaction rate data, *Chem. Eng. Sci.*, **24**, 1519–1527 (1970).

Franckaerts, J., and G. F. Froment, Kinetic study of the dehydration of ethanol, *Chem. Eng. Sci.*, **19**, 807–818 (1964).

Hill, W. J., Statistical techniques for model building, Ph. D. thesis, University of Wisconsin–Madison (1966).

Hill, W. J., and W. G. Hunter, A note on designs for model discrimination: Variance unknown case, *Technometrics*, **11**, 396–400 (1969).

Hill, W. J., W. G. Hunter, and D. W. Wichern, A joint criterion for the dual problem of model discrimination and parameter estimation, *Technometrics*, **10**, 145–160 (1968).

Press, W. H., S. A. Teukolsky, W. T. Vetterling, and B. P. Flannery, *Numerical Recipes in FORTRAN*, 2nd edition, Cambridge University Press (1992).

Reilly, P. M., Statistical methods in model discrimination, *Can. J. Chem. Eng.*, **48**, 168–173 (1970).

Stewart, W. E., M. Caracotsios, and J. P. Sørensen, Parameter estimation from multiresponse data, *AIChE J.*, **38**, 641–650 (1992); Errata, **38**, 1302 (1992).

Stewart, W. E., T. L. Henson, and G. E. P. Box, Model discrimination and criticism with single-response data, *AIChE J.*, **42**, 3055–3062 (1996).

Stewart, W. E., Y. Shon, and G. E. P. Box, Discrimination and goodness of fit of multiresponse mechanistic models, *AIChE J.*, **44**, 1404–1412 (1998).

Tschernitz, J., S. Bornstein, R. B. Beckmann, and O. A. Hougen, Determination of the kinetics mechanism of a catalytic reaction, *Trans. Am. Inst. Chem. Engrs.*, **42**, 883–903 (1946).

# Author Index

Abrams, D. S., 133
Agarwal, A. K., 133
Agin, G. L., 174
Aitchison, J., 92
Aitken, A. C., 172
Aluko, M., 29, 31
Amiridis, M. D., 32
Amundson, N. R., 37, 62
Anderson, D. F., 133
Anderson, E., 62, 133, 188, 216
Anderson, T. F., 133
Anderson, T. W., 172
Aris, R., 29, 30, 32, 33, 62, 64
Arrhenius, S., 30, 133
Asbjornsen, O. A., 136
Ashmore, P. G., 30
Astarita, G., 30
Athalye, A. M., 133
Atkinson, A. C., 133, 135
Avery, N. R., 30
Aziz, R. A., 63

Bacon, D. W., 138, 139, 172, 174
Bader, G., 31
Bai, Z., 62, 133, 188, 216
Bain, R. S., 172
Bajramovic, R., 133
Balse, V. R, 31
Baltanas, M. A., 30
Bard, Y., 133, 172, 257
Bates, D. M., 134, 172
Bayes, T. R., 76, 92, 172
Bazaire, K., 64
Beall, C. E., 31
Beckmann, R. B., 139, 257
Bell, N. H., 137

Benson, S. W., 30
Berger, J. O., 92
Bernstein, R. B., 34
Berry, D. A., 92
Biegler, L. T., 139, 172, 216
Bird, R. B., 62, 216
Bischof, C., 62, 133, 188, 216
Bischoff, K. B., 32, 62, 136
Blakemore, J. E., 30
Blakemore, J. W., 134, 257
Blau, G. E., 172, 174, 216
Boag, I. F., 172
Bock, H. G., 134
Bodenstein, M., 30
Booker, A., 136
Booth, G. W., 134
Bornstein, S., 139, 257
Boudart, M., 30, 32
Bowen, J. R., 30
Box, G. E. P., 30, 76, 92, 134,
    139, 172, 174, 257
Box, M. J., 134, 172, 257
Bradshaw, R. W., 30, 134, 257
Brenan, K. E., 216
Briggs, G. E., 30
Brisk, M. L., 133
Broadbelt, L. J., 31
Brønsted, J. N., 31
Brown, L. F., 62
Bunch, J. R., 135, 173, 188
Bunting, P. S., 34
Burke, A. L., 134
Butt, J. B., 31, 174
Buzzi-Ferraris, G., 135, 173

Cadle, P. J., 63

Campbell, E. H., 31
Campbell, S. L., 216
Canu, P., 135, 173
Capsakis, S. C., 31
Caracotsios, M., 135, 173, 174, 216
Carberry, J. J., 31
Cardona-Martinez, N., 32
Chambers, J. M., 135
Chang, H.-C., 29, 31
Chou, M. Y., 31
Chow, G. C., 173
Christiansen, J. A., 31
Clarke, G. P. Y., 135
Cleveland, W. S., 135
Conway, B. E., 31
Cooper, S. L., 137
Corcoran, W. H., 30
Cornfield, J., 92
Coutie, G. A., 134
Cullinan, H. T., 62
Curtiss, C. F., 31, 62
Cutlip, M. B., 31

Dalal, V. P., 36
Damiano, J. J., 172, 216
Daniel, C., 135
Daniel, J. W., 188
Dartt, S. R., 37
Davidson, B., 30, 134, 257
Davies, C. W., 31
de Groot, S. R., 31
De Moivre, A., 76
Debye, P., 31
Delahay, P., 31
Deming, W. E., 135
Demmel, J., 62, 133, 188, 216
Dennis, J. E., Jr., 135
Deuflhard, P., 31, 137
Djéga-Mariadassou, G., 30
Donaldson, J. R., 135
Dongarra, J. J., 62, 133, 135, 173, 188, 216
Douglas, J. M., 137
Downie, J., 138, 139, 172, 174

Draper, N. R., 134, 135, 137, 139, 172, 173, 257
Du Croz, J., 62, 133, 188, 216
Duever, T. A., 134
Dullien, F. A. L., 63
Dumesic, J. A., 31, 32
Dumez, F. J., 135
Dunsmore, I. R., 92

Eckert, C. A., 32
Edelson, D., 32
Edgar, T. F., 137
Efroymson, M. A., 135
Eigen, M., 32
Eisenhart, C., 76
Ekerdt, J. G., 63
Erjavec, J., 34, 137, 172, 257
Espie, D. M., 135
Etzel, M. R., 216
Evans III, R. B., 62, 63, 174
Evans, M. G., 32
Eyring, H., 32, 37

Fariss, R. H., 135
Farrow, L. A., 32
Fedorov, V. V., 133, 135
Feinberg, M., 32
Feller, W., 76
Feng, C. A., 35
Feng, C. F., 62, 135, 173
Fisher, R. A., 92, 135, 173
Fjeld, M., 32
Flannery, B. P., 138, 188, 257
Forzatti, P., 135, 173
Franckaerts, J., 32, 257
Frane, J. W., 135
Frank-Kamenetskii, D. A., 32
Fredrickson, A. G., 32
Fristrom, R. H., 31
Froment, G. F., 30, 32, 62, 135, 136, 257
Fuguitt, R. E., 173

Gamson, B. W., 62
Garrett, A. B., 137
Gates, B. C., 32

Gauss, C. F., 76, 92, 136, 173
Gavalas, G. R., 30
Gay, D. M., 135
Geankoplis, C. J., 64
Gear, C. W., 216
Geisser, S., 92
Georgakis, C., 32
Gibbs, J. W., 32
Gibbs, S. J., 133
Goddard, S. A., 32
Gola-Galimidi, A. M., 63
Golden, D. M., 33
Golub, G. H., 188
Gorman, J. W., 175
Gossett, W. S. ("Student"), 136
Gotoh, S., 64
Graham, W. R. C., 33
Greeley, J., 33
Green, D. W., 63
Greenbaum, A., 62, 216
Grens, E. A., II, 133
Greppi, L. A, 31
Grieger, R. A., 32, 36, 137
Gring, J. L., 173
Guay, M., 173
Guertin, E. W., 63
Gunn, R. D., 62
Guttman, I., 136

Haldane, J. B. S., 30
Halsey, G. D., Jr., 33
Hammarling, S., 62, 133, 188, 216
Hanson, R. J., 137
Happel, J., 33
Harold, M. P., 33
Hartley, H. O., 136
Hase, W. L., 33
Hawkins, J. E., 173
Haynes, H. W., 62
Heinemann, R. F., 33
Henri, V., 33
Henson, T. L., 134, 137, 139, 257
Herschbach, D. R., 33
Hertzberg, T., 136
Herzfeld, K. F., 33

Hettinger, W. P., Jr., 173
Hill, C. G., 33
Hill, W. J., 134, 136, 137, 172, 257
Himmelblau, D. M., 136
Hinshelwood, C. N., 33
Hirschfelder, J. O., 31, 62
Ho, T. C., 31, 33
Hoerl, A. W., 134, 257
Hoffman, R., 37
Hofmann, H., 35
Holbrook, K. A., 35
Hooke, R., 173
Horak, J., 62
Horiuti, J., 33
Hosten, L. H., 135, 136
Hougen, O. A., 2, 33, 37, 62, 64, 139, 175, 257
Hsiang, T., 136
Hubbard, A. B., 173
Huber, P. J., 136
Hunter, J. S., 76, 134, 136
Hunter, W. G., 76, 134, 136–138, 172, 173, 257
Hutchison, P., 33
Hwang, S. T., 62

Jackson, R., 62
Jansen, A. R., 63
Jeeves, T. A., 173
Jeffreys, H., 93, 174
Jennrich, R., 138
Johansen, L. N., 33
Johnson, M. F. L., 62
Johnson, R. A, 137
Johnston, H. S., 33
Jones, C. R., 136

Kabel, R. L., 33
Kalogerakis, N., 137
Kammermeyer, K., 62
Kanemasu, H., 134
Kataoka, T., 31
Kee, R. J., 33
Keith, C. D., 173

Kenney, C. N., 31
Kim, I.-W., 137
Kimball, G. E., 32
King, C. J., 62
Kirkby, L., 172
Kittrell, J. R., 34, 137
Klaus, R., 174
Kleiner, B., 135
Kondratiev, V. N., 36
Kostrov, V. V., 62, 135, 173
Krambeck, F. J., 34
Kreevoy, M. M., 34
Krug, R. R., 137
Kuo, J. C. W., 37
Kustin, K., 32

Laidler, K. J., 34
Langmuir, I., 34
Lapidus, L., 62, 133, 138, 172
Laplace, P. S., 76, 93
Laurence, R. L., 35
Law, V. J., 135
Lawson, C. L., 137
Lee, P. M., 93
Lee, S. H., 137
Lee, W. M., 34
Legendre, A. M., 137, 174
Levenberg, K., 137
Levenspiel, O., 36
Levich, V. G., 34
Levine, R. D., 34
Lightfoot, E. N., 62, 133, 216
Lind, S. C., 30
Lindeberg, J. W., 76
Lindemann, F. A., 34
Lindgren, B. W., 92
Lobão, M. W., 138
Lu, Y. T., 62, 174
Lucas, H. L., 134, 172
Lumpkin, R. E., 137
Luss, D., 33
Lutkemeyer, H., 30
Luus, R., 137, 139
Lynch, D. T., 33

MacGregor, J. F., 172, 257
Machietto, S., 135
Mah, R. S. H., 30
Mahan, B. H., 34
Malinauskas, A. P., 62, 63
Maloney, J. O., 63
Manogue, W. H., 62
Marcus, R. A., 34
Marks, M., 172
Marquardt, D. W., 137
Marr, D. F., 63
Mason, E. A., 62, 63, 174
Mastenbrook, S. M., 139
Mavrikakis, M., 33
Maxwell, J. C., 63
Mazur, P., 31
McCabe, R. W., 34
McCune, L. K., 63
McKenney, A., 62, 133, 188, 216
McLean, D. D., 173, 174
Mead, R., 137
Mears, D. E., 63
Meeden, G., 93
Menten, M. L., 34
Meyer, R. R., 137
Mezaki, R., 136, 137, 174
Michaelis, L., 34
Miller, J. A., 33
Milligan, B. A., 31
Mimashi, S., 36
Missen, R. W., 36
Moler, C. B., 135, 173, 188
Monteiro, J. L., 138
Moore, J. W., 34
Morton, W., 31
Mukesh, D., 31, 34
Muller, M. E., 134
Murray, G. D., 93

Nam, T. T., 63
Nelder, J. A., 137
Neufeld, P. D., 63
Newman, J. S., 34
Newman, M. S., 137
Noble, B., 188

Nørskov, J. K., 33
Nowak, U., 31, 137
Noyes, R. M., 34

O'Connell, J. P., 63
Okamoto, D. T., 137
Ostrouchov, S., 62, 133, 188, 216

Pacheco, M. A., 34
Pearson, R. G., 34
Pease, R. N., 35
Pelzer, H., 35
Penlidis, A., 134
Perry, R. H., 63
Petersen, E. E., 34
Peterson, T. I., 134, 137
Petzold, L. R., 35, 216
Pfaendiner, J., 31
Pfeffer, R., 63
Pinchbeck, P. H., 138
Pinto, J. C., 138
Plackett, R. L., 138
Polanyi, M., 32, 35
Poling, B. E., 35, 63
Poore, A. B., 64
Powell, M. J. D., 138, 174
Prater, C. D., 37, 175
Prausnitz, J. M., 63
Press, W. H., 138, 188, 257
Pritchard, D. J., 138, 174
Pritchard, H. O., 35
Prober, R., 63

Ralston, M. L., 138
Ramaswami, D., 37, 64
Rase, H. F., 63
Ratkowsky, D. A., 138
Rawlings, J. B., 63
Ray, W. H., 35, 64, 138
Razón, L. F., 35
Reilly, P. M., 133, 136, 138, 257
Reiner, A. M., 137
Reklaitis, G. V., 35
Rekoske, J. E., 32
Rideal, E. K., 35

Rippin, D. W. T., 138, 174
Robinson, P. J., 35
Robinson, W. E., 173
Rogers, G. B., 63, 174
Root, T. W., 137
Rosenbrock, H. H., 138, 174
Roth, P. M., 137
Rothfeld, L. R., 63
Rudd, D. F., 31
Rupley, F. M., 33

Sarnowski, K. T., 31
Satterfield, C.N., 63
Saxena, S. C., 63
Schipper, P. H., 35
Schmitz, R. A., 35
Schnabel, R. B., 135
Schneider, D. R., 35
Schneider, P., 62
Schubert, E., 35
Schwedock, M. J., 138
Scott, D. S., 63
Seinfeld, J. H., 35, 138
Sellers, P. H., 33
Shabaker, R. H., 63, 174
Shampine, L. F., 216
Shannon, C. E., 138
Sheppard, N., 30
Shinnar, R., 35
Shon, Y., 174, 257
Sinfelt, J. H., 35
Slin'ko, M. G., 35
Slin'ko, M. M., 35
Smith, H., 135
Smith, J. M., 35, 64
Smith, P. J., 36
Smith, W. D., Jr., 137
Smith, W. R., 36
Smoluchowski, M. V., 36
Smoot, L. D., 36
Sobol, M., 174
Sohn, H. Y., 36
Solari, M. E., 174
Somorjai, G. A., 36
Sorensen, D., 62, 133, 188

Sørensen, J. P., 36, 63, 64, 138, 139, 174
Sowden, R. G., 35
St. John, R. C., 139
Stefan, J., 63
Steinberg, D. M., 138
Steiner, E. C., 174
Stephenson, J. L., 63
Stewart, G. W., 135, 139, 173, 188
Stewart, W. E., 36, 62–64, 135, 139, 173–175, 216, 257
Stigler, S. M., 76, 93, 139
Suen, S. Y., 216
Szekely, J., 36
Szepe, S., 36

Tan, H. S., 139
Taylor, G. I., 64
Taylor, T. I., 36
Tee, L. S., 64
Teeter, B. C., 64, 139
Teter, J. W., 173
Teukolsky, S. A., 138, 188, 257
Thodos, G., 62
Thomas, J. M., 36
Thomas, W. J., 36
Tiao, G. C., 76, 92, 134, 172
Tjoa, I.-B., 139
Toor, H. L., 64
Trautz, M., 36
Trevino, A. A., 31
Trommsdorf, E., 36
Trotman-Dickenson, A. F., 35
Truhlar, D. G., 34, 36
Tschernitz, J., 139, 257
Tsuchida, E., 36
Tsuchiya, H., 32
Tukey, P. A., 135

Uppal, A., 64

Van Loan, C. F., 188

van t'Hoff, J. H., 36
Varma, A., 64
Vetterling, W. T., 138, 188, 257
Villadsen, J. V., 64
Voevodsky, V. V., 36

Wadsworth, M. E., 36
Waissbluth, M., 36
Wakao, N., 64
Walter, J., 32
Wang, B-C., 139
Wang, J. C., 64
Watson, C. C., 137
Watson, K. M., 33
Watts, D. G., 134, 172
Weekman, V. W., 36
Wei, J., 37, 175
Weidman, D. L., 64, 175
Welsch, R. E., 135
Westerterp, K. R., 37
Westmoreland, P. R., 37
Whitwell, J. C., 37
Wichern, D. W., 136, 257
Wigner, E., 35
Wilhelm, R. H., 63
Wilke, C. R., 64
Wilkes, S. S., 136
Wilkinson, F., 37
Williamson, J. E., 64
Wilson, E. J., 64
Windes, L. C., 138
Wood, F. S., 135
Woodward, R. B., 37
Wynne-Jones, W. F. K., 37

Yang, K. H., 2, 37, 175
Yoshida, F., 37, 64
Young, T. C., 139, 175

Zelen, M., 76
Zellner, A., 93
Zeman, R. J., 37
Ziegel, E. R., 175

# Subject Index

Activated complex, 13, 14, 17, 25
Activity
  catalyst, 8, 53, 99, 120, 164, 233
  coefficient, 8
  thermodynamic, 8
Athena Visual Studio, ii, 1, 2, 39, 141, 159, 162

Balances
  mass, 40
  material, 40
  mechanical energy, 42
  momentum, 41
  total energy, 41
Batch reactor, *see* Reactor
Bayesian estimation, 74, 77–93, 96, 110, 142, 165, 219
Boundary layer, 48, 50

Catalyst activity, *see* Activity
Chemical equilibrium, *see* Equilibrium
Chemical species, *see* Species
Chilton-Colburn $j$-factor, 47
Collision frequency, 12, 16, 17, 24, 26
Collocation, iii, 162
Conditional density, *see* Conditional probability
Conditional probability, 67, 78, 80, 105, 112, 114, 157
CSTR, 44

D-optimal experimental design, 115

DDAPLUS (Differential-algebraic sensitivity analysis code), 1, 16, 101, 165, 171, 189–216
Detailed balance, 12
Dusty-gas model, 55, 58

Elementary step, *see* Reaction
Energy balance, *see* Balances
Enthalpy, 10, 41, 47
Enthalpy of adsorption, 29
Entropy, 9, 117
Entropy of adsorption, 29
Equilibrium
  chemical, 3, 7, 9, 11, 12, 14, 15, 17, 20, 23, 24, 28, 29, 208, 233
  quasi-, 18, 22–26, 40
Estimation
  Multiresponse, 72, 95, 125, 141, 162, 166, 169, 244, 248, 249
  Single-response, 95, 141, 218
Ethylene, 3, 7
Ethylene oxidation, 3
Ethylene oxide, 7
Exothermic reaction, 24, 43
Experimental design, 65, 66, 77, 87, 95, 96, 100, 110, 113, 115, 117–119, 142, 156, 157, 160, 164

$F$-distribution, 156
$F$-test, 107, 154
Fixed-bed reactor, *see* Reactor
Fluid mechanics, 50
Frequency factor, 12, 14

Gaussian elimination, 6, 177, 185–187

Gibbs energy, 9
Goodness of fit, 2, 96, 107, 113, 114, 122, 141, 154, 159, 161, 217, 219, 229, 231, 234, 249
GREGPLUS (Generalized REGression package), 1, 2, 66, 90, 91, 96, 98, 99, 101, 102, 104, 106, 107, 109, 110, 112–114, 116–121, 128, 129, 131–133, 141, 144–149, 151, 153–155, 159, 160, 162, 164, 170, 184, 217–257

Heat capacity, 10, 11
Helmholtz energy, 9
Highest-posterior-density, 110, 115, 129, 131–133, 142, 153, 160, 170, 171, 217, 229, 231, 246
Hougen-Watson kinetics, *see* Langmuir-Hinshelwood-Hougen-Watson kinetics
HPD, *see* Highest-posterior-density

Inference, *see* Statistical inference
Internal energy, 10, 41

Jeffreys noninformative prior, *see* Prior

Kinetics
    chemical, 3, 9, 14–16, 23, 24, 26, 102, 120, 159, 233

Lack of fit, 107, 114, 120, 154
Langmuir-Hinshelwood kinetics, *see* Langmuir-Hinshelwood-Hougen-Watson kinetics
Langmuir-Hinshelwood-Hougen-Watson kinetics, 22, 29, 58
Line search, 104, 105, 129, 133, 152, 195

Mass balance, *see* Balances
Material balance, *see* Balances

Matrix
    inversion, 181, 187
    partitioned, 177, 180, 181
    rank, 6, 7, 26, 40, 73, 102, 162, 164, 179, 180, 183, 190
Mechanical energy balance, *see* Balances
Molecularities, 11
Momentum balance, *see* Balances

Neptune, 125
Nonideal solution, 14, 27
Normal distribution, 2, 69, 71–73, 75, 79, 82, 84, 91, 92, 95, 107, 108, 117, 132, 141–143, 154, 155, 218
Normal equations, 127, 128

Outliers, 125

Parametric sensitivities, *see* Sensitivities
Partitioned matrix, 177, 180, 181
Perry's handbook, 53
PFR, *see* Reactor
Planck's constant, 14
Plug-flow fixed-bed reactor, *see* Reactor
Plug-flow homogeneous reactor, *see* Reactor
Pluto, 125
Polymerization, 5, 16, 26
Prandtl number, 48
Prior
    information, 77–79
    informative, 79, 80
    noninformative, 2, 79, 84–89, 142, 165
Probability, 65–76
Production rate, 3, 5–7, 16, 23, 26, 28, 46, 58

Quadratic programming, 105, 150, 152, 219
Quantum mechanics, 14
Quasi-equilibrium, *see* Equilibrium

Quasi-steady-state approximation,
   16, 18–20

Radicals, 16, 26
Random variable, 68, 69, 71, 72,
   75, 76
Rank, 6, 7, 26, 40, 73, 102, 162,
   164, 179, 180, 183, 190
Rate expression, 3, 5, 7, 12, 13,
   15–18, 20, 21, 23, 24, 27, 28, 58,
   101, 122
Reaction
   elementary, 11, 14, 18, 22, 24,
      27
   first-order, 16, 17, 119
   gas-liquid, 10
   gas-solid, 10
   heterogeneous, 5, 10, 20, 22, 26,
      27, 39, 40, 46, 101
   homogeneous, 5, 26, 40, 43–45
   network, 15, 16, 27
   second-order, 17
   third-order, 24
Reaction rate, 3, 5–7, 12, 13, 15,
   16, 20, 27, 28, 53, 58, 99, 101,
   122, 164
Reactor
   batch, 26, 43, 119, 159, 208,
      210, 211
   perfectly-stirred tank (CSTR),
      44
   plug-flow fixed-bed, 46
   plug-flow homogeneous, 45
Replication, 113, 114, 157
Residuals, 65, 66, 74, 106, 107,
   114, 125, 129, 131–133, 153,
   161, 218, 229, 233

Reynolds number, 47
Robust estimation, 125

Sensitivities, iii, 1, 100, 101, 126–
   128, 162, 165, 190–192, 195–
   198, 200, 205, 208, 214, 231,
   248, 250
Species
   chemical, 4, 5, 8, 11, 13, 14, 16–
      18, 23, 25–29, 40, 43, 44, 46,
      49, 50, 55, 57, 58, 60, 161,
      164, 210
Spectroscopic data, 11
Standard state, 8–10, 13
Statistical inference, 2, 74, 77, 78,
   80, 108, 110, 112, 116, 154, 156,
   161
Statistical mechanics, 11, 14
Stoichiometric coefficient, 4, 11,
   24, 25, 27, 48
Stoichiometric matrix, 3, 6, 7, 26,
   40, 73, 75
Stoichiometry, 3, 26, 58
Successive quadratic programming,
   150, 219
Surface
   diffusion, 16, 21, 55–57
   Langmuir, 22

$t$-distribution, 109, 110
Tortuosity, 57, 58
Trace criterion, 116, 170
Trace intermediates, 25
Trace of a matrix, 143
Transition
   phase, 11
   state, 14, 24